Semi-Solid Processing of Aluminum Alloys

Shahrooz Nafisi • Reza Ghomashchi

Semi-Solid Processing of Aluminum Alloys

Springer

Shahrooz Nafisi
University of Alberta
Edmonton, AB, Canada

Reza Ghomashchi
The University of Adelaide
Adelaide, SA, Australia

ISBN 978-3-319-82083-5 ISBN 978-3-319-40335-9 (eBook)
DOI 10.1007/978-3-319-40335-9

Softcover reprint of the hardcover 1st edition 2016

Printed on acid-free paper

This Springer imprint is published by Springer Nature
The registered company is Springer International Publishing AG Switzerland

Foreword

With the publication of this book, it is just 45 years since the seminal doctoral thesis of David B. Spencer laid the foundation for semi-solid metal (SSM) processing. Research interest remains high in developing the fundamentals of semi-solid forming, and industrial applications of these fundamentals continue to grow.

The authors of this book have had a long and distinguished career in the field of solidification processing and have collaborated very closely since 2003 on the solidification and metallurgy of aluminum alloys. This book is an outgrowth of such fruitful collaboration and a very detailed study on the semi-solid processing of aluminum alloys.

An important strength of the book is its emphasis on fundamentals, including on how SSM microstructure evolves and how stirring and thermal variables can alter the structure and alloy distribution. The fundamentals of solidification are used to interpret nucleation, growth, and disintegration of primary phase particles. Important points are made regarding characterization of SSM structures including the roadmap, "methodology," and necessary tools and equipment to characterize such a structure. The book is mainly on rheocasting, but thixocasting also forms a chapter. SSM processes available or in the process of being developed are described.

Industrial applications are outlined in a chapter by Dr. Steve Midson, based on his decades of experience in the field, most recently in China, where SSM research and development is currently particularly strong. The time also appears ripe for an expansion of SSM activities in other parts of Asia and in the West. However, it must be recognized that major innovations, such as SSM, become adopted only slowly by industries where capital costs are high, profit margins are modest, and failure to meet customer requirements carry a high penalty.

The still relative newness of industrial applications of the technology makes it evident that increased industrial utilization of the process will depend on continuing close collaboration between researchers and production personnel.

Cambridge, MA
June 2016

Merton C. Flemings

Preface

Numerous manufacturing techniques have been implemented for the production of near net shape components of metals and their alloys. In principle, these techniques could be categorized as liquid or solid based. Casting is a manufacturing route initiated from liquid, while forging, extrusion, and rolling are some examples of the solid-based fabrication techniques. A new category of metallic alloys processing was introduced during the last quarter of the twentieth century combining the two principles and inheriting both liquid and solid-based know-how. The new processing route, the so-called semi-solid processing of alloys, claimed improved quality at lower cost. The lower cost is mainly associated with tooling cost and reduced energy consumption while achieving enhanced properties is due to better control over microstructural evolution during the course of alloy processing and parts fabrication.

Since the introduction of semi-solid metal (SSM) processing as a possible and viable fabrication route in the early 1970s at MIT, a great deal of research has been carried out worldwide resulting in industrialization of the process. The process is now gaining momentum as an effective alternative to the classical manufacturing processes of casting and forging. Knowledge of different routes and understanding the formation of slurries, including the application of solidification science during SSM processing, are necessary for further development and also selection of the optimum and most appropriate process, including process parameters, for any application. This book provides a vast amount of knowledge and know-how for SSM processes which is mandatory for any researcher, SSM technologist, and educator. It is also useful for both undergraduate and graduate teaching of solidification and casting of metals and their alloys. This book demonstrates the methodology and necessary tools and equipment to characterize semisolid metallic alloys. It also provides the reader with the fundamental knowledge on semi-solid metal processes.

After the introductory Chap. 1 on aluminum industry and its role in economic development along with some useful statistics, Chap. 2 discusses the semi-solid metal technologies available to date, while Chap. 3 concentrates on the solidification, stirring, and alloy distribution during SSM processing of mainly Al–Si alloys. Chap. 4 gives a detailed account of the techniques used to characterize SSM billets which include metallographic and flow behavior, i.e., rheological characteristics. Chap. 5 is concentrated on rheocasting of Al–Si alloys where the findings on SSM research are presented for rheocast billets produced by low pouring temperature, electromagnetic stirring, EMS, and Swirled Enthalpy Equilibration Device, SEED, processes. Chap. 6 explains the melt treatment currently employed for SSM billets, while Chap. 7 reports on the recent findings of research activities on the thixocast billets' production routes. In Chap. 8, we have invited Dr. Steve Midson, Midson group-Colorado School of Mines, to highlight some industrial applications of SSM processes in the production of high-quality engineering parts.

The book is based on our research on SSM processes over the last 13 years including one of our latest publications which was recognized "2013 winner of the Buehler Technical Paper Merit Award for Excellence" by The Awards Committee of the International Metallographic Society and Metallography. As well, the book covers the work of many researchers and other well-respected research centers worldwide. We are grateful to many of our colleagues and collaborators as well as other researchers who have kindly provided us with their results and permission to include them in this book. We particularly wish to acknowledge the following individuals whose assistance and comments made this work possible and helped us to overcome many hurdles:

- Dr. Omid Lashkari, Canada
- Dr. Alireza Hekmat, Canada
- Prof. Eugenio Jose Zoqui, University of Campinas, Brazil
- Dr. Frank Czerwinski, CanmetMATERIALS, Natural Resources Canada/Government of Canada
- Prof. Jessada Wannasin, Prince of Songkla University, Thailand
- Dr. Peyman Ashtari, Canada
- Prof. Magnus Wessén, Jönköping University, Sweden
- Dr. Jayesh Patel, Zyomax Ltd, UK
- Dr. Jim Yurko, USA
- Drs. Shan Jin, and Hai-Lin Chen, Thermo Calc, Sweden
- Pascal Coté, STAS Inc., Canada
- Djibril René, European Aluminium Association, Belgium
- Véronique Bouchard, Trans-Al Network, Canada
- Matt Meenan, The Aluminum Association, USA
- Rosanna Boyd, Australian Aluminium Council, Australia

An enormous task of finalizing the book was about acquiring permission for materials we used in this context and we would like to acknowledge the assistance of the following:

- Prof. Plato Kapranos, Sheffield University, UK
- Prof. David St. John, Queensland University, Australia

- Prof. Hao Wang, University of Southern Queensland, Australia
- Prof. Merton Flemings, MIT, USA
- John Jorstad, USA
- Prof. Annalisa Pola, Università degli Studi di Brescia, Italy
- Stephen LeBeau, Thixomat, Inc., USA
- Prof. Xiangjie Yang, School of Mechanical Engineering, Nanchang University, China
- Dr. K.M. Kareh, Imperial College, London, UK
- Jon Olson, Phillips-Medisize Corporation, USA
- Prof. Nobutaka Yurioka, Japan
- Prof. S.J. Hong, Division of Advanced Materials Science, Kongju National University, South Korea
- Prof. Donggyu Kim, Dong-A University, South Korea

TTP and Scientific Net (Dr. Andrey Lunev), ASM International, NADCA, AFS, TMS, Elsevier, Maney (Routledge, Taylor & Francis Group), Springer, and many others who are missed in this list.

We would like to acknowledge the financial support for some parts of our work on semi-solid processing of aluminum alloys which was provided by Alcan International (Rio Tinto-Alcan), Natural Science and Engineering Research Council of Canada, CQRDA, and the University of Quebec at Chicoutimi, UQAC. In this regard, we should acknowledge the contribution of Prof. Andre Charrette, UQAC, Prof. Frank Ajersch of Ecole polytechnic de Montreal, Prof. Hojatollah Vali of McGill University, Dr. Bahadir Kulunk of STAS Inc., and special thanks to Dr. Joseph Langlais of Alcan for continued support.

We are also grateful to CanmetMATERIALS, Natural Resources Canada/Government of Canada (Drs. Jennifer Jackman, Mahi Sahoo, Daryoush Emadi, and Mahmoud T. Shehata), for supporting our work on EMS process by providing access to research facilities at CanmetMATERIALS.

Finally, the authors would like to acknowledge Prof. Jalal Hedjazi of Iran University of Science and Technology who was instrumental in motivating the authors on solidification and casting.

Edmonton, AB, Canada — Shahrooz Nafisi
Adelaide, SA, Australia — Reza Ghomashchi

Contents

Chapter 1
Introduction

Abstract An overall view of aluminum production and application is presented with the latest statistics. The challenges facing the aluminum industry and major aluminum producers worldwide are mentioned. Guidelines are given for future development of aluminum shaping with main emphasis on casting and foundry technology. Semi-solid metal casting is introduced as an alternative to reduce environmental effects of aluminum shape casting through reduction of energy consumption and generation of pollutants.

Since its first commercial production in 1854, aluminum has proved to be one of the most versatile and useful materials discovered and extracted from the earth. It is the material of choice in a wide range of design and engineering applications because of its light weight, formability, corrosion resistance, and high specific strength; and is the world's second most used metal after iron (steels). Moreover, it is one of the most readily recyclable materials in use for packaging, medical and electrical applications, and motor and automotive manufacturing. Nearly three-quarters of the aluminum ever produced in the world remains in use today because by its nature, it is endlessly recyclable.

The aluminum industry is vital to the world economy with a production rate of 57.5 Mt in 2015 with China (31.2 Mt), Middle East (5.1 Mt), North America (4.5 Mt), Europe[1] (4.4 Mt), CIS (3.9 Mt) and Rest of Asia (3.4 Mt) being the top six aluminum producing regions in the world [1]. As an example of the importance of aluminum to individual country's economy, aluminum contributed to Australian economy with nearly A$4.4 billion worth of export in 2014 [2]. For Canada as the third producer of aluminum, exportation of aluminum represents more than 10 % of the overall value of goods exported by Canada in 2013 [3]. The world production of aluminum since 1950 is given in Fig. 1.1 and the pie chart for 2015 production in Fig. 1.2 shows China producing more than half of the world's aluminum [1].

Given its inherent properties and potential as a sustainable resource, new ways of using aluminum in manufacturing processes and manufactured products are currently being investigated worldwide. The roadmap on aluminum [4, 5] identifies specific areas where research is urgently needed to bring aluminum to its full potential.

[1]EU + EFTA + Bosnia and Montenegro

S. Nafisi, R. Ghomashchi, *Semi-Solid Processing of Aluminum Alloys*,
DOI 10.1007/978-3-319-40335-9_1

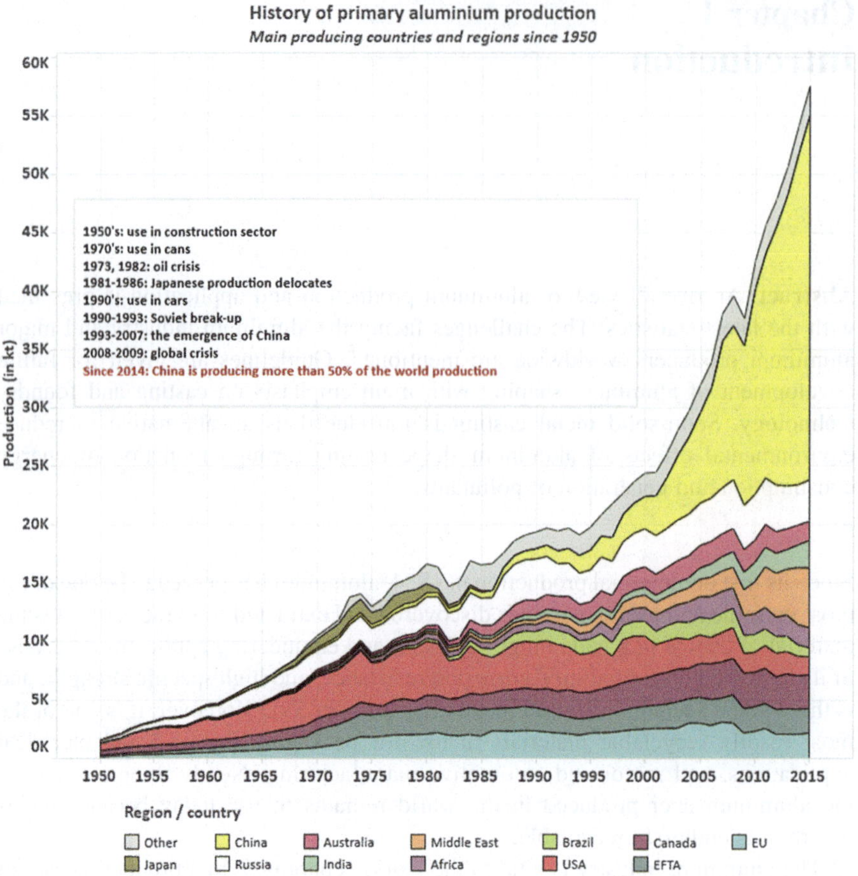

Fig. 1.1 The production of primary aluminum worldwide from 1950 to 2015 [1]

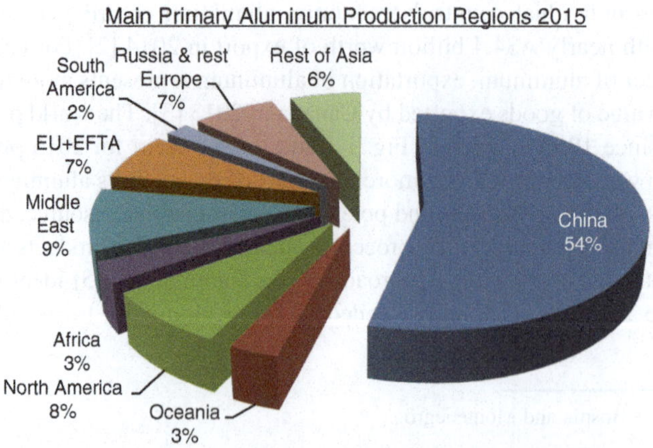

Fig. 1.2 The world production of aluminum for 2015 (reproduced from Ref. [1])

Solidification and recycling, fabrication, and finished products are particularly detailed. Prediction of structure, stress, and strain; understanding of relationships between structure and materials performance; development of new manufacturing processes; and forming techniques such as semi-solid casting are among the most important research avenues identified. As an example, the latest Canadian roadmap identifies the main challenges faced by the aluminum industry [5] to be:

(a) The need for the development of new aluminum products having superior performance, produced using more efficient processes.
(b) Better access to predictive and actual product performance testing facilities so as to meet the most demanding needs of the transportation, construction and energy industries.

For development of new products, various forms of casting, in particular, offer enormous possibilities since this production method is predominantly suited to the use of recycled aluminum. However, the casting processes need to be innovative and cost-effective. It is also highlighted that new applications must be found rapidly and product-oriented research should be prioritized.

A close look at the nature of aluminum research over the last few decades shows that efforts have chronologically been concentrated on: primary production and related technology development with emphasis on environment and energy efficiency, forming technologies, foundry and wrought alloys, and composite materials (e.g., [6–9]). Although such intensive research activities have resulted in better understanding of aluminum products, they equally generated new questions and opened the way to new research avenues.

Renewed interest in aluminum intensive vehicles (AIV) and aluminum space frames for passenger cars has resulted in a new focus on aluminum wrought and foundry alloys casting, including the introduction of novel melt treatment and alloy manipulation techniques such as semi-solid metal (SSM) processes. The near-net shape casting and forming techniques such as conventional high pressure die casting (HPDC) are inadequate due to large defect content [10, 11] and therefore need to be customized to produce parts with load bearing capabilities. Equally important are new alloys and melt treatment techniques to be developed to enhance the quality of as-cast products. It is of particular interest in the context of the current book that the first item identified in the Canadian roadmap [5] for shape casting of aluminum alloys related to the *alloy development for semi-solid rheocasting of structural components,* one of the principle objectives of the research carried out by the authors, as well as new techniques for the production of thin wall cast parts.

One of the important issues facing the world community in general and the developed industrialized countries in the twenty-first century, in particular, is the impact of technology on the environment. This is particularly highlighted in technologies that have been the backbone of industrialization age in nineteenth and twentieth centuries, namely raw materials conversion such as steel making and aluminum production with high greenhouse gas emissions and fabrication routes for the shaping of the base metals into engineering components such as casting.

Although casting is the most economical route to convert raw materials into readily usable engineering components, its full potential has not yet been exploited simply due to its impact on the environment and, of course, the integrity of the product itself. SSM processes appear to be the answer for the development and fabrication of cast products, cheaper with superior quality and less energy intensive. The issue of less energy consumption in the context of SSM processes is in line with the current debates on environment and therefore should help the manufacturing industry to progress toward the "*green manufacturing*".

The applications of less energy intensive light weight as-cast products in automotive and aerospace industry will further mitigate the effect of pollution on our environment as a result of less fuel consumption.

1.1 Near Net-Shaped Casting, N2SC

In order to reduce energy consumption in manufacturing engineering components, it is necessary to approach this goal from two fronts:

1. Fabrication route, that is, more direct transformation of raw materials to readily usable engineering parts
2. Feedstock for the fabrication route

For casting, the above-mentioned approach can be highlighted into combining near net shape casting route with Semi-Solid Metal (SSM) processing as the source of feedstock.

As shown in Fig. 1.3, shape casting involves so many steps but generally could be divided into design and production stages [12, 13]. The effective design time has been reduced through application of Computer Aided Design (CAD) and Computer Aided Manufacturing (CAM) software packages. As for production, the recent surge in casting simulation software packages created a great impact on the implication of our understanding of solidification principles on casting design and operation. Such predictive tools have enabled foundry engineers to increase the chances of fabricating sound cast parts by upfront evaluation. The casting engineers are now able to combine experience with visual analysis tools to identify the key parameters in reducing rejection rates and improving quality in shorter time. This eventually reduces cycle time, improves quality, and most importantly boosts productivity.

There is ample information in open literature on the principles of modeling and simulation of solidification and casting [14, 15] and a wide range of commercial software packages available in the market for the modeling of shape casting. The packages are able to provide detailed information on heat flow, fluid flow, mold filling, radiation, thermomechanical stresses, and microstructural prediction and evolution. The design and creation of such predictive tools as well as their application require good knowledge of solidification principles and the physical characteristics of the alloy to be cast.

The term near net shape casting covers the casting techniques where the as-cast product does require none or at least a minimum amount of post-solidification/casting treatments, that is, the red broken line rectangle segment in Fig. 1.3. These are the following operations:

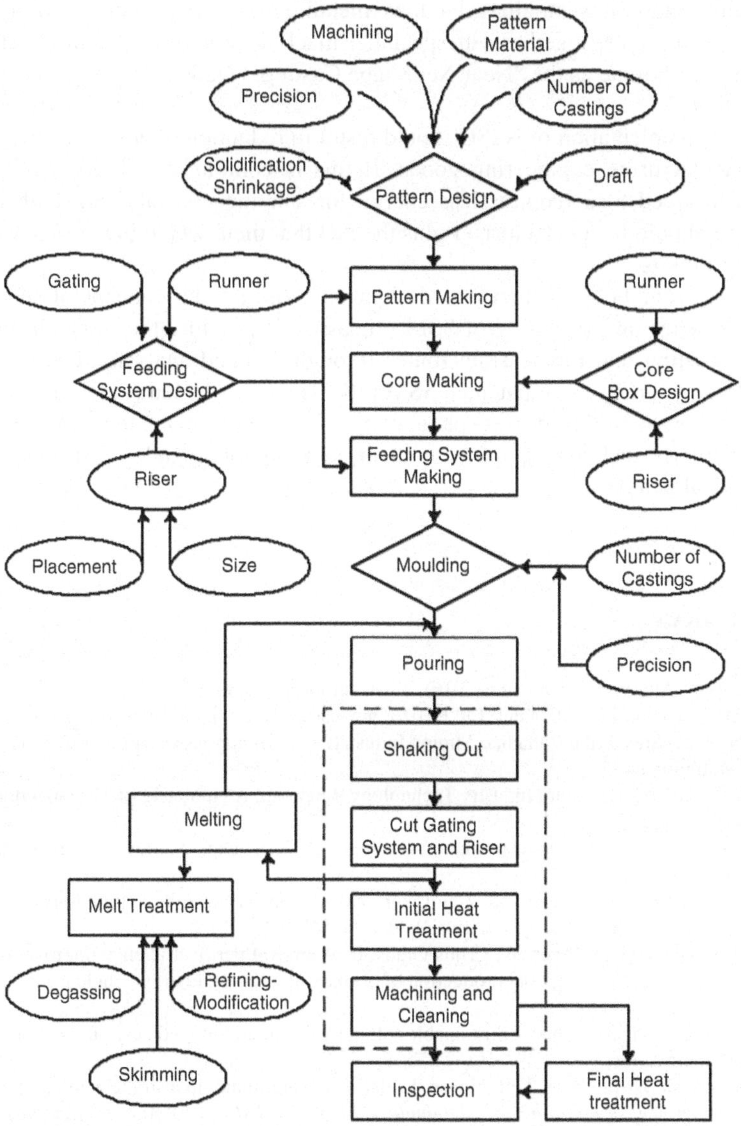

Fig. 1.3 Schematic diagram showing the many steps to be considered for production of an engineering as-cast product (reproduced from [12])

- Removal of feeding and gating systems
- Cleaning and removal of cores
- Machining
- Heat treatment

In this regard, it is anticipated to have metallic molds in most cases, for near net shape casting routes together with specialize machines and/or procedures to classify a casting technique as the "Near Net Shape Casting," N2SC. The key word here is "*precision*."

The implementation of N2SC should result in reduction of energy consumption for manufacturing engineering goods. If this is combined with SSM processes where the feedstock requires less energy for shaping, the fabrication of as-cast products should be much cheaper plus the fact that there would lower maintenance cost for tooling.

The current book is therefore designed to discuss the concept of feedstock through better understanding of SSM processes. It provides the research findings on rheocasting and thixocasting routes through detailed analysis of SSM billets produced by a number of patented SSM processes. This includes melt treatment of Al–Si alloys as well as process parameters effect in optimizing the SSM feedstock characteristics such as solid phase content, its morphology, alloy distribution, and rheological behavior.

References

1. European Aluminium Statistics (2016), www.european-aluminium.eu
2. Australian Aluminium Council Ltd (2016), www.aluminium.org.au/statistics-trade
3. Facts & Figures of the Canadian Mining Industry, The Mining Association of Canada (2014), www.mining.ca
4. The Canadian Aluminium Industry Technology Roadmap, An initiative of The Government of Canada (2000)
5. Canadian Aluminium Transformation Technology Roadmap, National Research Council Canada (2006)
6. W. Zhang, H. Li, B. Chen, CO2 emission and mitigation potential estimations of China's primary aluminum industry. J. Clean. Prod. **103**, 863–872 (2015)
7. K. Kermeli, P.H. ter Weer, W. Crijns-Graus, E. Worrell, Energy efficiency improvement and GHG abatement in the global production of primary aluminum. Energy Efficiency **8**, 629–666 (2015)
8. Aluminum Industry Vision—Sustainable solutions, US aluminum industry publication (2001), www.energy.gov
9. R. Love, M. Skillingberg, T. Robinson, Updating the aluminum industry technology roadmap. in *Conference: Symposium on Aluminum held at the TMS 2003 Annual Meeting*, ed. by S.K. Das (San Diego, 2003), Aluminum, 215–223 (2013)
10. R. Ghomashchi, Die filling and solidification of Al-Si alloys in high pressure die casting. Scandinavian J. Met. **22**(2), 61–67 (1993)
11. R. Ghomashchi, High pressure die-casting: effect of fluid flow on the microstructure of LM24 die-casting alloy. J. Mat. Proc. Tech. **52**, 193–206 (1995)
12. R. Ghomashchi, Solidification and foundry technology, Metallurgical Engineering Course, University of South Australia, Adelaide, Australia (1997–2000)
13. R. Ghomashchi, Process control and optimization of near net-shaped aluminum-silicon alloys premium cast products. in *Modeling, Control and Optimization in Nonferrous and Ferrous Industry*, ed. by F. Kongolie. Invited paper, Process Control and Optimization in Ferrous and Non-Ferrous Industry (TMS, Chicago, 2003)

14. Modeling of casting, welding and advanced solidification processes. in *Proceedings of the Tenth International Conference*, ed. by D.M. Stefanescu, J.A. Warren, M.R. Jolly (Destin, 2003)
15. B.L. Ferguson, R. Goldstein, S. MacKenzie, R. Papp (eds.), Thermal process modelling. in *Proceedings from the Fifth International Conference on Thermal Process Modeling and Computer Simulation* (Orlando, 2014)

Chapter 2
Semi-Solid Metal (SSM) Technologies

Abstract The concept of semi-solid casting is discussed with main emphasis on rheocasting along with a brief account of the mechanisms responsible for the microstructural development during rheocasting. A state of art of the available SSM processes discussed to highlight the engineering features of SSM processes already developed as well as those still at the development stage. After a detailed description of rheo-routes, thixo-routes are also mentioned to provide a concise view of SSM processes available to aluminum casting and foundry industry.

Semi-Solid Metal (SSM) processing is gaining momentum as an effective alternative to the classical manufacturing processes of casting and forging. This category takes advantages of both liquid and solid forming processes. In the absence of shear forces, the semi-solid slug is similar to a solid, for example, self-standing, while with application of shear, the viscosity is reduced appreciably and the material flows like a liquid, that is, thixotropic behavior; "*dual characteristics where the SSM billet has solid-like characteristics, but flows like a liquid under shear stress.*" Such unique characteristics have made SSM routes attractive alternatives to conventional casting on an industrial scale.

Knowledge of different routes and understanding the formation of slurries is a necessity for choosing the optimum and most appropriate process for any application. The current chapter deals with the principles in the SSM studies including an overview of different processing routes.

Technologies for SSM processing are generally structured into two basic groups (Fig. 2.1):

- Rheo-routes
- Thixo-routes

The rheo-route involves the preparation of SSM slurry of alloys from liquid phase and its direct transfer into a die or mold for component shaping. The term "Slurry-on-Demand", SoD, has been coined in industry to describe slurry making operations that take place in the cast shops, thus providing a constant supply of slurry for shaping operations.

The thixo-route is basically a three-step process, involving preparation of a feedstock material of appropriate length and weight having an equiaxed or globular, or indeed having the potential to transform into spherical structure on further

S. Nafisi, R. Ghomashchi, *Semi-Solid Processing of Aluminum Alloys*,
DOI 10.1007/978-3-319-40335-9_2

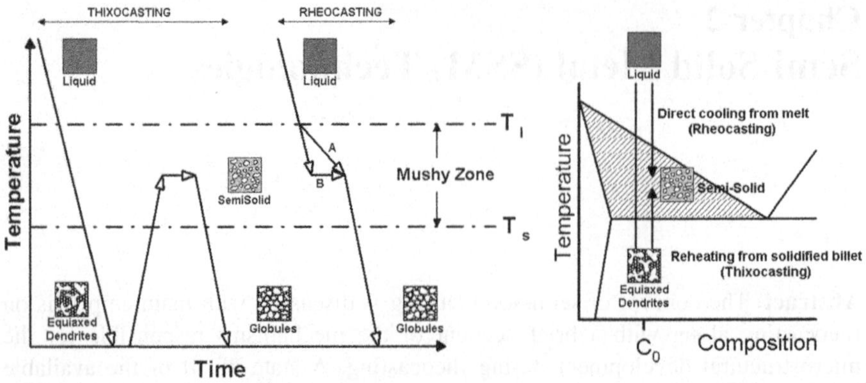

Fig. 2.1 Schematic representations of SSM processes [1] (reprinted by permission of Taylor & Francis Ltd)

processing. The second step is reheating of the feedstock material to temperatures between solidus and liquidus (mushy zone) to generate a semi-solid structure. The final step is shaping of the mush with thixotropic characteristics. The feedstock for thixo processes may include a variety of sources: rheocast billets, pellets or chips (e.g., Mg casting), spray-formed solids, grain-refined ingots, compacted powders, or mechanically deformed bars [2, 3]. Thixocasting was the first commercially viable source for semi-solid processing. However, due to various concerns such as billet preparation cost, limited production facilities, restricted alloys, in house recycling difficulties, and overall high cost, competitiveness of the process has been decreased.

When rheocasting was first discovered, it was believed mechanical disintegration of dendritic structure during solidification is the main cause of forming the globular structure. However, with more research and innovation in the SSM field, it is believed that there are two distinct mechanical and thermal-based processes for SSM preparation. In other words, globularization is achieved by mechanical means, such as an impeller or vigorous melt flow by employing, for example, electromagnetic force and through close control of heat balance, respectively. In these processes, the dominant globularization mechanisms are different. The following mechanisms are responsible for microstructural evolution of SSM products as schematically illustrated in Fig. 2.2 [1].

- *Mechanical and thermo-mechanical fragmentation of dendrites.*
  The concept of dendritic structure transformation during solidification was proposed by Flemings [3, 4], Doherty [5], and later elaborated by Hellawell [6]. The same concept is applicable to SSM processes. Following the nucleation and initial growth, dendrites are broken mechanically or by localized remelting, also known as dendrite root remelting, due to direct stirring by an impeller, or indirect stirring, electromagnetic stirring (EMS). With continuous shearing of the melt, dendrites morphology changes to "rosettes" and/or "globules." With further stirring, particle ripening becomes dominant, driven by the reduction of interfacial surface energy.

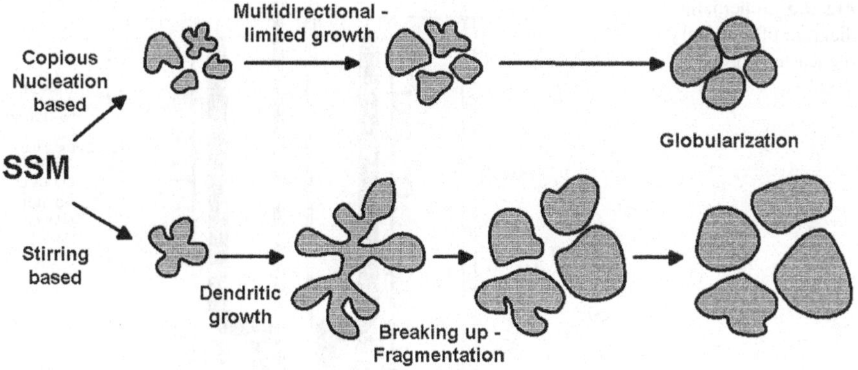

Fig. 2.2 Two different mechanisms for rheocasting process [1] (reprinted by permission of Taylor & Francis Ltd)

- *Creating multiple nucleation near the liquidus temperature with limited and multidirectional growth, excessive nucleation*
  This is a thermally activated mechanism where localized undercooling is artificially generated within the bulk liquid to accelerate nucleation, leading to "excessive nucleation" analogs to that of the "big bang" theory or "copious nucleation" initially proposed by Elliot and Chalmers [7, 8]. From the solidification point of view, if the mean free path between the nuclei becomes small due to excessive nucleation, grains grow slowly due to limited constitutionally supercooled boundary layer and multidirectional heat flow. Such solidification conditions should eventually lead to the formation of more or less spherical primary phase particles.

As illustrated in Fig. 2.2, excessive nucleation-based processes are preferred because of elimination of at least one step and subsequently shorter processing time and better globule morphology.

2.1 Rheo-Route Techniques

As mentioned earlier, there is a wide range of processes patented or under investigation worldwide. In spite of technical and technological differences amongst the available rheo-routes, the most significant processes are as follows:

2.1.1 Mechanical Stirring

The implication of stirring during solidification of alloys was initially originated at the Massachusetts Institute of Technology, MIT [9]. Melt agitation is commonly generated by means of augers or screw [10–12], impellers, paddles, or some special

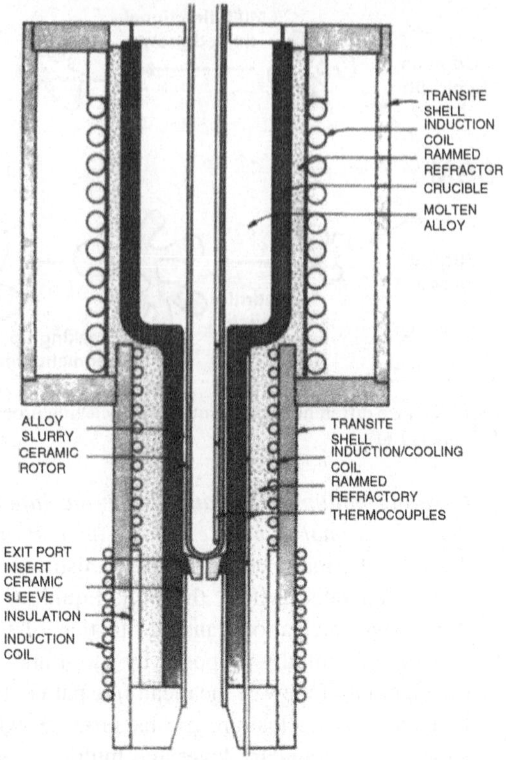

Fig. 2.3 Schematic diagram of a primary rheocaster [9]

kind of agitators [13–16]. The applied shear forces by melt agitation during solidification create nondendritic structures. The process was advanced from batch into a continuous process. As shown in Fig. 2.3, in the simple continuous process, superheated liquid within the holding vessel flows into a gap between the stirrer and the outer cylinder where it is simultaneously agitated and cooled. Slurry flows from the bottom of the rheocaster either to be cast directly to shape (rheocasting) or to be solidified as feedstock material for subsequent reheating and thixoforming.

In spite of mechanical stirring being the first SSM processing technique, it has found limited applications on industrial scale due to its drawbacks such as erosion of the stirrer (particularly with more chemically aggressive alloys), the contamination of the slurry by oxides and dross, gas entrainment, low productivity, and the difficulty in process control [2, 17, 18]. In addition, slurries produced by such processes tend to contain larger, ripened rosette particles and less homogeneity than other processes (Fig. 2.4). Such morphologies immobilize and entrap more liquid, adversely affect the rheological behavior of the slurry by reducing the effective liquid fraction.

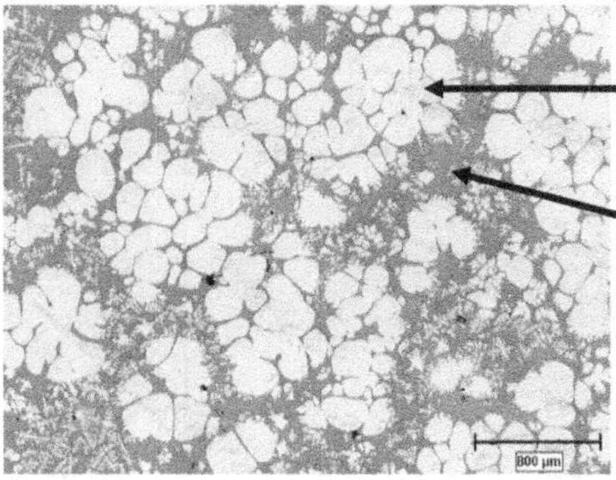

Fig. 2.4 Conventional mechanical stirring of 356 alloy (Al-7Si-0.35Mg), quenched from ~587 °C

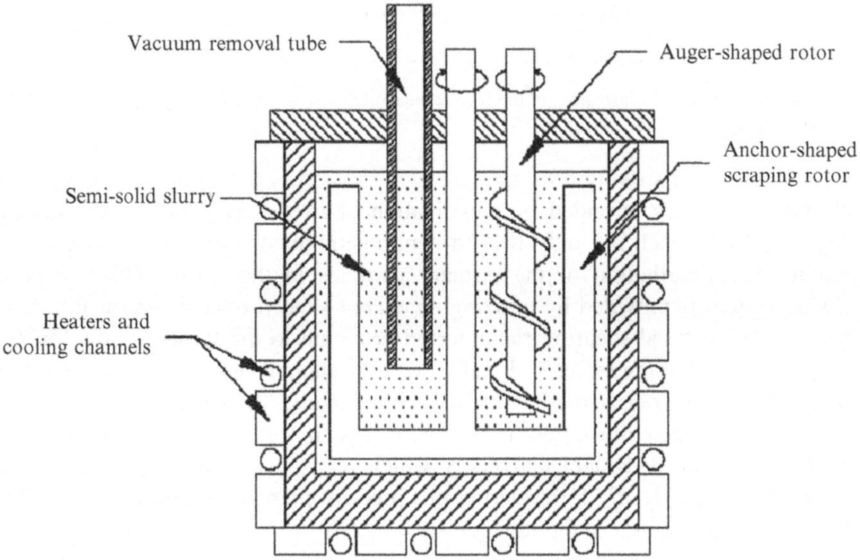

Fig. 2.5 Schematic of direct slurry formation process and machine components [19]

The aforementioned shortcomings have resulted in introducing new processes using helicoidal screws and in most cases; shearing and solidification are carried out in different sections of the rheocaster. The main reason for the modification seems to be the improvement in the microstructural uniformity of semi-solid billets.

Figure 2.5 shows an example of a rheocasting process named DSF, "Direct Slurry Formation" [19, 20]. The process is divided into three steps; formation of semi-solid metal, maintaining the slurry, and transportation of the slurry to the diecasting machine. Based on the production rate, superheat, and desired fraction solid, the amount of the heat extraction is calculated. In the next step, adequate

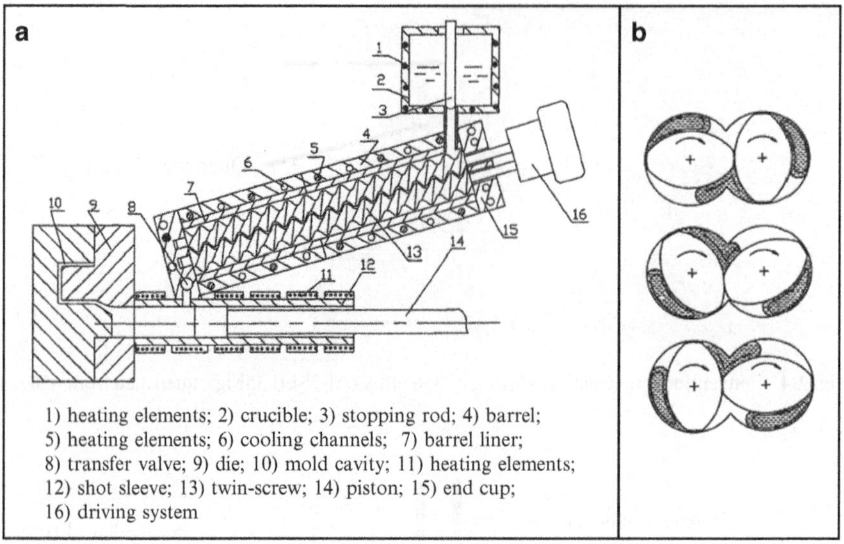

Fig. 2.6 (**a**) Schematic illustration of the Rheo diecasting twin screw technology. (**b**) Flow pattern on the cross section of the extruder [10]

shearing forces are provided by combination of two screws (auger and anchor shaped rotors) which not only lead to dendrite fragmentation but also assist the homogenous distribution of the primary particles in the slurry. This complex mixing system is provided by utilizing an anchor-shaped rotor for vertical homogeneity with vertical shearing rods located proximal to the furnace walls for the scraping of solidifying material. Finally, with the support of a vacuum system, the prepared slurry is transferred directly to a cold chamber die-casting press.

Figure 2.6a also illustrates a successful design for rheocasting of Al- and Mg-based alloys by Brunel University which referred as "rheo diecasting twin screw technology, RDC" [10]. During this process, prepared liquid is fed into the device and the liquid is rapidly cooled down to a desired temperature while mechanically sheared under cooling at a controlled rate by a pair of closely intermeshing screws. Combination of these screws, which is claimed, results in a fluid flow in the form of "8" and therefore pushing the liquid along the axial direction and as a consequence liquid undergoes multidirectional forces (Fig. 2.6b). Figure 2.7 compares a cross-sectional view from the edge to the center of a tensile sample produced by HPDC and rheo die casting process (twin screw technology). As seen, the microstructure produced by mechanical stirring is more uniform through the cross section.

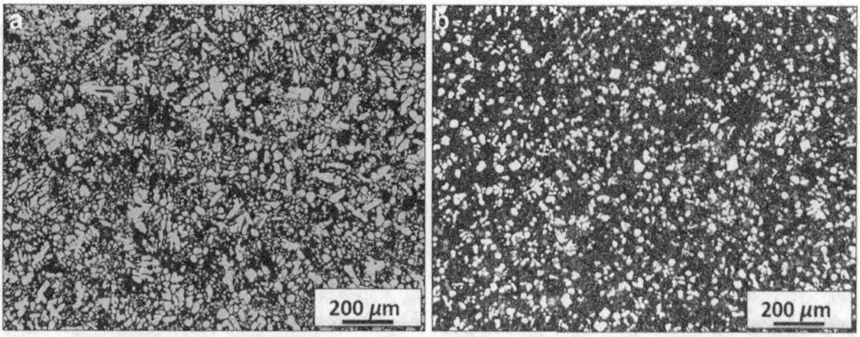

Fig. 2.7 Cross-sectional views of the microstructures from area between the edge and the center of the tensile samples (6.4 mm Dia.) of A380 (Al−9Si−3Cu) produced by: (**a**) HPDC and (**b**) RDC (Courtesy of Zyomax) [21]

2.1.2 Magneto Hydro Dynamic (MHD) Stirring or Electro Magnetic Stirring (EMS)

It was claimed that electromagnetic stirring process was developed to overcome the associated complications with the direct mechanical stirring [22, 23]. Indeed, the application of EMS in casting process is not new and it was borrowed from steel continuous casting process. It was shown that depending on the EMS unit location on the caster (e.g., mold, below the mold, or before final solidification point), it has several advantages including elimination of centerline segregation, converting columnar structure to equiaxed, removing inclusions from melt, elimination of mold powder entrapment, hotter steel in meniscus, etc. [24].

The local shear is generated by a dynamic electromagnetic field and the solidifying metal is the rotor. The stirring action resulted from liquid movement shears the dendrites causing the dendrites' fragmentation during the process. This ultimately generates a material flow between solidification front and bulk liquid. The new fragmented equiaxed particles and wall crystals come close to the solidification front and recirculated toward the liquid phase during crystal formation and growth. Recirculation of equiaxed crystals results in partially remelting of dendrite arms and after several recirculations, the globules are created. The stirring is deep within the treated (filtered and degassed) liquid; therefore, the contamination is almost eliminated. Such strong agitation breaks up the dendrites and transfers dendrite branches as well as newly formed grains into the bulk, improving distribution of particles and creating a uniform temperature throughout the slurry.

Figure 2.8 shows billets cast with and without application of EMS with a very large superheat of about 230 °C. By continuous and uniform cooling parallel to dendrites fragmentation and spheroidization during EMS process, it seems that the process is reasonably capable for structural evolution. It has been reported [25] that MHD casting process could be able to deliver the desired solidified microstructure

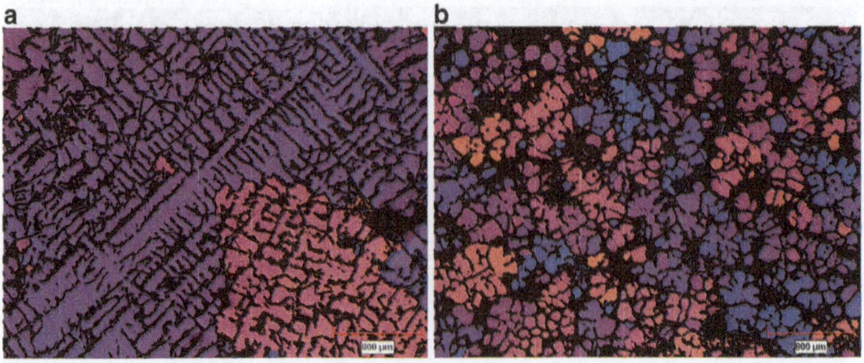

Fig. 2.8 Polarized light micrographs showing the effectiveness of EMS in refining the structure of A356 alloy cast at 850 °C (**a**) without stirring and (**b**) EM stirred

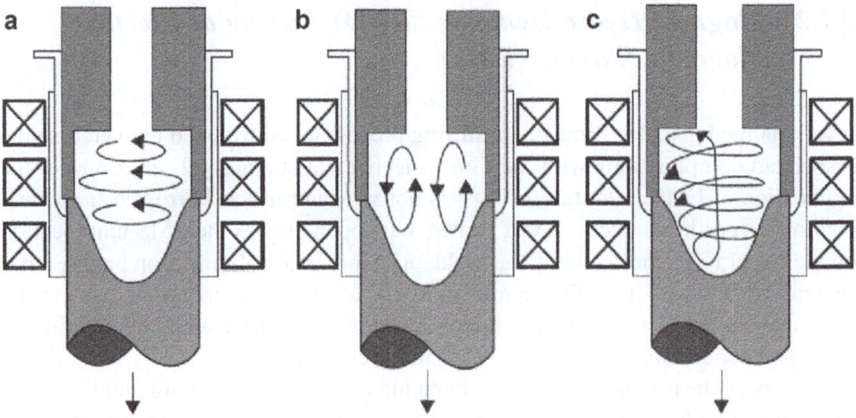

Fig. 2.9 Schematic of electromagnetic coils for MHD stirring and solid particle flow pattern in the mushy zone (**a**) due to rotational inductive coils, (**b**) due to linear inductive coils, and (**c**) helicoidal stirring (reproduced from [26])

with a grain size that is normally about 30 µm. This compares with the 100–400 µm grain size produced by mechanical stirrers.

There are three modes of electromagnetic stirring capable of achieving vertical, horizontal, and helical fluid flow. The helical mode is a combination of the vertical and horizontal modes. In the horizontal flow mode, solid particles are moved in a quasi-isothermal plane so that mechanical shearing is more likely to be the leading mechanism for dendrite disintegration (Fig. 2.9a). In the case of vertical flow mode, dendrites are almost equally fragmented at the solidification front. However, dendrites are recirculated to the hotter region of the stirring zone with possible partial remelting. Hence, thermal processing is dominant over mechanical shearing (Fig. 2.9b).

The continuous casting is made through either a vertical or horizontal arrangement, which to be determined by the casting direction in relation to the gravity. Vertical stirring has been employed in vertical caster only, while horizontal stirring has been used in both vertical and horizontal casters. In addition, the microstructures produced from these feedstocks are affected by the design of the inductive coils. Furthermore, electromagnetic force fields in the mushy zone are not uniform which may result in billets having different degrees of structural modification in the radial direction.

According to Niedermaier et al. [26], the main advantages of horizontal stirring include cost-effectiveness and continuous production; however, the billet quality is affected by gravity vector. From a technological point, the essential dissimilarity lies in the gravity and casting direction interrelationship. In contrast to the horizontal EMS stirring, the vertical EMS casting takes advantage from symmetrical solidification and with almost no restriction of the billet diameter. However, the vertical system has disadvantages such as discontinuous production, high investment, and production costs (however it could be used in combination with high pressure diecasting for parts production [27]). It is claimed that EMS application not only breaks down dendrites to encourage globularization, but also could have beneficial effects on secondary phases, such as intermetallic particles and eutectic silicon in Al–Si foundry alloys [28, 29].

MHD process was the first commercial route for semi-solid metal processing and has been the most effective and most common method for production of thixo billet for decades.

2.1.3 New Rheocasting NRC™ or UBE Process

The UBE process was developed by UBE industries—Japan [30] for production of aluminum and magnesium alloys slurries. This is regarded as the slurry on demand "SoD" process and relies on thermal treatment of liquid metal instead of stirring. The sequence of processing steps leading to the formation of SSM slurries is (Fig. 2.10) as follows:

- Molten alloy is prepared conventionally with a specific superheat (step 1)
- The liquid prepared in step 1 is poured into a thermally insulated vessel. The vessel or cup is coated and cleaned between each cycle to maintain a consistent process. The transfer of molten alloy could be direct or indirectly through a jig, that is, cooling slope (step 2). If a cooling slope is used, then it acts as a nuclei generator as discussed in Sect. 2.1.4 (it is important to note that grain refiner addition and also the amount of superheat in step 1 depend on the application of jig).
- In step 3, the alloy is held within the vessel for a specific time within the mushy zone to obtain a specific fraction of solid appropriate for the subsequent forming

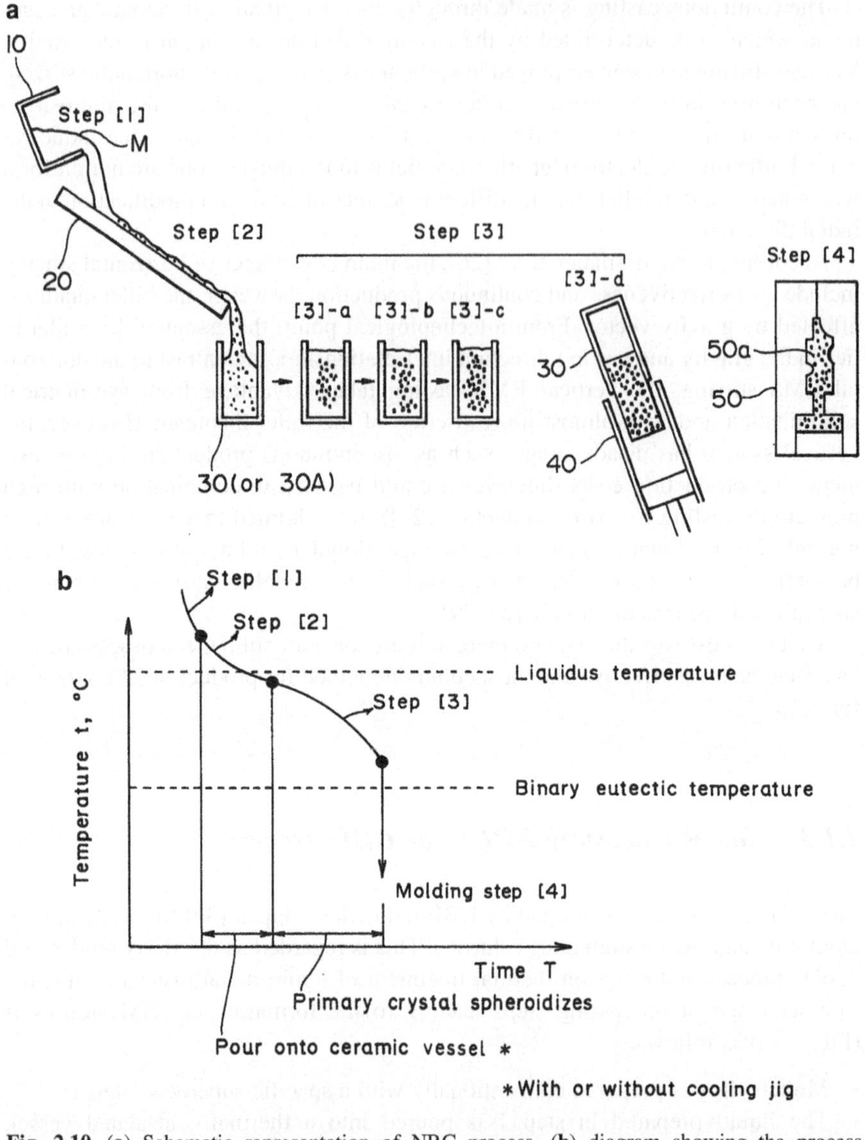

Fig. 2.10 (**a**) Schematic representation of NRC process, (**b**) diagram showing the process sequence for hypoeutectic Al–Si alloys (10) ladle; (20) cooling jig; (30) ceramic container; (40) injection sleeve; and (50) mold [30]

process. In this step and with a uniform thermal gradient within the vessel, it is claimed that fine and equiaxed dendrites grow as the fraction solid increases with decreasing temperature.

- In step 4, the as-cast SSM billet is pressure formed in a permanent mold.

2.1.4 *The Cooling Slope Process (CSP)*

Application of cooling slope is a widespread practice for semi-solid slurry production. This process is based on the simple pouring the melt down on a cooling slope with subsequent solidification in a die which could be used directly for rheo-processing or indirectly to thixo-processing (e.g., Fig. 2.11). Refinement of the solid particles within the slurry could be simply performed by directing the molten metal through a water/air cooled tube [32–34] or by other means or even the cooling slope could be linked to a vibrator [35]. Therefore, the key variables of this process are the length and angle of the cooling slope, slope material, and superheat of the molten metal. Cooling slope practice may result in oxide film formation and gas pickup which should be considered during commercialization process.

Semi-solid strip casting using an unequal diameter twin roll caster is another case of using cooling slope. As depicted in Fig. 2.12, the semi-solid slurry is made by a cooling slope and then rolled in an unequal diameter twin roll caster. A conventional twin roll caster is not suitable for rheocasting as the SSM slurry easily solidifies in the nozzle. It was stated that A356 Al–Si alloy could be produced by this method for use in press forming [36].

The mechanism of semi-solid globule formation in this process was studied by Motegi et al. [37]. They proposed "crystal separation theory" explaining that the nucleation and growth start on the chill mold wall (inclined cooling slope) and its detachment due to the fluid motion (Fig. 2.13) is responsible for globule primary particles formation. The detachment was claimed to be more efficient by application of vibration [35].

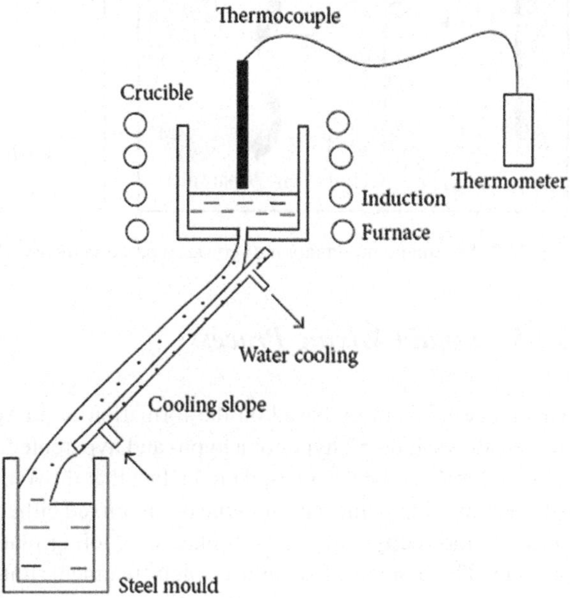

Fig. 2.11 A cooling slope casting system [31]

Fig. 2.12 Schematic of unequal diameter twin roll caster (adapted from [36])

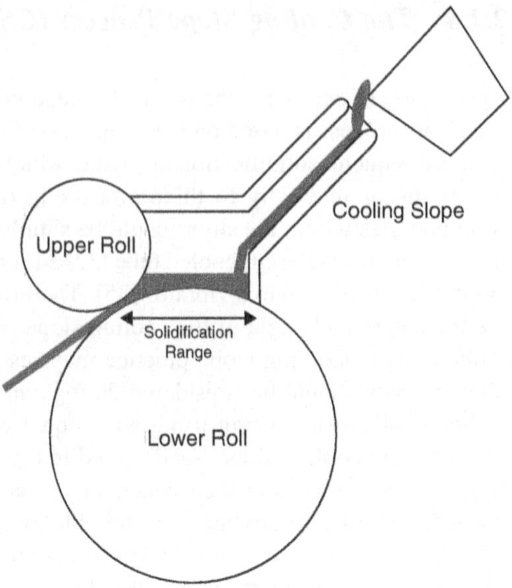

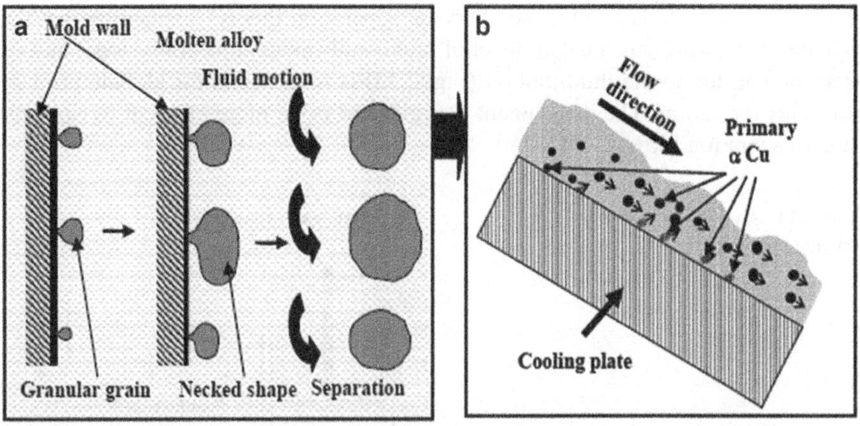

Fig. 2.13 Schematic presentation of crystal separation theory [37]

2.1.5 *Liquid Mixing Process*

This process is simply based on the formation of a new alloy from mixing of two molten alloys, either 2 hypo- or a hypo- and hypereutectic Al–Si alloy [38, 39]. The rational for this process is based on the fact that if two or more molten alloys having different melting points and superheats are mixed either directly within an insulated vessel or indirectly with a first contact on cooling plates, they create nuclei in the mixture. The amount of superheat, melt treatment, holding time within the vessel,

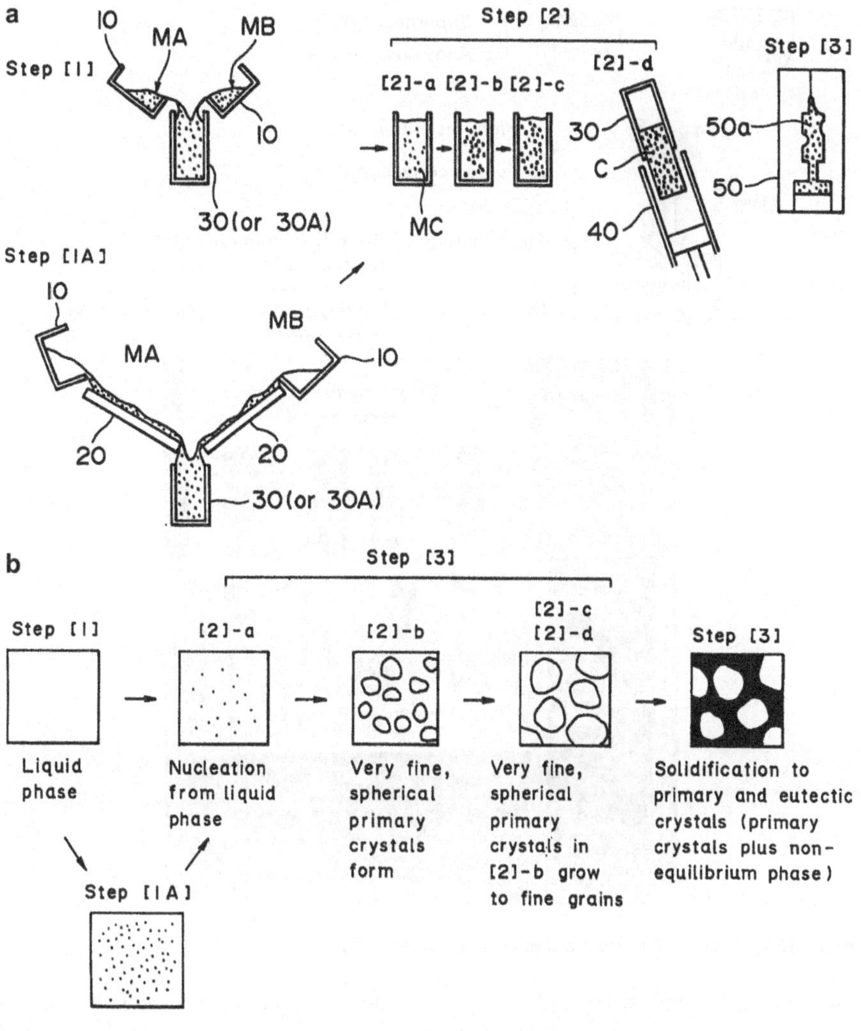

Fig. 2.14 (**a**) Schematic representation of one mixing process; (**b**) a diagrammatic representation showing the metallographic structure of a shaped part (10) ladle; (20) cooling jig; (30) ceramic container; (40) injection sleeve; and (50) mold [30]

weight ratio of two molten alloys, and the mixing method are the main factors that must be fully controlled, see Fig. 2.14 [30].

Another version of the liquid mixing was introduced by Apelian et al. [40, 41] dubbed "Continuous Rheoconversion Process, CRP". In this process, two melts (either from the same alloy or two different alloys) are mixed within a reactor. The reactor provides heat extraction, copious nucleation, and convection during initial stage of solidification leading to the formation of thixotropic structure (Fig. 2.15). It was claimed that CRP is a flexible process and could be used for thixo and

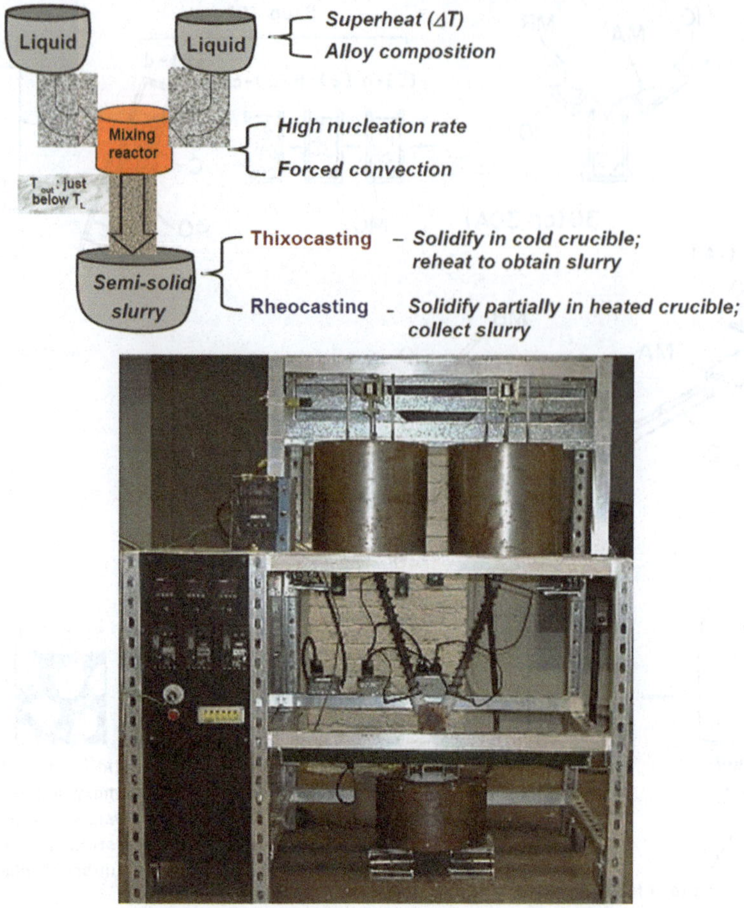

Fig. 2.15 Schematic and a picture of CRP apparatus [40]

rheo-routes. For industrial applications, the reactor was optimized and simplified in such a way that only one melt is involved with a cooling system incorporated into the reactor (Fig. 2.16). The CRP reactor could be mounted above the shot sleeve of die-casting machine (the melt could be pumped from the holding furnace to the reactor). CRP can be used to produce wrought aluminum alloys via shaped casting to achieve high strength of wrought alloys in cast parts [40, 41].

2.1.6 Semi-Solid Rheocasting SSR™ Casting

This is a process proposed initially by MIT [42–45]. It has been claimed to produce fine SSM structures with no entrapped liquid. The process, Fig. 2.17, consists of the following steps:

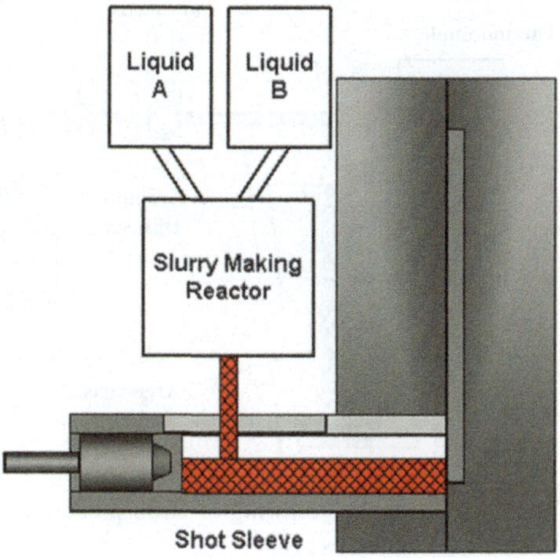

Fig. 2.16 Schematic of a CRP within a diecasting machine (adapted from [41])

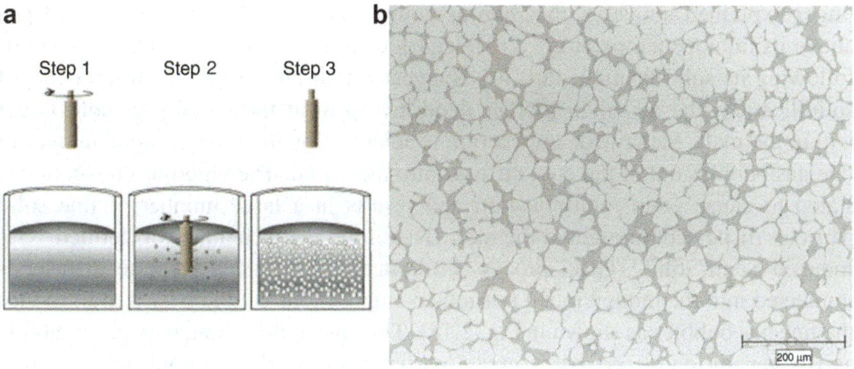

Fig. 2.17 (**a**) Sequence of the new MIT process and (**b**) micrograph for 356 cast alloy [45]

- Melt agitation over a very short period of time at temperatures within mushy zone close to liquidus using a so-called Spinning Cold Finger or Cold Diffuser such as copper or graphite rod.
- Localized heat extraction.
- A short period of slow cooling or isothermal holding within the mushy zone.

It is claimed the combined stirring/cooling action around the liquidus temperature of an Al–Si alloy results in copious nucleation of primary α-Al particles in the melt. The SSR process is shown to be compatible to both low and high fraction solid castings. It is worth noting that the low fraction solid slurry could be handled like a liquid and therefore there is no need for modification of the conventional diecasting machine, for example, the stroke of the machine does not need to be lengthened. Figure 2.17b shows a typical SSM structure produced by this method.

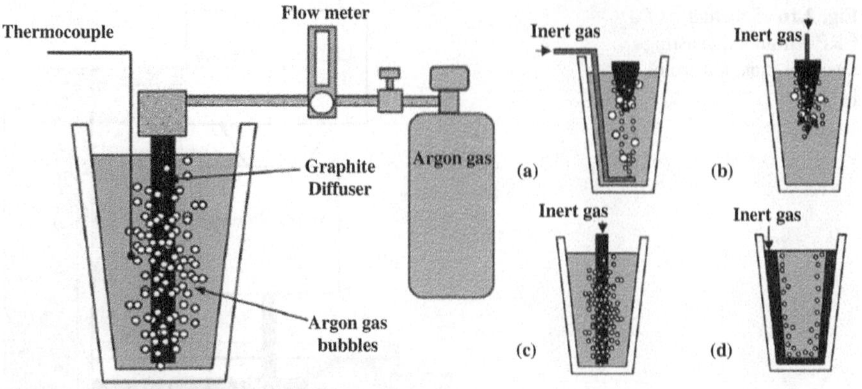

Fig. 2.18 Schematic drawing of GISS process [46] and examples of possible configurations to introduce gas bubbles [47]

Wannasin et al. [46] have shown that if a combination of localized chill (cold finger explained above) with vigorous convection is applied to a melt held just above its liquidus temperature; a nondendritic structure can form in a few seconds following its solidification start. It is suggested that during the initial moments of solidification, the so-called mother dendrites form at the vicinity of cold finger within the melt. Meanwhile convection is achieved by the flow of large number of fine gas bubbles out of the cold diffuser into the liquid. The vigorous convection is stated to cause grain multiplication and results in a large number of fine solid particles dispersed throughout the melt. Therefore, the technique was termed "Gas Induced Semi-Solid", GISS process [46]. In this process, a graphite diffuser is immersed into the molten metal (various configurations could be used to introduce the fine gas bubbles as shown in Fig. 2.18). The convection created by gas bubbling assists generation of secondary nuclei particles and leads to a nondendritic structure, Fig. 2.19. The first step in this process is to determine the liquidus temperature and then select process temperature of a few degrees above its liquidus. The rheocast time is about 5–30 s. Figure 2.19 shows an example of a rheocast Al7% Si billet structure [48]. This technique has been developed and commercialized in Thailand [49].

Based on the SSR casting, other similar processes were also reported. In one case [50], molten metal is directed toward a hollow vertical rotating stainless steel rod having its temperature controlled by a cooling system. The rotating speed is adjusted to achieve the optimum structure. The rotating prepared slurry is then poured into a graphite crucible (Fig. 2.20).

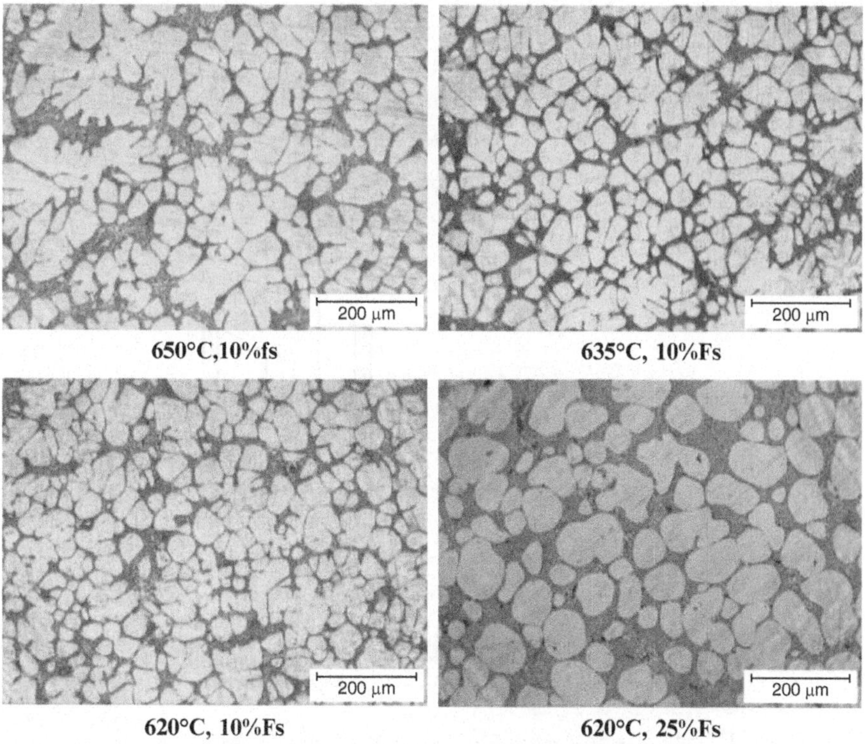

Rheocasting temperature (°C)	Rheocasting time (s)	Holding time (s)	Fs before casting (%)
650	25	-	10
635	15	-	10
620	10	-	10
620	10	70	25

Fig. 2.19 Microstructures of semi-solid cast for Al7Si–0.32 Mg0.46Fe alloy by the GISS process using different temperature and solid fractions [48]

2.1.7 RheoMetal Process (Also Called Rapid Slurry Forming Process, RSF)

The RheoMetal process is based on the enthalpy exchanges between alloy systems in controlling the final solid fraction of the slurry. In this process, at least two predetermined alloy systems having different heat content (normally two, one liquid, and one solid) are mixed together through stirring to produce a new alloy with the required enthalpy and solid fraction. This is essentially different from most other SSM processes where heat extraction (outer cooling) and temperature control of the melt are the key issues for controlling the solid fraction of the slurry.

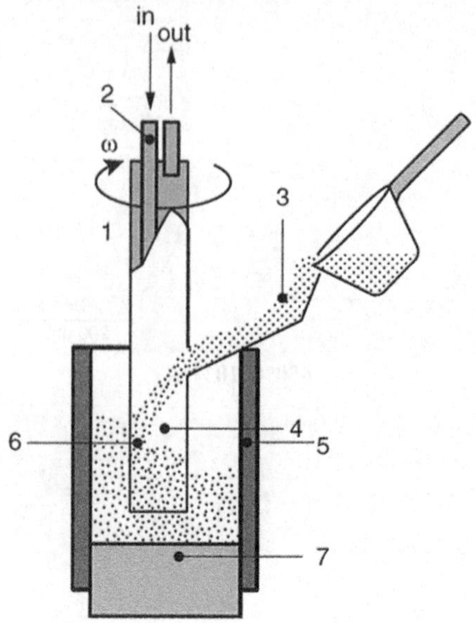

Fig. 2.20 Rotating rod equipment; 1. connecting to electric motor, 2. cooling system, 3. pouring system, 4. self-rotating rod, 5. temperature control system of slurry prepared chamber, 6. melt film, and 7. semi-solid slurry (reproduced from [50])

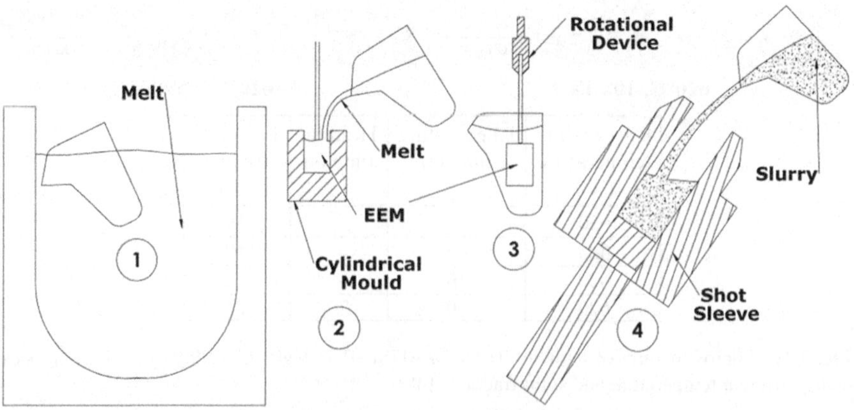

Fig. 2.21 Schematic diagram of the RSF process (reproduced from [51]) (1) extract the melt, (2) pour into a mold to make EEM, (3) prepare a slurry by means of stirring and EEM melting, and (4) pour the slurry into the shot chamber

Schematic presentation of the process is given in Fig. 2.21 [51]. A liquid metal is poured into an insulated vessel or container. Then a predetermined amount of solid alloy, which is attached to a stirrer, is added to the melt and start stirring. The solid material is comparatively cold having a relatively low enthalpy. Accordingly, it absorbs the heat from the melt and exchanges enthalpy with melt. Gradually it will partially or totally melt away and at the end will mix evenly with the original melt to form a new alloy system with desired enthalpy and solid fraction. The added solid

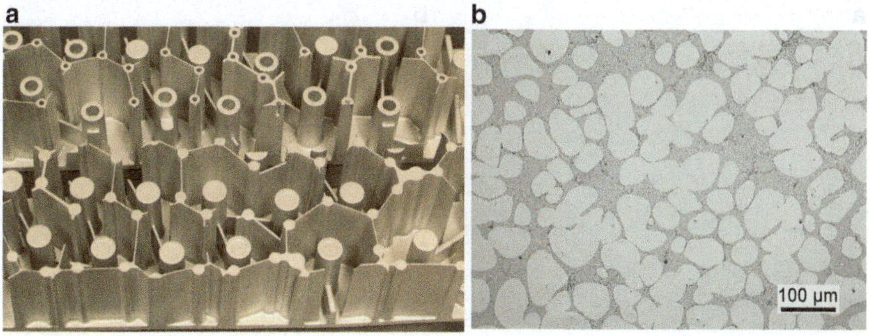

Fig. 2.22 (**a**) A section of cavity filter for telecom industry with wall thickness less than 0.5 mm, AlSi6Cu2.5 [52]. (**b**) microstructure of A356 slurry quenched in water, RSF process [53]

material is also called "Enthalpy Exchange Material", that is, EEM. Evidently, the starting temperature and composition of the melt and the EEM as well as the added percentage of the solid alloy will determine the final solid fraction of the slurry. The slurry can be directly used for rheocasting, or cast as feedstock material which could be reheated into a semi-solid state for thixoforming. This process is commercialized by RheoMetal AB in Sweden; an example is presented in Fig. 2.22a [51–54]. It was claimed that the grains obtain from this process for A356 alloy have globular morphology of about 50–100 μm depending on different parameter settings, Fig. 2.22b [53].

2.1.8 Ultrasonic Treatment

It is well established that application of ultrasonic treatment to a liquid metal at a starting temperature above its liquidus can effectively produce a fine and nondendritic microstructure, which is suitable for subsequent reheating and thixoforming operations. According to the literature, introduction of high power ultrasonic into a liquid can lead to two basic physical phenomena [55]: cavitations and acoustic streaming.

Cavitation includes the formation, growth, pulsation, and collapsing of tiny bubbles within the liquid. It is stated that the compression rate of these unsteady bubbles can be so high that hydraulic shock waves could be generated by their collapsing. Primary dendrites are broken by these waves, and therefore producing potential nuclei. The transmission of high-intensity ultrasonic waves also includes the initiation of steady state acoustic streaming within the melt. The various kinds of streams results in vigorous mixing and homogenization of the melt [55].

Hydraulic and acoustic streams create shock waves initiating from collapsing of cavitation bubbles in order to break down the dendrites. In addition, acoustic streams generated by ultrasonic treatment will homogeneously distribute these fine solid particles. By application of any means of vibration, structural evolution

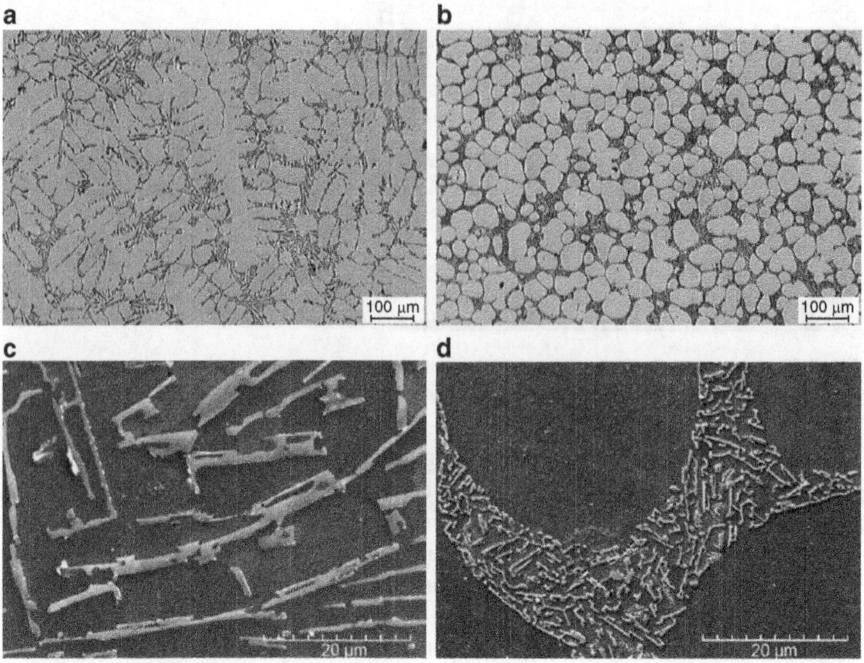

Fig. 2.23 Structural evolution of A356 alloy (**a**) (**c**) without ultrasonic treatment and (**b**) (**d**) with ultrasonic treatment [59]

occurs including grain refinement, suppression of columnar grain structure, increased homogeneity, and reduced segregation [55].

Abramov et al. [55–57] investigated the possibility of obtaining a thixotropic structure by an ultrasonic treatment in Al–Si-based alloys and found it as an effective way of treating major commercial Al–Si alloys. By ultrasonic vibration, most of the silicon particles were broken and strength of the alloy increased. Ultrasonic vibration at 20 kHz was introduced into A356 alloy as it was cast into a copper mold at the temperature of 630 °C [58, 59]. It was found that not only a globular/nondendritic microstructure was obtained, but also the morphology of eutectic silicon was altered from a coarse acicular plate-like form when no ultrasonic vibration was used, to a finely dispersed rosette like form when ultrasonic treatment was employed, Fig. 2.23. Other researchers [60–62] had similar findings and emphasized that an optimized combination of process parameters such as treatment time and alloy temperature are key factors. As well, it was found that ultrasonic vibration can not only refine the primary α-Al and primary Si in Al–Si alloys, but also could refine the intermetallic compounds such as Fe containing intermetallics.

In a different commercial trial, billets were prepared by mechanical vibration in the ultrasound range between 10 and 100 kHz [63] (note that the produced billets could be used as thixo feedstock). As shown in Fig. 2.24, a vibration producer is

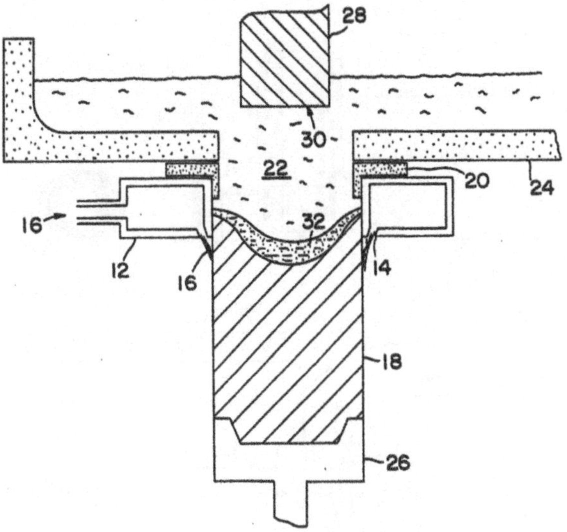

Fig. 2.24 Exemplary embodiment of the vibration continuous cast [63]. 12) Cooled die, 14) Gap, 16) Coolant, 20) Fire-resistant insert, 26) Start up base, 28) Vibration producer, 32) Mushy zone

immersed into the liquid metal exactly above the billet production unit which eventually results to the formation of a fine globular structure upon solidification. It is worth noting that this stirring effect is maintained up to the end of solidification in contrast with the electromagnetic stirring which is dependent on the viscosity of the solidifying metal. Experiments have shown that in 356 alloy, the grain size was reduced almost by 50 % [63]. Pola et al. [60] successfully used the ultrasound system effectively in production of 7 in. A356 billet in a direct chill pilot plant.

2.1.9 Hitachi Process

In this process, the liquid is directly transferred into the shot sleeve using an electromagnetic pump [64, 65], Fig. 2.25. The application of electromagnetic field over the water cooled shot sleeve causes the stirring and continuous cooling of the melt within the shot sleeve to produce the required SSM slurry before injection. As a result of cooling from shot sleeve as well as induction heating from electromagnetic field, it was claimed that the temperature of the liquid in the shot sleeve decrease uniformly. This process is stated to have the following benefits:

- By direct casting from liquid state, it offers the rheocasting advantages.
- By EMS application, uniform temperature distribution in the shot sleeve is guaranteed.
- Electromagnetic pumping results in oxide inclusions reduction and pouring weight is accurate.

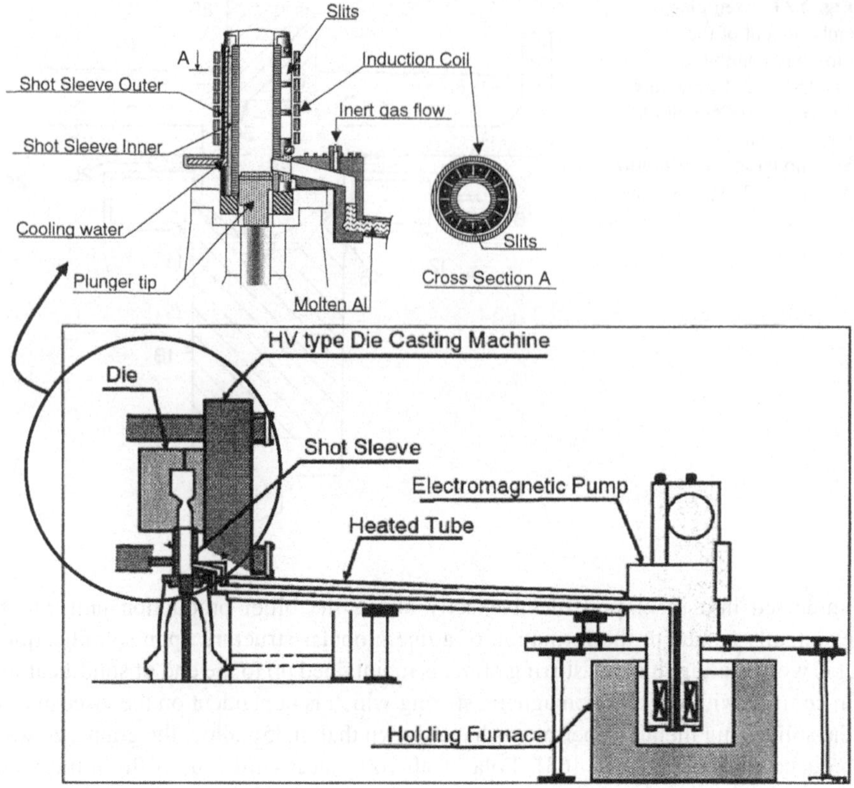

Fig. 2.25 Hitachi process with focusing the electromagnetic stirring in the shot sleeve [65]

- Due to less oxide contamination and protected atmosphere as well as vigorous agitation, mechanical properties of the casting prepared by this process are better than those of squeeze casting and are almost the same as those of conventional SSM process.

It seems that this process has been used by Hitachi Metals Inc on a limited basis [65].

2.1.10 Low Pouring Temperature or Superheat Casting

It has long been realized by foundrymen and ingot casters that low pouring temperature not only causes the formation of equiaxed grains, but also reduces casting defects, such as segregation and gas and shrinkage porosity. In semi-solid processing, the lower pouring temperature or in general the reduced temperature provides longer tool life, better dimensional tolerance due to less expansion and

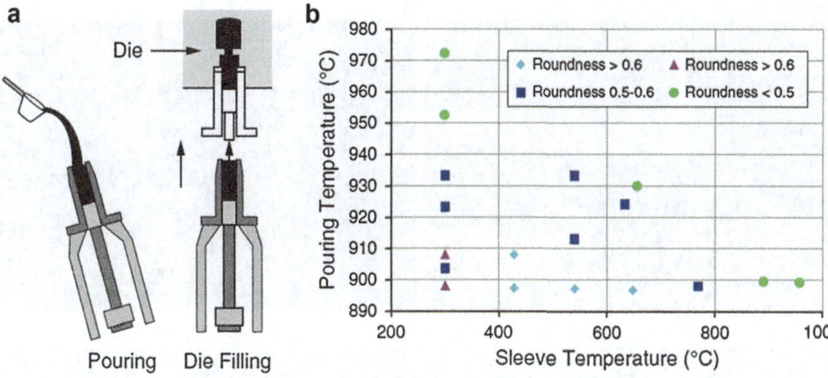

Fig. 2.26 (**a**) Schematic of casting method and (**b**) effects of pouring and sleeve temperature on roundness factor, A356 alloy (reproduced from [66])

shrinkage, and related defects. Convection during mold filling is an important factor as solidification begins. For low pouring temperatures, the natural convection due to pouring redistributes the nuclei formed, whereas at high pouring temperatures, the natural convection subsides before solidification even begins and all the formed nuclei remelt leaving the solidification to start with a relatively motionless melt.

Shibata et al. [66] as well as the authors [67] have shown that spherical primary solid phase can be easily obtained by low pouring temperature method. As shown in Fig. 2.26, the melt was poured into the shot sleeve and injected into the die cavity after a defined cooling period. The effect of pouring and sleeve temperatures on the roundness (roundness was defined as ratio of alpha particles over imaginary circle having the same circumference) of α-Al particles is shown in Fig. 2.26b. Higher roundness of primary α-Al particles were obtained at the lower pouring/sleeve temperatures. This phenomenon appears to be based on equiaxed grain formation initiated by instantaneous nucleation within the bulk liquid, which could be considered as similar to copious nucleation mechanism [8].

Wang et al. [68, 69] produced AlSi7Mg0.35 alloy with different pouring temperatures in a stepped die using a vertical injection squeeze casting machine. In their experiments, pouring temperature was decreased from 725 to 625 °C. With decreasing pouring temperature, the as-cast microstructures changed from coarse dendritic (725 °C) to fine rosette-like (625 °C). Also with reheating the billets at 580 °C and quenching, particle size decreased with decreasing pouring temperature (Fig. 2.27). Table 2.1 summarizes their results. The variation of shear stress with displacement after isothermal holding at 580 °C for 15 min is also reported [69]. The 725 °C cast samples show high shear strength in excess of 50 kPa, whereas for the 675 °C material, the shear strength is reduced significantly to 20 kPa. When the pouring temperature is reduced to 650 °C, the shear resistance of the material is very low at about 5 kPa and the shear stress does not drop significantly after the maximum for 650 °C graph in Fig. 2.28.

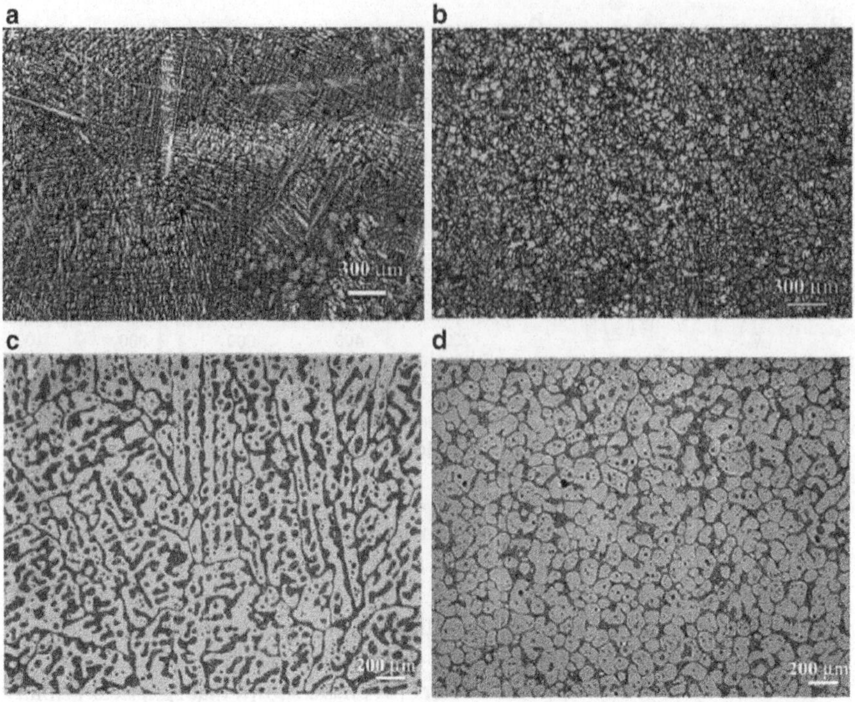

Fig. 2.27 Microstructures of (**a**) as-cast at 725 °C, (**b**) as-cast at 625 °C, (**c**) and (**d**) partially remelted and isothermally held for 15 min from specimens with initial microstructure exhibited in (**a**) and (**b**), respectively [69]

Table 2.1 Summary of the microstructural parameters measured in the as cast and reheated billets at 580 °C for 15 min [69]

As cast microstructure			Reheated microstructure	
Casting condition	Structure	Grain size (μm)	Particle size (μm)	Morphology
725 °C	Coarse-grained, dendritic	900	310	Solid network
675 °C	Medium-grained, dendritic	350	160	Irregular globular
650 °C	Fine-grained, dendritic	200	102	Spherical globular
625 °C	Fine rosette-like	180	100	Spherical globular

Interestingly, it was found that pouring technique influences the grain size which was shown for the first time by Chalmers [8, p. 267], Dahle et al. [71] repeated the practice by using an experimental arrangement as shown in Fig. 2.29. A thin cylinder of metal screen was placed into a mold and then molten metal was poured both from the wall and center of the mold. One can see that the grain size in the center is much finer when liquid is poured from the cold wall because of the large amount of nuclei produced in the first contacts.

Fig. 2.28 Variation of shear stress with displacement for the 356 alloy cast at different temperatures and isothermally held at 580 °C (reproduced from [70])

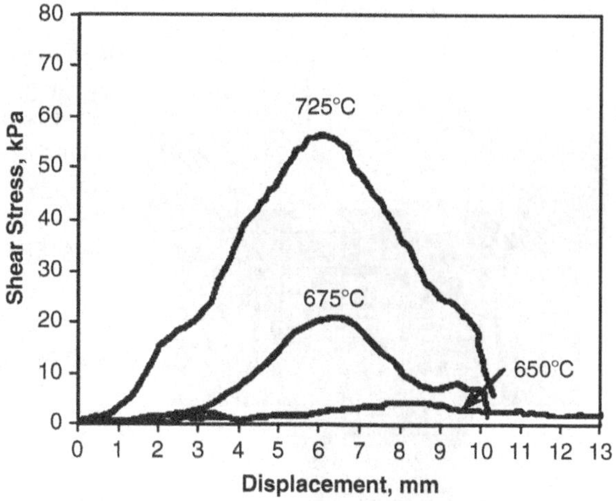

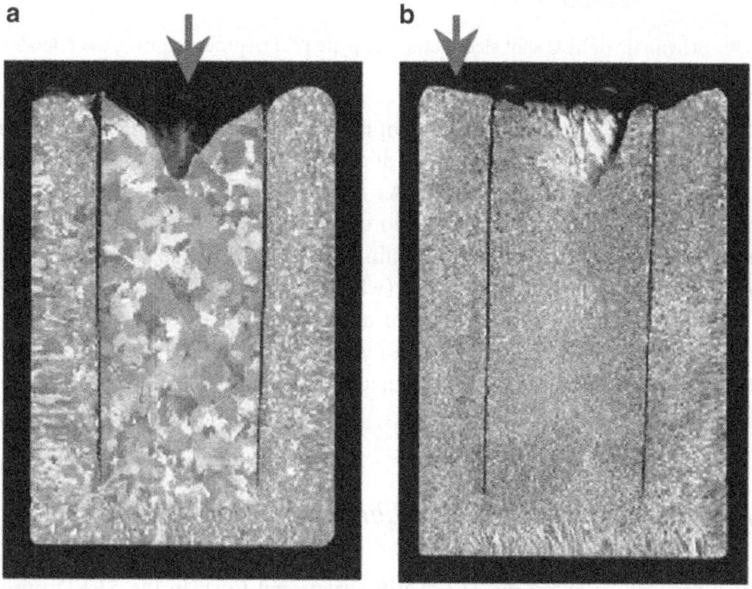

Fig. 2.29 Macrostructures produced by different pouring techniques (a) central pouring and (b) wall pouring (each casting had 50 mm width, 70 mm height) [71]

2.1.11 Sub Liquidus Casting (SLC®)

The SLC® process considered another simple slurry production approach to SSM processing introduced by THT Presses, Inc., in 2000. This process involves pouring grain refined and/or modified melt at very low or near liquidus temperatures into a

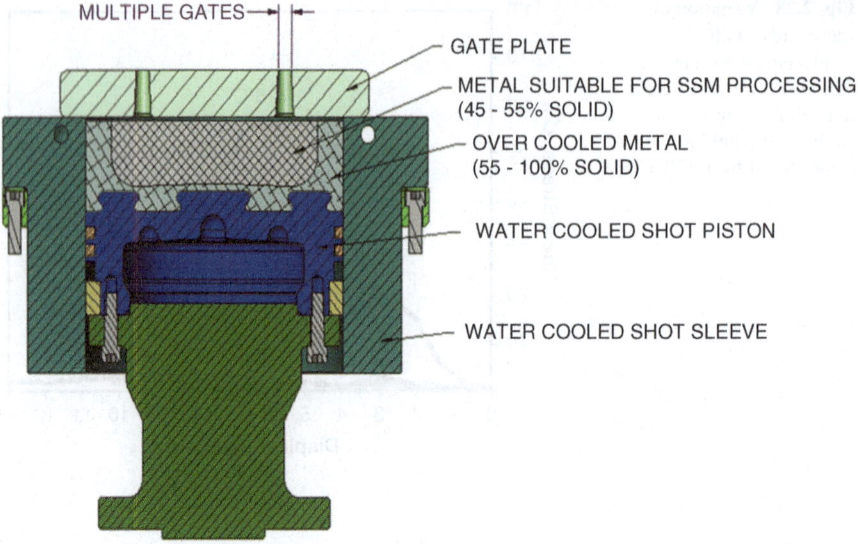

Fig. 2.30 Schematic of SLC shot sleeve and gate plate [72] (reproduced from the Foundry Trade Journal)

vertical shot sleeve, controlled cooling of metal in the shot sleeve and injection into the die cavity, following the formation of desired fraction solid and slurry ripening within the cold chamber (Fig. 2.30). As shown, slurry is formed within the shot sleeve. A gate plate transfers slurry into the die cavity. The large diameter of the shot sleeve results in the use of melt within central part of the sleeve, avoiding the use of material closer to the mold wall (which has the higher solid fraction). It has been reported to yield a globule size of approximately 75 μm. Unlike other SoD routes, the SLC® process requires no slug (billet) preparation equipment or processing time outside of the casting machine [72–76].

2.1.12 Swirled Enthalpy Equilibration Device (SEED)

This is a process patented by ALCAN International [77]. In the SEED process, a superheated alloy is poured into a cylindrical mold. The mold is then rotated off-center at certain RPM. Duration of this stage depends on the dimension of the mold and the mass of charge, but typically between 30 and 60 s was reported [78]. In the next step, the swirling motion is stopped, and after a brief pause of 5–10 s the bottom plug is removed to allow some remaining liquid to drain. The degree of superheat and the swirling and drainage time and speed are selected in such a way as to allow a solid fraction of 0.3–0.4 to form before drainage. This is based on heat exchange between the mold and molten alloy. After a specific time between 30 and 45 s, the prepared billet is unloaded and transferred into a high pressure diecasting machine to fabricate the finished products.

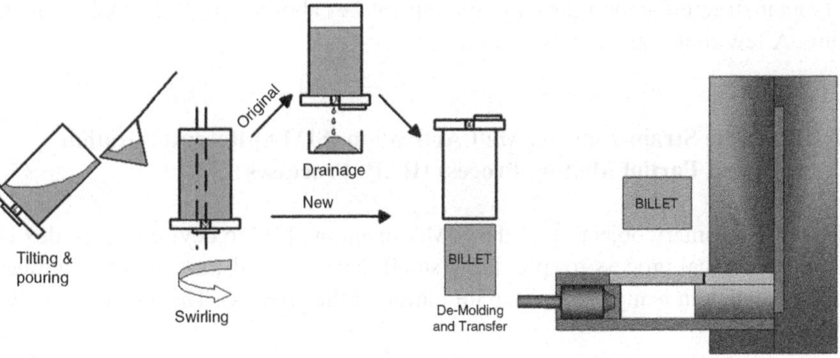

Fig. 2.31 Preparation procedure of slug in SEED process (note the new and original ones)

Fig. 2.32 Pre industrial SEED unit at the pilot plant as well as various sizes of crucible and slug (Courtesy of ttp Publications, Inc., [80])

Based on the feedback from industrial users, the process was further developed to have two alternative modes, with and without the drainage step; both reported to have similar structure and mechanical properties [79] (Fig. 2.31). It has been also claimed that the process is capable of handling a range of foundry and wrought alloy compositions such as 206, 319, 356/357, AA6061, and AA6082. In addition, the process can produce different slug dimensions and weights up to about 18 Kg, Fig. 2.32 [80]. At the present time, this process is being commercialized and specific information on the process is available in the open literature [78–82].

2.2 Thixo-Route Techniques

2.2.1 Thermomechanical Treatments

In the conventional deformation processes such as rolling and extrusion, dendritic microstructure is deformed and distorted. This heavily deformed structure is an excellent feedstock for thixo billet and could be easily changed to equiaxed and globular structure upon reheating to temperatures above recrystallization temperature. A few routes are addressed here.

2.2.1.1 The Strain-Induced Melt Activation (SIMA)/Recrystallization and Partial Melting Process (RAP) Processes

It was the primary objective of the SIMA inventors [23] to devise a more flexible and economical process for providing small diameter feedstock for some wrought alloys (which in many cases is the limitation of the process). This process involves (Fig. 2.33):

- Following melting, casting, and cooling to room temperature, the billet is reheated to recrystallization temperature and extruded. Then the billet is quenched and further cold worked.
- Reheating the cold worked billet to the semi-solid temperature range. In this step, following partial remelting, an extremely fine, uniform, and nondendritic spherical microstructure is generated (Fig. 2.34).
- Thixoforming the billet.

There is also another process similar to the SIMA referred as "Recrystallization And Partial melting", RAP process. In SIMA process, hot working is performed above the recrystallization temperature, whereas RAP process has a warm working step, Fig. 2.33.

If the feed stock is sufficiently deformed and recrystallized, reheating to temperatures above solidus will generate partial remelting and formation of ideal slurry composed of rounded solid primary particles within a liquid matrix. The initial deformation may be carried out above the recrystallization temperature (hot working) followed by cold working at room temperature [23, 25], or alternatively below

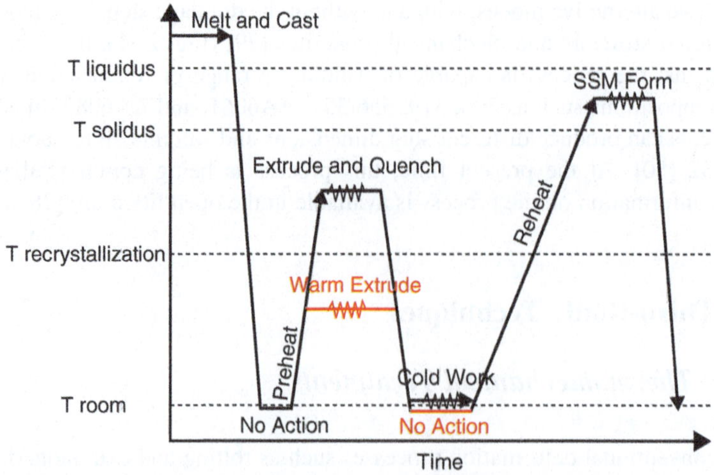

Fig. 2.33 SIMA and RAP processes (the difference for RAP is in *red*) (adapted from [83])

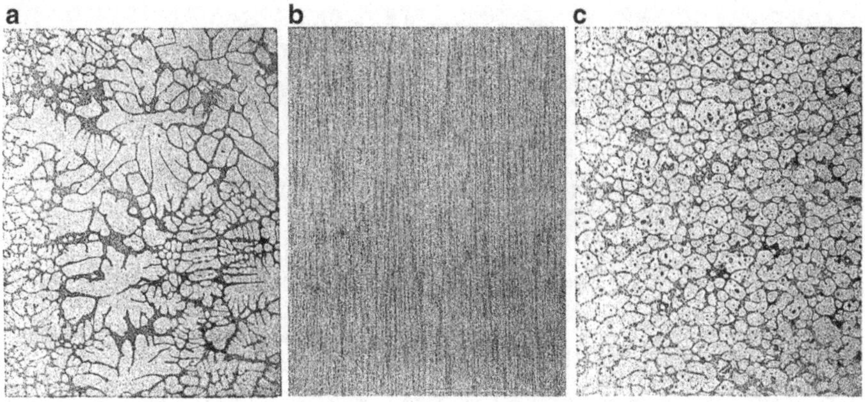

Fig. 2.34 SIMA processing route for semi-solid forming, 357 alloy (**a**) direct chill cast with 6 in. diameter, (**b**) longitudinal section of the extruded stretched bar, and (**c**) cross section of the reheated and quenched sample [23]

the recrystallization temperature (warm working) to ensure the maximum strain hardening, as also suggested by Kirkwood and coworkers [2, 84].

Sufficient cold deformation induces recrystallization during subsequent heating when strain is above a critical amount. Recrystallization leads to the formation of a large quantity of high angle grain boundaries, which are able to melt readily when heated above solidus (partial remelting). Therefore, greater degree of cold working results in the smaller grain size, that is, finer globules. This is shown in Fig. 2.35 for Al7%Si alloy cold rolled with various thickness reductions [85] (it is important to indicate that the strain may not be uniform across the billet).

2.2.1.2 Equal Channel Angular Pressing (ECAP)

Severe plastic deformation (SPD) techniques have been proved to be capable enough to produce thixo feedstock. In this method, a high plastic strain is applied to a solid feedstock without any significant dimensional change [86]. Equal Channel Angular Pressing (ECAP) gained attention as a viable source for thixo feedstock [87–89]. In this process, the SSM feedstock is pressed directly in two channels having a constant cross section which is bent at an angle (such as 90 or 120°). As the cross section of the sample remains constant, the process can be repeated for many passes till the optimum structure and properties achieved. The severity of plastic deformation is a key in the formation of globular structure. The more passes result in more deformation; however, it should be noted that there should be a balance between passes, that is, deformation, and reheating temperature and time to prevent the grain growth after reheating. An example of A356 alloy prepared with various processing methods is shown in Fig. 2.36 [87] which the ECAP sample has been proved to have the smallest grain and globule sizes and highest sphericity in the semi-solid state.

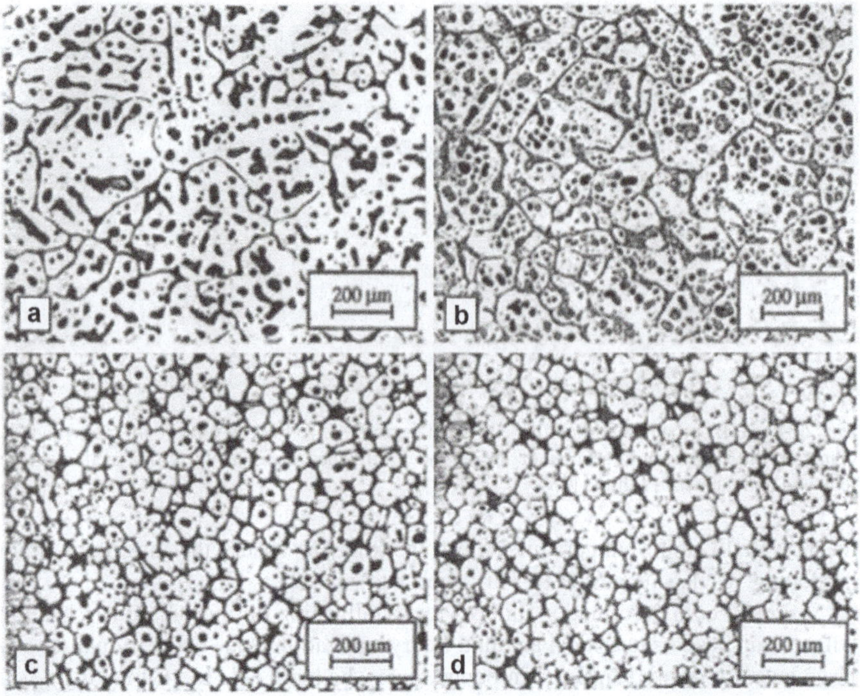

Fig. 2.35 Microstructure of conventionally chill cast AlSi7Mg0.6 cold rolled (**a**) 0 %, (**b**) 10 %, (**c**) 25 %, and (**d**) 40 % prior to isothermally held at 580 °C for 30 s for partial remelting. All samples were quenched to room temperature from 580 °C [85]

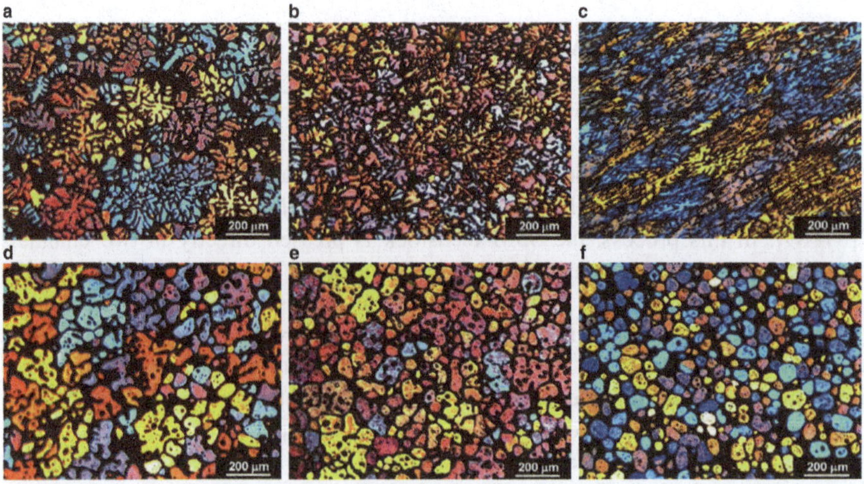

Fig. 2.36 Images of A356 alloy produced by (**a**) water-cooled mold cast, (**b**) water-cooled mold cast and enhanced by refining and electromagnetic stirring, and (**c**) water-cooled mold casting route with a single pass by ECAP. Images of partial melting at 580 °C, (**a**), (**b**), and (**c**) correspond to (**d**), (**e**), and (**f**) [87]

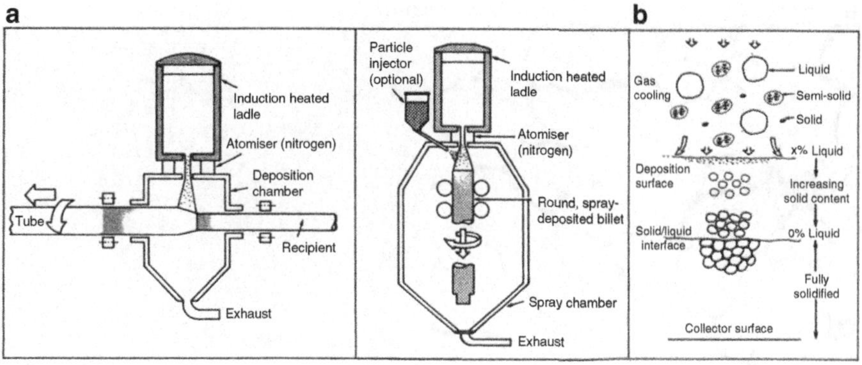

Fig. 2.37 (**a**) Two different arrangements for tube and billet production by Osprey process, (**b**) solidification mechanism of Osprey process [90] (reprinted by permission of Taylor & Francis Ltd)

2.2.2 Spray Casting (Osprey)

The process with two different billet manufacturing routes is shown in Fig. 2.37. Briefly, the specified alloy is induction melted in a crucible located on the top of the spray chamber (an option is linking the melting unit directly to a tundish). The liquid is directed through a nozzle into a gas atomizer with various flow rates for different alloys. This results in atomization of the liquid stream into different droplet sizes. The liquid droplets are cooled by the atomizing gas and subsequently interact with the substrate.

According to Mathur et al. [91], there are two stages for the process: One is when the droplets are in-flight and just interact with the atomizing gas, and the other is when they impact and interact with the substrate. In the first stage, the generated droplets are classified as fully liquid, semi-solid, or completely solid. Droplets are collected on a substrate and solidified to form a shaped casting as depicted in Fig. 2.37b. Following atomization, the droplets impinge, consolidate, and solidify on the substrate to form a homogenous structure.

In the case of aluminum alloys, Al4%Cu was chosen for spray forming since Al–Cu alloying systems are prone to micro/macrosegregation and porosity formation. Billets of about 30 kg weights were formed, and Fig. 2.38 compares structure produced by continuous casting and spray forming. Spray formed material has almost globular structure with homogenous dispersed precipitates, whereas dendrite formation is quite obvious in continuous cast sample. The distribution of the main elements Cu, Si, Mn, Mg, and Fe of the sprayed deposits was analyzed in axial and radial direction. No gradient in radial and axial direction has been reported for copper element [92].

The Osprey process has been applied to a wide range of alloys including aluminum alloys and its composites, high temperature alloys such as high speed steel, superalloys, and copper alloys [90–92].

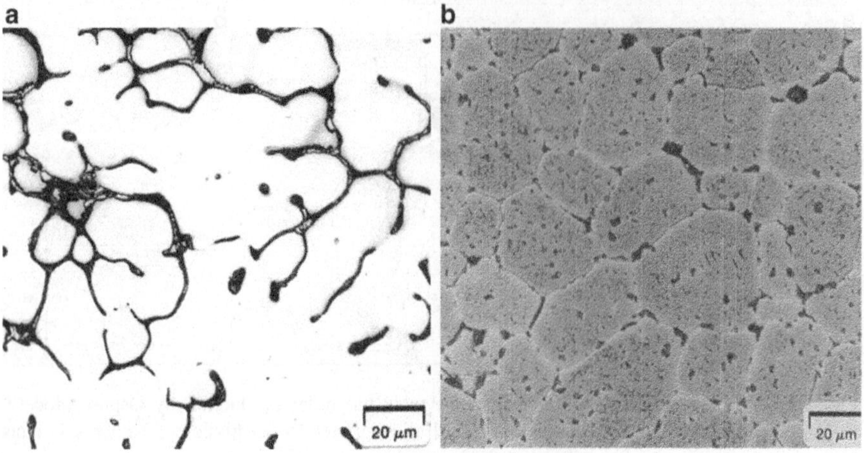

Fig. 2.38 Al4%Cu (**a**) continuous cast and (**b**) spray formed [92]

2.2.3 Liquidus or Low Superheat Casting

As described in Sect. 2.1.10, this process could be a potential source for the thixocasting process. With low pouring temperature, the resulting microstructure is usually fine and nondendritic. Partial remelting and isothermal holding of the sample will generate globular structure suitable for thixoforming. This technique has been reported for both cast and wrought aluminum alloys [2, 17].

Wang [68] performed experiments on AlSi7Mg0.35 billets prepared with different pouring temperatures. The produced billets were reheated to 580 °C by an induction furnace and injected into a stepped die using a vertical injection squeeze casting machine (plate 98 mm width, steps 20,15,and 10, and 5 mm thick and each one 30 mm long, Fig. 2.39a). It was shown that the casting made from billets with pouring temperature of 725 °C only filled half of the die cavity, the first two and half steps. Also X-radiography revealed presence of numerous pores within the cast piece. Lowering the pouring temperature of billets to 650 °C resulted in a significant improvement to the filling ability and the mold was filled completely except at one top corner. The internal defects were also significantly reduced. A further decrease in billet pouring temperature resulted in a completely filled casting and no defects were observed on cast piece, Fig. 2.39c–e.

This technique could be used for industrial applications, but the major obstacles may arise from difficulties related to the accuracy and uniformity of temperature control, and consistency and uniformity of resulting microstructure in large-scale production [17].

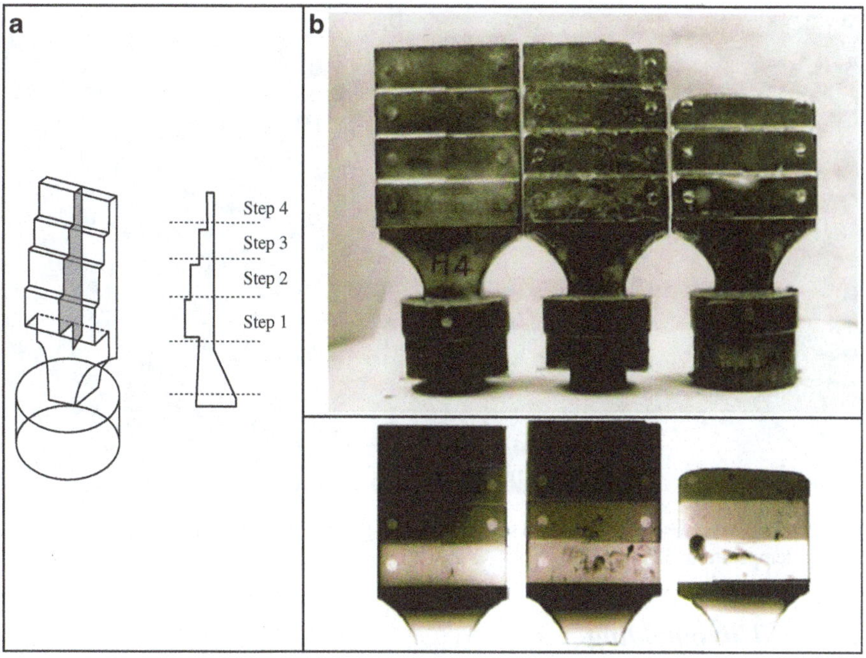

Fig. 2.39 (**a**) Schematic of stepped die and (**b**) cast pieces from different pouring temperatures (*above*) and their X-radiographs (*below*). Pouring temperatures are 725 °C (*right*), 650 °C (*middle*), and 625 °C (*left*) (mold was preheated to 150 °C. Billets were isothermally held at 580 °C for 5 min prior to casting) [68]

2.2.4 *Chemical Grain Refinement*

Chemical grain refinement is a routine practice in batch and continuous casting of aluminum alloys. This technique has also been considered for feedstock production for thixocasting [93–96]. In this method, globular SSM slurries can be obtained by simply reheating grain refined billets. Billets could be made or purchased from cast houses or primary producers. It is claimed that in this technique, the morphology of the primary α-Al particles is still globular or rosette-like, but the fraction of liquid entrapped within the primary particles is significantly larger than EMS slurries [97].

In another study, it was shown that the grain refinement approach is more flexible and cost-effective compared to the EMS technique [98]. It was stated that the type of grain refiner is a critical parameter in the establishment of the final microstructure and by using a new trade mark grain refiner, namely SiBloy (basically boron-based refiner), concluded that during reheating of refined billets, the B-refined billets have four times less entrapped liquid compared to TiB_2-refined billets (Fig. 2.40). Authors confirmed this finding and the details will be discussed later in Chap. 6.

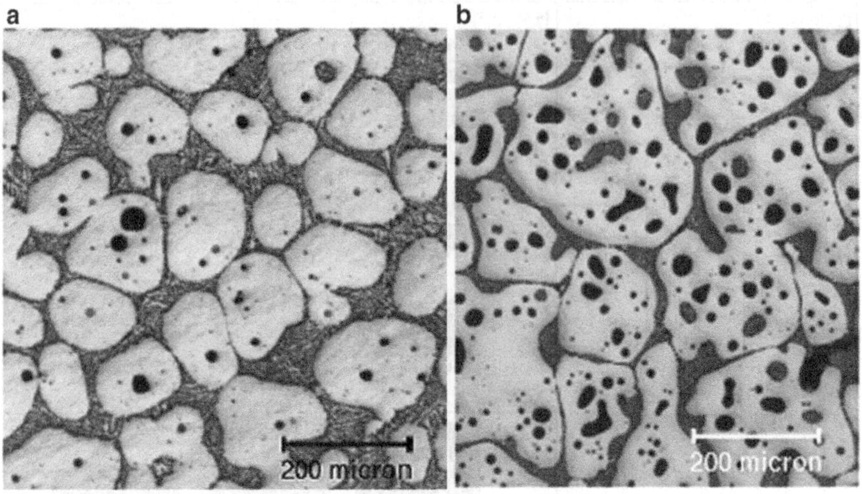

Fig. 2.40 A comparison of semi-solid structures of (**a**) B-refined billets and (**b**) TiB_2 grain refined (reheat temperature: 585 °C) [98]

2.2.5 *Thixomolding*

Thixomolding is a similar process to plastic injection molding which was developed especially for magnesium alloys (Fig. 2.41). The raw material for the process is neither liquid (as in rheocasting) nor solid billet (as in thixocasting), but rather solid particulates in a form of pellets, chips, or granules with a typical size of 2–5 mm obtained by machining, atomization, or other comminution methods [99]. During processing, alloy particulates are fed into a heated barrel of the injection molding system where they are partially melted, transformed into thixotropic slurry, and finally injected into a mold cavity. Both the mechanically comminuted chips and rapidly solidified granules possess unique microstructural features that allow them to transform into thixotropic slurry under the sole influence of heat. Thus, in contrast to rheomolding, no shear is necessary [100].

The core of the injection system is an Archimedean screw which performs both rotary and translational movements. To prevent the highly reactive magnesium feedstock from oxidation, an argon atmosphere is maintained within the machine barrel [101]. There are several processing differences between injection molding and diecasting. First, as opposed to furnace-environment melting in diecasting, during injection molding the slurry preparation is more complex and is controlled by temperature distribution along the barrel length and to lesser extent by the screw rotation. Moreover, due to injection of semi-solid slurry instead of overheated melt, there is a lower heat impact on the mold which can withstand higher number of shots than tools used in diecasting. Injection speed during injection molding depends additionally on the solid fraction and may be lower than that required in diecasting.

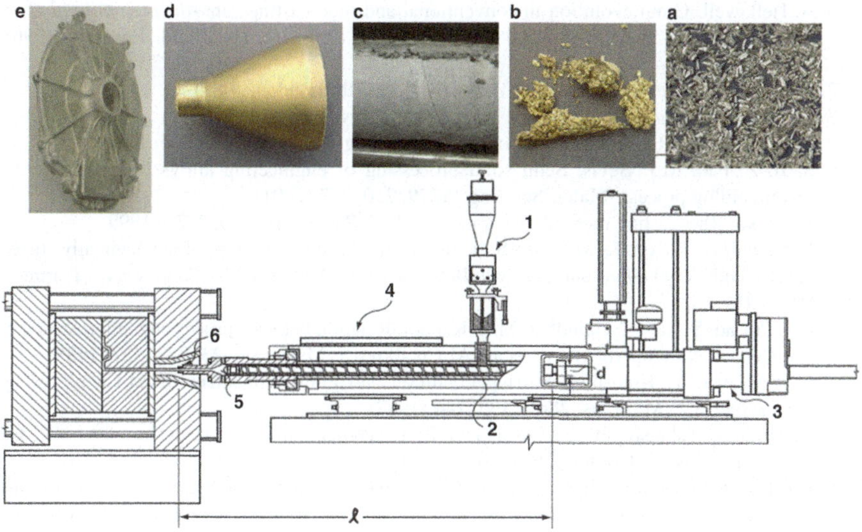

Fig. 2.41 The flow path (l) of magnesium alloy during thixomolding: 1—feeding system; 2—screw; 3—motor; 4—barrel along with heaters, 5—nozzle; and 6—mold. Feedstock transformation (**a**) comminuted chips; (**b**) initial stage of processing; (**c**) alloy from end screw; (**d**) alloy from nozzle area; and (**e**) final product [99]

In addition to general benefits associated with semi-solid processing, there are advantages unique to thixomolding, including environmental friendliness with no SF_6 requirement, no melt loss due to closed processing system, an ability to manufacture thinner parts with more complex shapes, and the capability to apply advanced techniques of slurry distribution to the mold, including hot runners [99]. However, despite substantial progress made during last decades, the technology still requires significant development, especially on the hardware side. While magnesium exhibits high affinity to oxygen, at temperatures of semi-solid or liquid states it is also highly corrosive toward materials it contacts, thus creating challenges on machinery components [102]. Due to processing nature of thixomolding, these thermal and corrosive challenges are particularly difficult to overcome.

References

1. S. Nafisi, R. Ghomashchi, Semi-solid metal processing routes: an overview. J. Canad. Metall. Quart. **44**(3), 289–303 (2005)
2. D.H. Kirkwood, Semi-solid metal processing. Int. Mater. Rev. **39**(5), 173–189 (1994)
3. M.C. Flemings, Behavior of metal alloys in the semi-solid state. Metal. Trans. A **22A**, 952–981 (1991)
4. M.C. Flemings, *Solidification Processing* (McGraw-Hill, New York, 1974)
5. R.D. Doherty, H.I. Lee, E.A. Feest, Microstructure of stir-cast metals. Mater. Sci. Eng. **A65**, 181–189 (1984)

6. A. Hellawell, Grain evolution in conventional and rheo-castings. in *4th International Conference on Semi-Solid Processing of Alloys and Composites* (Sheffield, England, 1996), 60–65
7. R. Elliot, *Eutectic Solidification Process* (Butterworth, London, 1983)
8. B. Chalmers, *Principles of Solidification* (Wiley, New York, 1964)
9. M.C. Flemings, R.G. Riek, K.P. Young, Rheocasting. Mater. Sci. Eng. **25**, 103–117 (1976)
10. S. Ji, Z. Fan, M.J. Bevis, Semi solid processing of engineering alloys by a twin screw rheomoulding process. Mater. Sci. Eng. **A229**, 210–217 (2001)
11. F. Asuke, Rheocasting method and apparatus, U.S. Patent 5865240, 2 Feb 1999
12. Y. Uetani, H. Takagi, K. Matsuda, S. Ikeno, Semi-continuous casting of mechanically stirred A2014 and A390 aluminum alloy billets. in *Light Metals 2001 Conference* (Toronto), 509–520
13. P.R. Prasad, S. Ray, L. Gaindhar, M.L. Kapoor, Relation between processing, microstructure and mechanical properties of rheocast Al-Cu alloys. J. Mater. Sci. **23**, 823–829 (1988)
14. K. Ichikawa, Y. Kinoshita, Stirring condition and grain refinement in Al-Cu alloys by rheocasting. J. I. M. Trans. **28**(2), 135–144 (1987)
15. M. Hirai, K. Takebayashi, Y. Yoshikawa, R. Yamaguchi, Apparent viscosity of Al-10% Cu semi solid alloys. ISIJ Inter. **33**(3), 405–412 (1993)
16. H.I. Lee, R.D. Doherty, E.A. Feest, J.M. Titchmarsh, Structure and segregation of stir-cast aluminum alloys. in *Proceedings of International Conference of Solidification* (Warwick, 1983), 119–125
17. Z. Fan, Semisolid metal processing. Int. Mater. Rev. **47**(2), 49–85 (2002)
18. A. Figueredo, *Science and Technology of Semi-Solid Metal Processing* (North American Die Casting Association, Rosemont, 2001)
19. C.S. Rice, P.F. Mendez, Slurry based semi-solid diecasting. Adv. Mater. Process. **159**(10), 49–53 (2001)
20. S.B. Brown, P.F. Mendez, C.S. Rice, Apparatus and method for integrated semi-solid material production and casting, U.S. Patent 5881796, 16 Mar 1999
21. Private communications with Jayesh Patel, Zyomax Ltd, UK (2016)
22. K.P. Young, D.E. Tyler, H.P. Cheskis, W.G. Watson, Process and apparatus for continuous slurry casting, U.S. Patent 4482012, 13 Nov 1984
23. K.P. Young, C.P. Kyonka, F. Courtois, Fine grained metal composition, U.S. Patent 4415374, 15 Nov 1983
24. L.G. Kun, *Continuous Casting; The Application of Electromagnetic Stirring (EMS) in the Continuous Casting of Steel*, vol. 3 (Iron & Steel Society, Warrendale, 1984)
25. M.P. Kenney, J.A. Courtois, R.D. Evans, G.M. Farrior, C.P. Kyonka, A.A. Koch, K.P. Young, *Semisolid Metal Casting and Forging, Metal Handbook, Vol. 15, "Casting"* (ASM Publication, Des Plaines, 2002) (Copyright 2002)
26. F. Niedermaier, J. Langgartner, G. Hirt, I. Niedick, Horizontal continuous casting of SSM billets. in *Fifth International Conference on Semi-Solid Processing of Alloys and Composites* (Golden, 1998), 407–414
27. T.W. Kim, C.G. Kang, S.S. Kang, Rheology forming process of cast aluminum alloys with electromagnetic applications. in *Ninth International Conference on Semi-Solid Processing of Alloys and Composites* (Busan, Korea, 2006) (published in Solid State Phenomena, vol. 116–117, 2006, 445–448)
28. S. Nafisi, D. Emadi, M. Shehata, R. Ghomashchi, A. Charette, Semi-solid processing of Al-Si alloys: effect of stirring on iron-based intermetallics. in *Eighth International Conference on Semi-Solid Processing of Alloys and Composites* (Limassol, Cyprus, 2004)
29. S. Nafisi, R. Ghomashchi, D. Emadi, M. Shehata, *Effects of Stirring on the Silicon Morphological Evolution in Hypoeutectic Al-Si Alloys, Light Metals 2005*, (TMS Publication), 1111–1116
30. M. Adachi, H. Sasaki, Y. Harada, Methods and apparatus for shaping semisolid metals, UBE Industries, European Patent EP 0 745 694 A1, 4 Dec 1996

31. D. Liu, H.V. Atkinson, P. Kapranos, W. Jirattiticharoean, H. Jones, Microstructural evolution and tensile mechanical properties of thixoformed high performance aluminium alloys. Mater. Sci. Eng. **A361**, 213–224 (2003)
32. Y. Uetani, R. Nagata, H. Takagi, K. Matsuda, S. Ikeno, Simple manufacturing method for A7075 aluminum alloy slurry with fine granules and application to rheo-extrusion. in *Ninth International Conference on Semi-Solid Processing of Alloys and Composites* (Busan, Korea, 2006) (published in Solid State Phenomena, vol. 116–117, 2006, 746–749)
33. T. Grimmig, A. Ovcharov, C. Afrath, M. Bunck, A. Buhrig-Polaczek, Potential of the rheocasting process demonstrated on different aluminum based alloy systems. in *Ninth International Conference on Semi-Solid Processing of Alloys and Composites* (Busan, Korea, 2006) (published in Solid State Phenomena, vol. 116–117, 2006, 484–488)
34. H. Guo, X. Yang, Continuous fabrication of sound semi solid slurry for rheoforming. in *Ninth International Conference on Semi-Solid Processing of Alloys and Composites* (Busan, Korea, 2006) (published in Solid State Phenomena, vol. 116–117, 2006, 425–428)
35. S. Saffari, F. Akhlaghi, New semisolid casting of an Al-25wt% Mg2Si composite using vibrating cooling slope. in *13th International Conference on Semi-Solid Processing of Alloys and Composites* (Muscat, Oman, 2014) (published in Solid State Phenomena, vol. 217–218, 2015, 389–396)
36. T. Haga, H. Inui, H. Watari, S. Kumai, Semisolid roll casting of aluminum alloy strip and its properties. in *Ninth International Conference on Semi-Solid Processing of Alloys and Composites* (Busan, Korea, 2006) (published in Solid State Phenomena, vol. 116–117, 2006, 379–382)
37. T. Motegi, F. Tanabe, New semi solid casting of copper alloys using an inclined cooling plate. in *Eighth International Conference on Semi-Solid Processing of Alloys and Composites* (Limassol, Cyprus, 2004)
38. M. Findon, A. de Figueredo, D. Apelian, M.M. Makhlouf, Melt mixing approaches for the formation of thixotropic semisolid metal structure. in *Seventh International Conference on Semi-Solid Processing of Alloys and Composites* (Tsukuba, Japan, 2002), 557–562
39. D. Saha, D. Apelian, R. Dasgupta, SSM processing of hypereutectic Al-Si alloy via diffusion solidification. in *Seventh International Conference on Semi-Solid Processing of Alloys and Composites* (Tsukuba, Japan, 2002), 323–328
40. Q.Y. Pan, M. Findon, D. Apelian, The continuous rheoconversion process CRP a novel SSM approach. in *Eighth International Conference on Semi-Solid Processing of Alloys and Composites* (Limassol, Cyprus, 2004)
41. Q.Y. Pan, S. Wiesner, D. Apelian, Application of the continuous rheoconversion process (CRP) to low temperature HPDC-part 1: microstructure. in *Ninth International Conference on Semi-Solid Processing of Alloys and Composites* (Busan, Korea, 2006) (published in Solid State Phenomena, vol. 116–117, 2006, 402–405)
42. R. Martinez, A. Figueredo, J.A. Yurko, M.C. Flemings, Efficient formation of structures suitable for semi-solid forming. in *Transactions of the 21st International Die Casting Congress* (2001), 47–54
43. M.C. Flemings, R. Martinez, A. Figueredo, J.A. Yurko, Metal alloy compositions and process, U.S. Patent 6645323, 11 Nov 2003
44. J.A. Yurko, R.A. Martinez, M.C. Flemings, Commercial development of the semi-solid rheocasting (SSR). in *Transactions of the International Die Casting Congress* (2003), 379–384
45. J.A. Yurko, R.A. Martinez, M.C. Flemings, SSR™: the spheroidal growth route to semi-solid forming. in *Eighth International Conference on Semi-Solid Processing of Alloys and Composites* (Limassol, Cyprus, 2004)
46. J. Wannasin, R.A. Martinez, M.C. Flemings, Grain refinement of an aluminum alloy by introducing gas bubble during solidification. Scripta Mater. **55**, 115–118 (2006)
47. J. Wannasin, S. Janudom, T. Rattanochaikul, R. Canyook, R. Burapa, T. Chucheep, S. Thanabumrungkul, Research and development of the gas induced semi solid process for

industrial applications. in *11th International Conference on Semi-Solid Processing of Alloys and Composites* (Beijing, China, 2010), 544–548

48. R. Burapa, S. Janudom, T. Chucheep, J. Wannasin, Effects of primary phase morphology on the mechanical properties of an Al-Si-Mg-Fe alloy in a semi solid slurry casting process. in *11th International Conference on Semi-Solid Processing of Alloys and Composites* (Beijing, China, 2010), 253–257
49. J. Wannasin, Applications of semi-solid slurry casting using the gas induced semi-solid technique. in *12th International Conference on Semi-Solid Processing of Alloys and Composites* (Cape Town, South Africa, 2012) (published in Solid State Phenomena, vol. 192–193, 2013, 28–35)
50. Z. Chen, L. Li, R. Zhou, Y. Jiang, R. Zhou, Study on refining of primary Si in semi-solid Al-25%Si alloy slurry prepared by rotating rod induced nucleation. in *13th International Conference on Semi-Solid Processing of Alloys and Composites* (Muscat, Oman, 2014) (published in Solid State Phenomena, vol. 217–218, 2015, 253–258)
51. M. Payandeh, A.E.W. Jafros, M. Wessén, Solidification sequence and evolution of microstructure during rheocasting of four Al-Si-Mg-Fe alloys with low Si content. Metallur. Mater. Trans. A **47**(3), 1215–1228 (2016)
52. M. Wessén, Rheogjutning av extremt tunnväggiga komponenter, Aluminium Scandinavia (2012), 16–17
53. M. Wessén, H. Cao, The RSF technology: a possible breakthrough for semi-solid casting processes. J. Metallur. Sci. Technol. **25**(2), 22–28 (2007)
54. http://www.rheometal.com
55. V. Abramov, O. Abramov, V. Bulgakov, F. Sommer, Solidification of aluminium alloys under ultrasonic irradiation using water-cooled resonator. Mater. Lett. **37**, 27–34 (1998)
56. V.O. Abramov, O.V. Abramov, B.B. Straumal, W. Gust, Hypereutectic Al–Si based alloys with a thixotropic microstructure produced by ultrasonic treatment. Mater. Design **18**, 323–326 (1997)
57. O.V. Abramov, Action of high intensity ultrasound on solidifying metal. Ultrasonics **25**, 73–82 (1987)
58. X. Jian, H. Xu, T.T. Meek, Q. Han, Effect of power ultrasound on solidification of aluminum A356 alloy. Mater. Lett. **59**, 190–193 (2005)
59. X. Jian, T.T. Meek, Q. Han, Refinement of eutectic silicon phase of aluminum A356 alloy using high intensity ultrasonic vibration. Scr. Mater. **54**, 893–896 (2006)
60. A. Pola, A. Arrighini, R. Roberti, Effect of ultrasounds treatment on alloys for semisolid application. in *10th International Conference on Semi-Solid Processing of Alloys and Composites* (Aachen, Germany, 2008) (published in Solid State Phenomena, vol. 141–143, 2008, 481–486)
61. S. Wu, J. Zhao, L. Zhang, P. An, Y. Mao, Development of non-dendritic microstructure of aluminum alloy in semi-solid state under ultrasonic vibration. in *10th International Conference on Semi-Solid Processing of Alloys and Composites* (Aachen, Germany, 2008) (published in Solid State Phenomena, vol. 141–143, 2008, 451–456)
62. W. Shusen, L. Chong, L. Shulin, S. Meng, Research progress on microstructure evolution of semi-solid aluminum alloy in ultrasonic field and their rheocasting. China Foundry **11**(4), 258–267 (2014)
63. J.P. Gabathuler, K. Buxmann, Process for producing a liquid-solid metal alloy phase for further processing as material in the thixotropic state, U.S. Patent 5186236, 16 Feb 1993
64. R. Shibata, T. Kaneuchi, T. Souda, Y. Iizuka, New semi solid metal casting process. in *4th International Conference on Semi-Solid Processing of Alloys and Composites* (Sheffield, England, 1996), 296–300
65. T. Kaneuchi, R. Shibata, M. Ozawa, Development of new semi-solid metal casting process for automotive suspension parts. in *Seventh International Conference on Semi-Solid Processing of Alloys and Composites* (Tsukuba, Japan, 2002), 145–150

66. R. Shibata, T. Kaneuchi, T. Souda, H. Yamane, Formation of spherical solid phase in die casting shot sleeve without any agitation. in *Fifth International Conference on Semi-Solid Processing of Alloys and Composites* (Golden, 1998), 465–469
67. O. Lashkari, S. Nafisi, R. Ghomashchi, Microstructural characterization of rheo-cast billets prepared by variant pouring temperatures. J. Mater. Sci. Eng. A **441**, 49–59 (2006)
68. H. Wang, Semisolid processing of aluminium alloys, Ph.D. Thesis (The University of Queensland, Australia, 2001)
69. H. Wang, C.J. Davidson, D.H.S. John, Semisolid microstructural evolution of AlSi7Mg alloy during partial remelting. Mater. Sci. Eng. **A368**, 159–167 (2004)
70. H. Wang, D.H.S. John, C.J. Davidson, M.J. Couper, Characterization and shear behavior of semisolid Al-7Si-0.35Mg alloy microstructures. Aluminum Trans. **2**(1), 56–66 (2000)
71. A.K. Dahle, J.E.C. Hutt, Y.C. Lee, D.H. St John, Grain formation in hypoeutectic Al-Si alloys, AFS Trans. (1999)
72. J. Jorstad, D. Apelian, Pressure assisted processes for high integrity aluminium castings-Part 2, Foundry Trade J. (2009), 282–287
73. J.L. Jorstad, M. Thieman, R. Kamm, SLC, the newest and most economical approach to semi-solid metal (SSM) casting. in *Seventh International Conference on Semi-Solid Processing of Alloys and Composites* (Tsukuba, Japan, 2002), 701–706
74. J.L. Jorstad, M. Thieman, R. Kamm, M. Loughman, T. Woehlke, Sub liquidus casting: process concept and product properties, AFS Trans. 80 (2003), paper 03-162
75. R. Kamm, J.L. Jorstad, Semi-solid molding method, U.S. Patent 6808004, 6 Oct 2004
76. A. Forn, S. Menargues, E. Martin, J.A. Picas, Sub liquidus casting technology for the production of high integrity component. in *10th International Conference on Semi-Solid Processing of Alloys and Composites* (Aachen, Germany, 2008) (published in Solid State Phenomena, vol. 141–143, 2008, 219–224)
77. D. Doutre, G. Hay, P. Wales, Semi-solid concentration processing of metallic alloys, U.S. Patent 6428636, 6 Aug 2002
78. D. Doutre, J. Langlais, S. Roy, The SEED process for semi-solid forming. in *Eighth International Conference on Semi-Solid Processing of Alloys and Composites* (Limassol, Cyprus, 2004)
79. P. Cote, M. Larouche, X.G. Chen, New developments with the SEED technology. in *12th International Conference on Semi-Solid Processing of Alloys and Composites* (Cape Town, South Africa, 2012) (published in Solid State Phenomena, vol. 192–193, 2013, 373–378)
80. J. Langlais, A. Lemieux, The SEED technology for semi-solid processing of aluminum alloys a metallurgical and process overview. in *Ninth International Conference on Semi-Solid Processing of Alloys and Composites* (Busan, Korea, 2006) (published in Solid State Phenomena, vol. 116–117, 2006, 472–477)
81. S. Nafisi, O. Lashkari, R. Ghomashchi, J. Langlais, B. Kulunk, The SEED technology: a new generation in rheocasting. in *CIM-Light Metals Conference* (Calgary, Canada, 2005), 359–371
82. D. Doutre, G. Hay, P. Wales, SEED: a new process for semi solid forming. in *Light Metals Conference* (CIM, Vancouver, Canada, 2003), 293–306
83. H. Atkinson, Alloys for semi-solid processing. in *12th International Conference on Semi-Solid Processing of Alloys and Composites* (Cape Town, South Africa, 2012) (published in Solid State Phenomena, vol. 192–193, 2013, 16–27)
84. D.H. Kirkwood, P. Kapranos, Semi-solid processing of alloys. Met. Mater. **5**, 16–19 (1989)
85. W.R. Loué, M. Suéry, Microstructural evolution during partial remelting of Al-Si7Mg alloys. Mater. Sci. Eng. **A203**, 1–13 (1995)
86. V.M. Segal, Materials processing by simple shear. Mater. Sci. Eng. **A197**, 157–164 (1995)
87. K.N. Campo, C.T.W. Proni, E.J. Zoqui, Influence of the processing route on the microstructure of aluminum alloy A356 for thixoforming. Mater. Charact. **85**, 26–37 (2013)

88. M. Moradia, M.N. Ahmadabadi, B. Poorganjic, B. Heidariana, M.H. Parsaa, T. Furuhara, Recrystallization behavior of ECAPed A356 alloy at semi-solid reheating temperature. Mater. Sci. Eng. **A527**, 4113–4121 (2010)
89. H. Meidani, S. Hosseini Nejad, M.N. Ahmadabadi, A novel process for fabrication of globular structure by equal channel angular pressing and isothermal treatment of semisolid metal. in *10th International Conference on Semi-Solid Processing of Alloys and Composites* (Aachen, Germany, 2008) (published in Solid State Phenomena, vol. 141–143, 2008, 445–450)
90. A. Leatham, A. Ogilvy, P. Chesney, J.V. Wood, Osprey process-production flexibility in material manufacture. Met. Mater. **5**, 140–143 (1989)
91. P. Mathur, D. Apelian, A. Lawley, Analysis of the spray deposition process. Acta Metall. **37** (2), 429–443 (1989)
92. H. Vetters, A. Schulz, K. Schimanski, S. Spangel, V. Uhlenwinkel, K. Bauckhage, Spray forming of cast alloys, an innovative alternative. in *Proceedings of the 65th world foundry Congress* (2002), 1089–1096
93. G. Wan, T. Witulski, G. Hirt, Thixoforming of Al alloys using modified chemical grain refinement for billet production. La Metallurgia Italiana **86**, 29–36 (1994)
94. J.P. Gabathuler, D. Barras, Y. Krahenbuhl, Evaluation of various processes for the production of billet with thixotropic properties. in *Second International Conference on Semi-Solid Processing of Alloys and Composites* (MIT, Cambridge, 1992), 33–46
95. G. Wan, P.R. Sahm, Particle growth by coalescence and Ostwald ripening in rheocasting of Pb-Sn. Acta Metall. Mater. **38**(11), 2367–2373 (1990)
96. H.P. Mertens, R. Kopp, T. Bremer, D. Neudenberger, G. Hirt, T. Witulski, P. Ward, D.H. Kirkwood, Comparison of different feedstock materials for thixocasting, EUROMAT 97. in *Proceedings of the 5th European Conference on Advanced Materials and Processes and Applications* (1997), 439–444
97. D. Apelian, Semi-solid processing routes and microstructure evolution. in *Seventh International Conference on Semi-Solid Processing of Alloys and Composites* (Tsukuba, Japan, 2002), 25–30
98. Q.Y. Pan, M. Arsenault, D. Apelian, M.M. Makhlouf, SSM processing of AlB2 grain refined Al-Si alloys. AFS Trans. (2004), Paper 04-053
99. F. Czerwinski, *Magnesium Injection Molding* (Springer, New York, 2008)
100. F. Czerwinski, On the generation of thixotropic structures during melting of Mg-9Al-1Zn alloy. Acta Mater. **50**(12), 3625 (2002)
101. F. Czerwinski, Controlling the ignition and flammability of magnesium for aerospace applications (A review). Corros. Sci. **86**, 1–16 (2014)
102. F. Czerwinski, Corrosion of materials in liquid magnesium alloys and its prevention. in: *Magnesium Alloys—Properties in Solid and Liquid States*, ed. by F. Czerwinski (INTECH, Rijeka, EU, 2014), 131–170

Chapter 3
Solidification and Alloy Distribution During SSM Processing

Abstract The process of semi-solid metal processing is also governed by solidification principles. However, the mechanisms controlling nucleation and growth of the primary phase and eutectic solidification are slightly different to conventional mechanism due to forced convection. The process of melt stirring influences alloy distribution and boundary layer characteristics. The mechanisms of nucleation and growth along with morphological evolution during melt stirring are discussed in this chapter.

3.1 Introduction

Solidification is governed by the nucleation and growth of the primary phase in an alloy system. In practice, growth does not take place with complete mixing of solute elements within the solidifying liquid. Therefore, the composition of liquid changes continuously during solidification, and the last solidified liquid is usually richer in solute than the first one. In other words, the resulting solid structure is chemically inhomogeneous both in macro and micro-scales. This is called "*segregation.*" The control of segregation in both micro and macro-scales is one of the most challenging tasks to improve the quality of as-cast products and their subsequent performance in service.

During solidification of alloy systems having partition coefficient[1] less than 1 ($k < 1$), the solute atoms are rejected out of the solidifying liquid into the solid–liquid interface to induce solute build up in the liquid adjacent to the solid. Therefore, three distinct zones are notable within such solidification system:

- Heterogeneous solid, solute distribution is controlled by its concentration gradient and heat flow
- Solid–liquid interface or boundary layer which is highly enriched due to the solute rejection from the growing solid phase
- Bulk liquid where solute is expected to distribute uniformly due to the convection and diffusion.

[1]The ratio of solute concentration in the solid to liquid at the interface is defined as the "partition coefficient or partition ratio."

S. Nafisi, R. Ghomashchi, *Semi-Solid Processing of Aluminum Alloys*,
DOI 10.1007/978-3-319-40335-9_3

The above-mentioned shortcomings due to the solidification could pose technological problems and reduce the likelihood of fabricating sound engineering parts. Consequently, the selection of an appropriate casting route and parameters are of great importance to mitigate such drawbacks.

In most SSM processes, induced fluid flow or in more general term "forced convection" is an integral part of the process. The effect of fluid flow has been widely studied on the micro- and macro-structure of solidified alloys independent of the process specifications. These studies have been mostly concerned with columnar-equiaxed transition and the process of macrosegregation. It is generally believed that fluid flow induces remelting, fragmentation, and normally disrupts dendritic solidification (e.g., [1–9]).

Apart from the aforementioned publications, there is not much research on the morphological evolution associated with the forced convection in the context of SSM processes. However, as will be discussed in the following sections, there are three mechanisms proposed for structural evolution during SSM casting. These include dendrite arm fragmentation, dendrite arm root remelting, and nucleation controlled mechanisms [10–15]. In addition to the proposed mechanisms, a further experimental-based discussion will be presented to specify how the microstructural evolution could be characterized during SSM processing of engineering alloys with emphasis on Al–Si casting alloys.

3.2 Solidification During Stirring

During an investigation on the hot tearing of alloys, it has been established that solidification in the stirred condition produces nondendritic structure [16]. Spencer et al. [16] measured the viscosity of Sn15%Pb alloy as a function of fraction solid during continuous shearing of the alloy by using a rotational rheometer. The results proved that while unstirred melts began to show strength around 0.2 fraction solid (20 %), by stirring the resulted slurries continued to behave like a liquid beyond 0.4 fraction solid (40 %) as seen in Fig. 3.1a. Joly and Mehrabian [17] carrying out more detailed study on the rheology of the same alloy showed that the viscosity is not only dependent on the fraction solid but also varies by cooling and shearing rates (Fig. 3.1b). Their experiment confirmed that increasing the shear rate induces morphological transition at shorter times and results in less liquid entrapped within the primary particles. The higher viscosity values for slowly stirred slurries were explained in terms of the buildup of clusters of solid particles at low stirring speeds and lower viscosity was due to breaking up of these clusters when the shear rate was increased.

The studies of Vogel et al. [18] on Al–Cu alloys showed that in the absence of stirring, the alloy has conventional dendritic structure but transforms to rosette-like particles when stirring is introduced using mechanical impeller. The rosette morphology will further transform to spherical particles as the rotational speed is

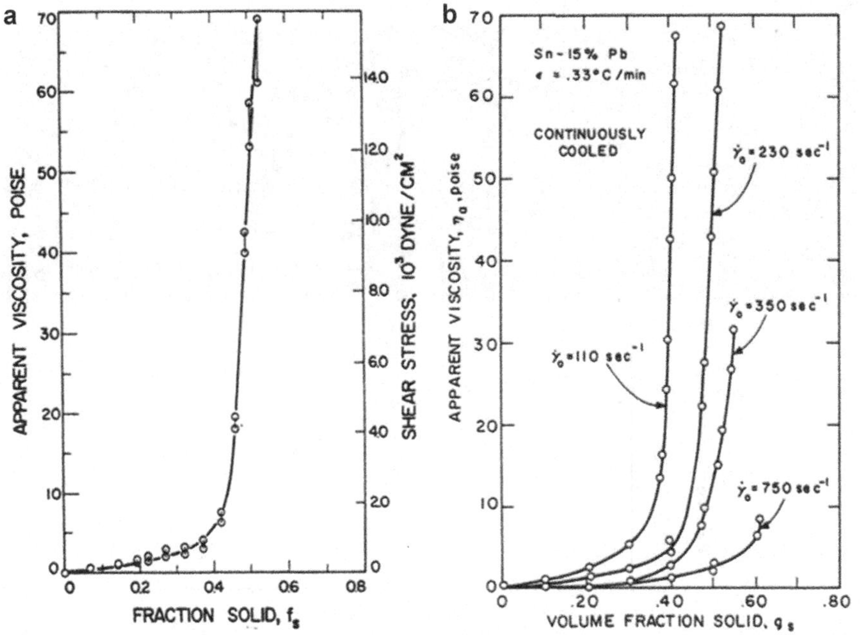

Fig. 3.1 Typical apparent viscosity versus fraction solid curve for Sn15%Pb alloy, cooled at ~0.33 °C min^{-1}; (**a**) with a constant shear rate of 200 s^{-1} [16], (**b**) sheared continuously at various shear rates [17]

increased, Fig. 3.2. It was also observed that the maximum particle size decreases by increasing the stirring speed (Fig. 3.2d).

Vogel et al. [18] studied the effect of stirring on the frequency of primary phase nucleation, using Al20%Cu in two different stirred and unstirred quenched conditions. They found the apparent increase in the number of primary particles between the unstirred and stirred materials and the increased density of particles during stir casting. In another test, the alloy was furnace cooled without stirring and quenched 1 min after the start of solidification (Fig. 3.3a). After structural examination, the same alloy was remelted and solidified again by furnace cooling while stirred at 1000 rpm (Fig. 3.3b). Their results suggested that stirring did not cause significant increase in primary nucleation during the first minute of solidification although close examination of their micrographs, shown in Fig. 3.3, may suggest that there is an increase in the number of primary particles with a reduced size confirming stirring somewhat suppressing the growth but increasing nucleation. Such conclusion appears appropriate considering that stirring reduces both thermal and compositional gradient enabling nucleation almost everywhere within the bulk liquid. This concept is more elaborated later in this chapter, see Sect. 3.3.1.

Vogel et al. [18] have also proposed a model to account for both the initial dendrite fragmentation and clustering reported by Joly-Mehrabian [17]. The model was based on the dendrite deformation followed by fragmentation and remelting of

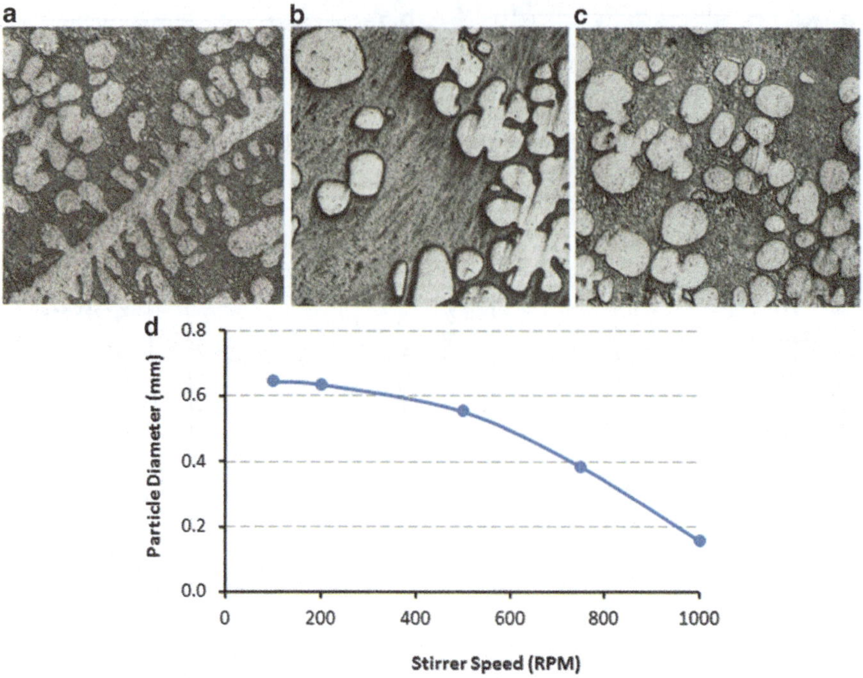

Fig. 3.2 Al20%Cu stir cast at: (**a**) without stirring, (**b**) 750 rpm, (**c**) 1000 rpm, and (**d**) particle size as a function of stirrer speed for Al24%Cu (reproduced from [18]) (reprinted by permission of Taylor & Francis Ltd)

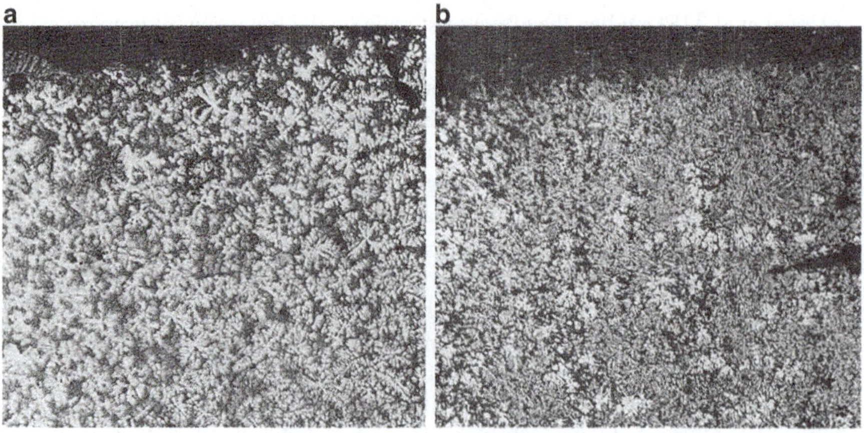

Fig. 3.3 Al20%Cu slow cooled and quenched 1 min after start of solidification; (**a**) without stirring, (**b**) same material stirred at 1000 RPM before quenching (×12) [18] (reprinted by permission of Taylor & Francis Ltd)

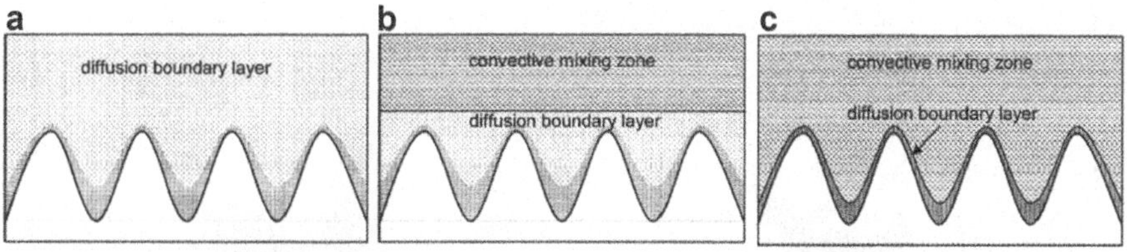

Fig. 3.4 A schematic illustration of the effect of fluid flow on the boundary layer; (**a**) infinite boundary layer under diffusive transport, (**b**) a finite boundary layer without interdendritic liquid penetration under laminar flow, and (**c**) extremely narrow boundary layer with interdendritic liquid penetration under turbulent flow [15]

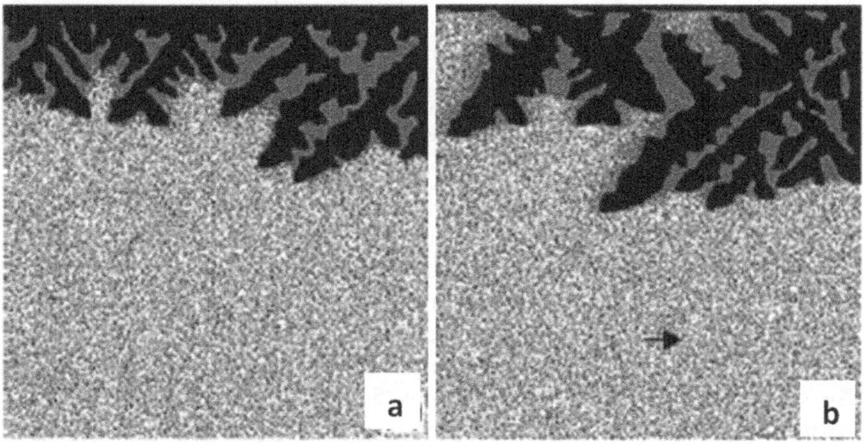

Fig. 3.5 Effect of fluid flow on the solidification morphology (*arrow* shows the direction of fluid flow); (**a**) purely diffusive flow, (**b**) forced flow [15]

grain boundaries. Das et al. [15] with computer simulation have shown that the growth morphology of dendrites is highly dependent on the nature of fluid flow as shown in Fig. 3.4.

In a pure diffusive flow, i.e., absence of flow in the liquid, solute transfer takes place by diffusion through approximately the entire volume of liquid, as if, there is an infinite diffusion boundary layer and the resulted growth structure is purely dendritic (Fig. 3.4a). At low and intermediate shear rates (Fig. 3.4b), the flow is laminar with a finite boundary layer existing around the growing particle beyond which the melt has homogeneity. It has been shown that the introduction of forced convection causes destabilization of the solid/liquid interface and enhances dendritic growth from a fixed substrate [15]. Das et al.'s result [15] is in agreement with the theoretical stability analysis of Vogel et al. [14] except that the model developed by Das et al. [15] predicts destabilization only under laminar flow condition and for solid particles growing from a substrate (Fig. 3.5).

Das et al. [15] have further suggested that lateral motion in the liquid prevents the formation of any effective concentration gradient ahead of the growing solid/

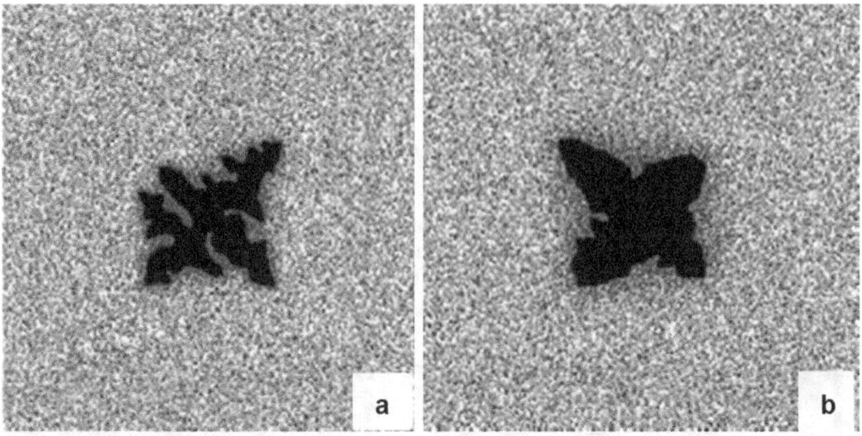

Fig. 3.6 Effect of rotation of an isolated particle in a forced liquid flow: (**a**) diffusive, (**b**) particle rotation with atoms movement allowed over three lattice distance [15]

liquid interface due to the rejection of solute. If their suggestion is indeed the case for laminar flow, then it is expected to have slow growth of the primary dendrites due to very limited constitutional supercooling at the interface. On the other hand, solute rejected within the interdendritic regions (lagging behind the growth front) could not be transported away due to weak laminar flow and therefore the buildup of enriched interdendritic boundary layer is expected to promote growth locally. The slow growth of primary and faster growth of secondary and tertiary dendrites eventually results in formation of bulky primary phase resembling that of rosette or globular morphologies. This was predicted in their simulation to have rosette growth under laminar flow. It was shown that isolated particles rotating in a laminar flow periodically stabilize and destabilize promoting a coarsening effect to form rosette morphology. Figure 3.6 shows the solidification structures formed under rotation of particles in a laminar flow.

At high shear rate, the flow characteristics change to turbulence and liquid penetrates into interdendritic regions (Fig. 3.4c). This would supply the solvent into the interdendritic regions and the rejected solute is quickly transported away, promoting compositional homogenization and discarding any possibility of constitutional undercooling. By uniform distribution of solvent, perturbation at the interface is not promoted and as a result, the solid grows more or less with a flat solid/liquid interface. This theory could also explain why spherical particles are normally observed from the initiation of solidification with a high shear rate.

Das et al. [15] obtained the growth data from the simulations of solidification for diffusive, laminar, and turbulent flows and presented the total number of atoms solidified as a function of time, see Fig. 3.7. The slope of the plot is an indication of the average growth rate at any instance. It is evident that under pure turbulent flow after an initial high growth rate, progressive growth retardation occurs as compared to pure diffusive or laminar flow.

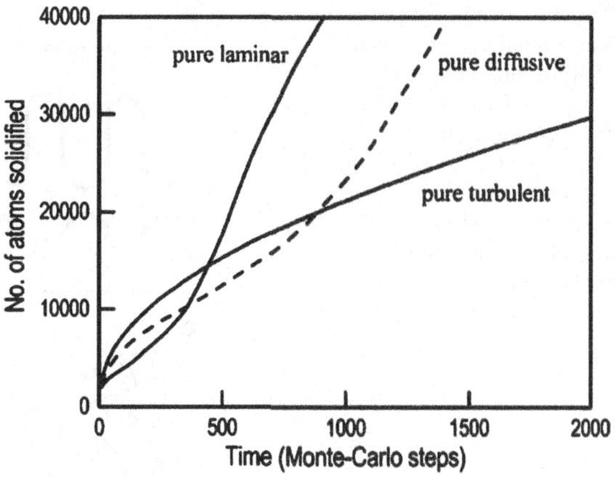

Fig. 3.7 Number of atoms solidified as a function of time under different fluid flow conditions [15]

In conclusion, intense shearing could homogenize the entire liquid and the whole liquid is undercooled prior to solidification. Therefore, the chance of nuclei survival is high throughout the volume of the melt and nucleation is expected to occur instantaneously in the bulk liquid [15]. After an initial high growth rate during turbulent flow (high shear rate), retarded growth rates result in formation of finer primary particles during SSM processes at high shear rate.

3.2.1 Morphological Evolution During Melt Stirring

Several mechanisms have been proposed for structural evolution under forced convection. These include dendrite arm fragmentation [10, 14, 18], dendrite arm root remelting [1, 12, 19, 20], and nucleation controlled mechanisms [6, 13], which further discussed in this section.

3.2.1.1 Dendrite Arm Fragmentation

The initial evidence of fragmentation is related to the work of Spencer et al. [16] where shearing was employed after partial solidification. In this kind of fragmentation, dendrite arms break off mechanically from the root due to the shear forces. Vogel and Cantor [14] developed a model to investigate the effects of stirring on the solidification of a particle from the melt under influence of thermal or solute flow. Numerical analysis verified that stirring destabilizes the solidifying interface and leads to a decrease in the critical radius for relative stability. Doherty and Vogel [10, 18] have proposed a dendrite arm fragmentation mechanism for grain multiplication, as schematically presented in Fig. 3.8 [10]. It was suggested that dendrite arms bend plastically under the shear force created by melt stirring. Plastic bending

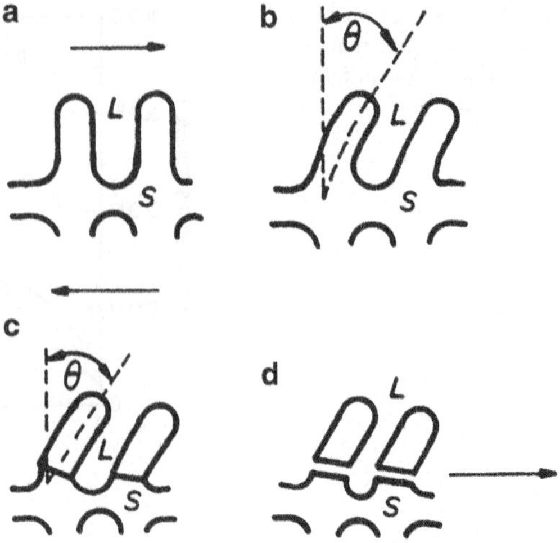

Fig. 3.8 Dendrite arm fragmentation mechanism [10]; (**a**) un-deformed dendrite, (**b**) after bending, (**c**) reorganization of the lattice bending to give grain boundaries, (**d**) fragmentation through wetting of grain boundary by liquid metal

introduces large misorientations into the dendrite arms in the form of "geometrically necessary dislocations." Such dislocations if randomly distributed will have a high elastic energy which can be reduced by dislocations migration to form grain boundary. Grain boundaries with misorientations greater than about 20° have an energy that is significantly greater than twice the solid/liquid interfacial energy ($\gamma_{gb} > 2\gamma_{sl}$ in which γ_{gb} and γ_{sl} are grain boundary and solid–liquid interphase boundary energies, respectively). If such high energy boundaries are formed then γ_{gb} dictates that the grain boundary will be replaced by a thin layer of liquid and thus the dendrite will break apart along the prior boundary [10].

The authors verified the aforementioned concept in the electromagnetically stirred Al7%Si alloy [21]. Electron backscatter diffraction (EBSD) analysis of the stirred and unstirred cast billets has shown that stirring generates more local misorientation within individual solid particles. This is an indication of plastic deformation during the course of SSM slurry preparation and it plays a key role on thixocasting where the reheating time is strongly dependent on the type and quantity of grain boundaries (see Sect. 5.4.1 for further information).

3.2.1.2 Dendrite Arm Root Remelting

Jackson et al. [19] by solidifying cyclohexanol and addition of fluorescein as impurity reached to the conclusion that dendrites in an alloy differ from those in a pure material. In a relatively pure material, the diameters of main stem and its branches are similar while with increasing impurity level, the radius of dendrite tips becomes much smaller than the width of the main stem. Figure 3.9a shows large branches attached by narrow necks to the stem. It was revealed that by holding the sample at a constant temperature and taking photo at a time interval, many of the

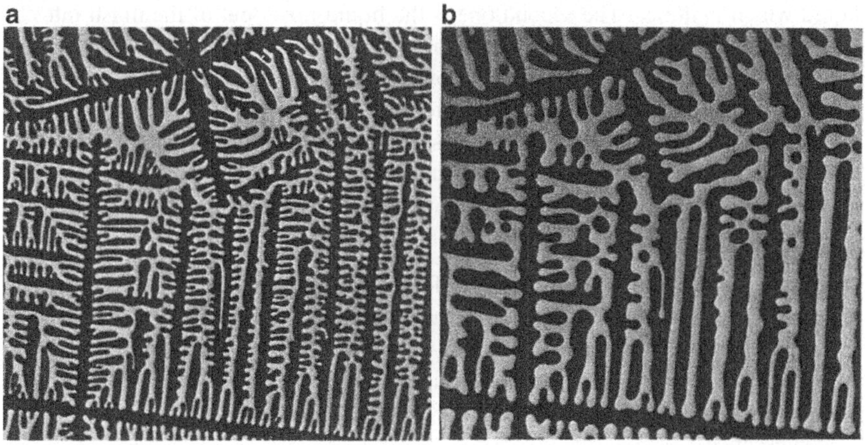

Fig. 3.9 Dendrites in cyclohexanol with addition of fluorescein, ×150; (**a**) just after dendrites grew, (**b**) 20 min after growing [19]

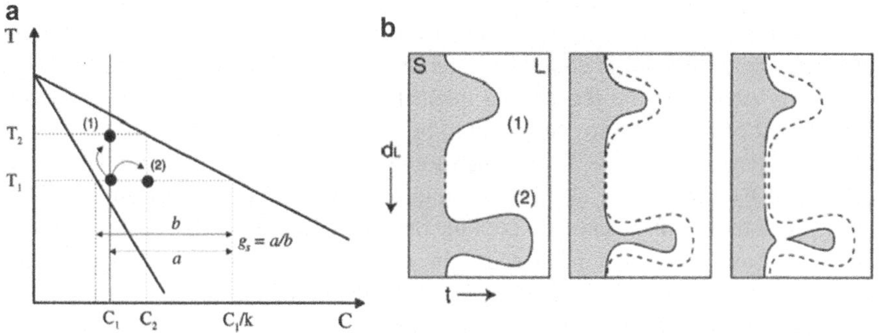

Fig. 3.10 (**a**) Schematic representation of dendrite arm remelting on a binary phase diagram (fraction solid given by lever rule: a/b), (**b**) root remelting mechanism [22]

secondary and tertiary branches have become detached. This detachment was claimed to be due to remelting of dendrites which has occurred either by recalescence of whole system or by local recalescence due to fluctuation caused by convective mixing or stirring (recalescence was defined as variation in growth rate). It was also reported that secondary or tertiary dendrite arms can detach from their root due to solute enrichment and thermo-solutal convection during solidification [1, 12]. Dendrite root remelting was schematically presented by Campanella et al. [22]. By a simple binary diagram, it was shown that local remelting can be achieved by two different mechanisms (Fig. 3.10a). Fraction of solid decreases if the temperature raises (path 1) or the average concentration (path 2) increases. The first assumption corresponds to fluid flow entering the mushy zone from the bulk liquid which is hotter. The second one is the liquid flows out of the mush into the bulk liquid which is enriched in solute due to the segregation in the direction of the dendrite tips growth. Since thermal flow is much faster than solutal diffusion, reheating is the predominant fragmentation mechanism (Fig. 3.10b).

Simultaneously with detachment of secondary and tertiary branches, dendrite arm coarsening could also take place. Dendrite arm coarsening is a surface tension-driven phenomenon, which has been shown to take either of two forms [20];

- Melting back of smaller dendrite arms which deposit onto larger arms, leading to an overall reduction in the number of dendrite arms present. This results to an increase in secondary dendrite arm spacing (named ripening)
- Agglomeration of secondary dendrite arms due to material depositing preferentially at regions of strong negative curvature along the solid/liquid interface, leading to the coalescence of several dendrite arms into large entities (named coalescence)

The driving force is reduction of total solid–liquid interface area, i.e., reduction in interfacial energy.

To explain the crystal multiplication, Hellawell [12] suggested that the refinement of the microstructures in stir cast processes may be caused by temperature fluctuation rather than by purely mechanical interactions. He stated that close to the melting point, the solid is entirely ductile and dendrites can be bent elastically or plastically depending on the stirring intensity but not fractured and therefore the detachment of the side arms supposed to be a local remelting phenomenon.

Using synchrotron fast X-ray microtomography, an experimental setup was developed by Limodin et al. [23] and microstructural evolution of an Al10%Cu alloy was studied at a constant cooling rate of 3 K min^{-1}. 3D observations of dendrites have generated further insight on dendrites morphological evolution during solidification, Fig. 3.11. On the left-hand side of the dendrite in Fig. 3.11, the roots between adjacent arms (1–5) progressively filled with solid and the tips of the adjacent arms grow until they touch and join each other. This could be the result of coalescence mechanism as depicted by Chen et al. [24] and Mortensen [20] for coarsening during isothermal holding. Another phenomenon is dissolution of arms 10, 12, and 15 from the tip toward the roots to the benefit of larger adjacent arms, i.e., remelting mechanism.

3.2.1.3 Nucleation-Controlled Mechanism

This is the “big-bang” or copious nucleation mechanism [6] where fine crystallites originated at the mold surface, the chill zone, detach and float or carried away by convection currents to the center of the casting to act as nucleation sites upon solidification. In another way, during the continuous cooling under forced convection, heterogeneous nucleation takes place simultaneously throughout the bulk liquid.

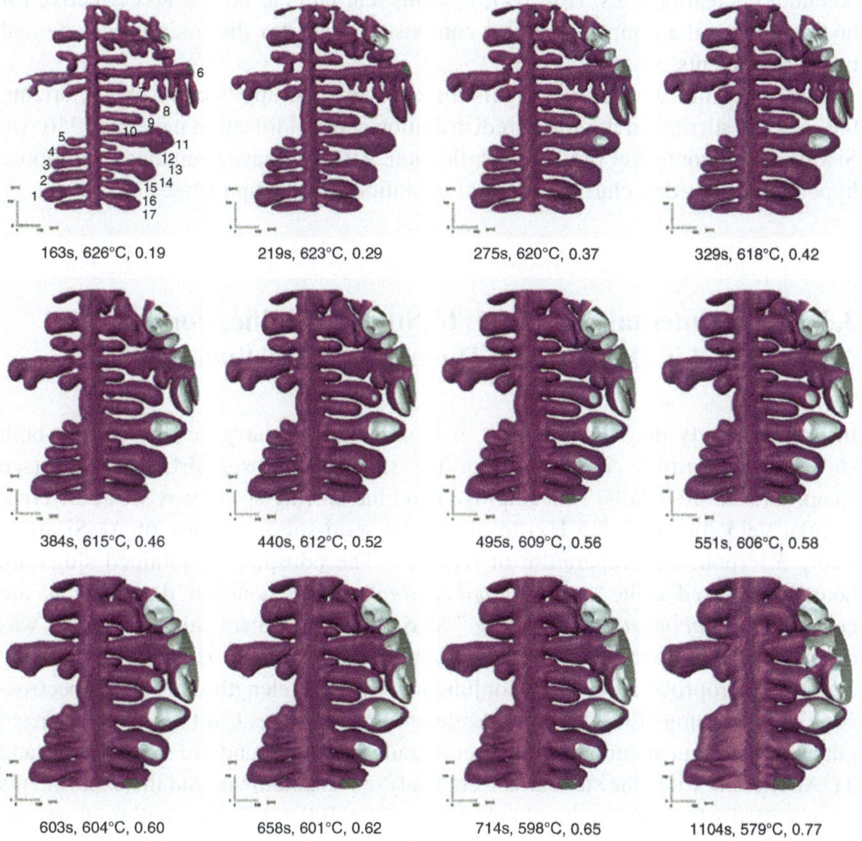

Fig. 3.11 In situ synchrotron X-ray microtomography of a dendrite with solidification time, Al10%Cu [23]

A continuous nucleation might take place in the absence of a distinct recalescence. The vigorous stirring prevents the establishment of stable diffusion fields necessary for dendrites evolution. Also multiple nucleations exist due to periodic passage through different temperature zones. Eventually, smooth rounded shapes are expected out of solidification.

Microscopic observation by Jackson et al. [19] showed that fluctuation in the growth rate of dendrites creates many isolated crystals of solid in the melt by remelting dendrite arms and as a result of convection there is a large number of fresh nuclei in the liquid. It was also claimed elsewhere [13] that comparing to conventional solidification, the actual nucleation rate may not increase but all the nuclei formed survive due to the uniform temperature field resulting in an increased effective nucleation event. In addition, the intensive mixing action should disperse the clusters of potential nucleation agents, giving rise to an increased number of potential nucleation sites. However, it seems that laminar flow is less effective for homogenizing the temperature and composition and for dispersing the potential nucleation agents.

The following sections highlights the experimental approach in characterizing the effect of stirring on the solute redistribution and solidification pattern of 356 Al–Si alloy in the context of process specification. Attempts have been made to propose hypotheses to further clarify structural evolution for SSM processes.

3.3 Experimental Approach to Study Solidification and Alloy Distribution During SSM Casting

In order to study the stirring effect, it is necessary to carry out the tests at both stirred and unstirred conditions. For the stirring purpose, SEED process (see Chap. 2 for more details) was employed and the stirring speed was set at different angular velocities (1.5–2.5 Hz) to obtain the semi-solid structure of Al–Si alloy, Table 3.1 (pouring temperature of 645 °C). The nonspun, nondrained slugs are hereafter referred as the "*conventional casting*" and the nonspun, drained slugs are called the "*semiconventional casting*." Variation of the chemical composition was checked by point analysis which performed across primary α-Al particles using electron microprobe analysis in conjunction with wavelength dispersive spectroscopy (WDS) using JEOL JXA8900L electron microprobe. Calibration was carried out with different natural and synthetic standard compounds of Al, Si, Mg, and Ti. After calibration, the standards were analyzed as unknowns and the experiments were started when the results of calibration were $100 \pm 1\,\%$. The optimum operating conditions, which obtained after preliminary trials, were 10 kV acceleration voltage, 20 nA beam current, 1 μm beam probe size, and 20 s counting time on the peaks. For the best quantitative results, analyses were carried out on the polished and unetched metallographic samples to increase the precision of the quantitative analytical results. For better conductivity, all the samples were coated with a thin layer of carbon. WDS analyses were done on the similar size primary α-Al globules with a ~100 μm circular diameter. The starting and finishing points of each line scan were set in the eutectic region and therefore with passing the beam through α-Al particles, sudden changes in chemical composition could be observed [25].

Table 3.1 Chemical analysis of 356 melts (wt%)

Sample	Process	Pouring Temp (°C)	Quenched Temp (°C)	Si%	Mg%	Fe%	Mn %	Ti%	Al
A1	Conv.	645	594	7.14	0.38	0.04	0.0027	0.0057	Bal.
A2	SEED			7.18	0.39	0.04	0.0028	0.0058	Bal.
B1	SEED		593	6.9	0.32	0.08	0.001	0.124	Bal.
B2	Semi Conv.			6.91	0.32	0.08	0.001	0.124	Bal.

3.3.1 Solidification During Stirring

3.3.1.1 Stir-Cast Nucleation

The nucleation (heterogeneous) which is the active mechanism to initiate primary α-Al usually takes place on foreign particles such as refiner, inclusions, oxides, and especially mold wall where the molten alloy was poured onto the mold wall to reduce turbulence. Due to effective stirring, the columnar dendrites growing off the mold surface are expected to fragment and the broken pieces are transported into the bulk liquid as a result of forced convection. The detachments of dendrites arms could also occur by dendrite arm root remelting mechanism [1, 12, 20] due to solute enrichment and thermal–solutal convection.

With considering heterogeneous nucleation, it is important to consider the size of nuclei and boundary layer. According to hypothesis of Hellawell [12], the critical size of nucleus will be far below the thickness of boundary layer around the moving particles. With stirring, these particles will be swept away with their small boundary layers and these nuclei alter the recalescence during solidification and it causes continuous fluctuations in the local temperature as suggested in Fig. 3.12a. Figure 3.12b shows typical cooling curves for samples cast with and without stirring which appears to support the aforementioned hypothesis. As shown, the assumption of Hellawell based on continuous nucleation appears to be correct. The authors just want to add that in the stirred sample, nucleation of primary particles is not only continuous, but also with stirring nucleation temperature shifts up. As indicated by arrows, forced convection increases the nucleation temperature by more than 6 °C. This value in the case of alloy distribution is quite important since by increasing the solidification commencement temperature, the liquidus curve is also moved up which could lead to a different composition as will be described later. It is interesting that there is no registered recalescence exists which is believed to be associated with improved heat transfer with agitation. Therefore, a longer nucleation event could mean more nuclei forming within a definite time, i.e., copious nucleation, resulting in a refined structure.

As a proof for better heat transfer, typical cooling curves of stirred and unstirred samples are shown in Fig. 3.13. For the exact same condition, by stirring, total solidification time decreases which is associated to the better contact between the mold and molten metal with less air gap and thus better heat transfer and higher cooling rate plus thermal homogenization through the bulk of liquid.

Microscopic observation by Jackson et al. [19] showed that fluctuation in the growth rate of dendrites creates many isolated primary particles within the melt and therefore a large number of nuclei exists in the liquid due to convective mixing or stirring. Also the concept of more continuous nucleation was suggested by Fan [13] as another nondendritic solidification mechanism named “growth controlled mechanism.” It was stated that in comparison to the conventional solidification, the actual nucleation rate may not be increased; however, the majority of nuclei would survive due to the uniform temperature field leading to an increased effective

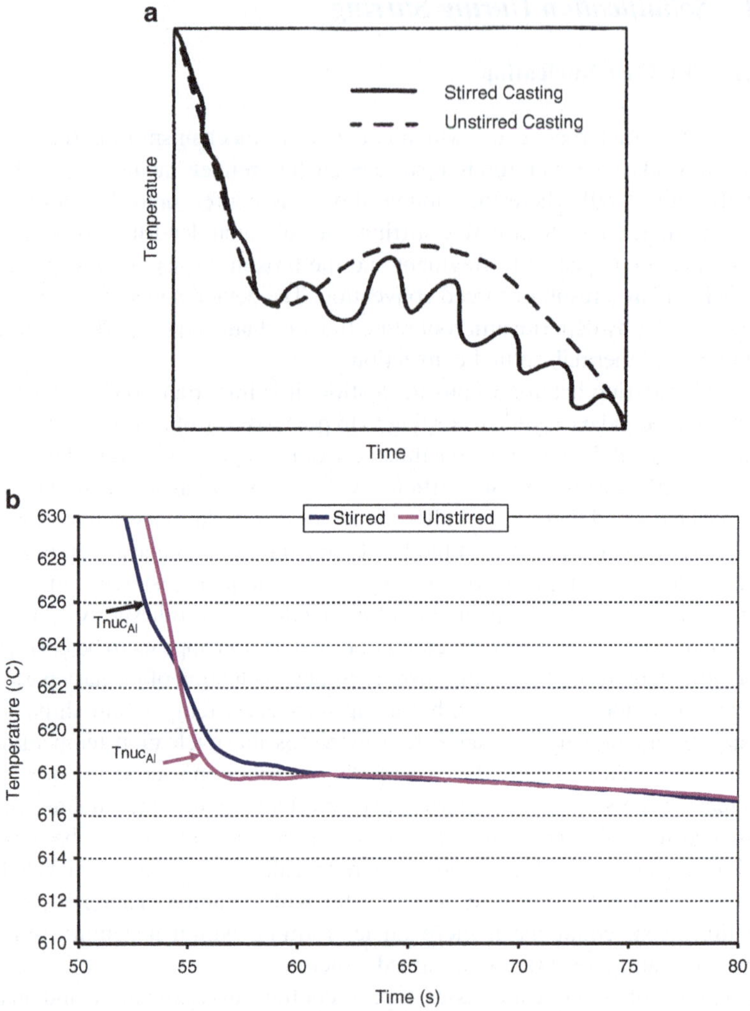

Fig. 3.12 (**a**) Schematic cooling curve expected for stirred and unstirred melt [12], (**b**) typical experimental results from continuous stirring, poured at 690 °C (*arrows* showing nucleation temperatures in both cases) [25]

nucleation rate. In addition, in the case of intensive mixing, the clusters of potential nucleation agents are dispersed, providing an increased number of potential nucleation sites.

3.3.1.2 Stir-Cast Growth

In the case of natural convection (unstirred melt), the solute distribution in the liquid is determined only by diffusion field and nucleation sites grow dendritically via columnar or equiaxed morphology. However, this is not the case for stirred

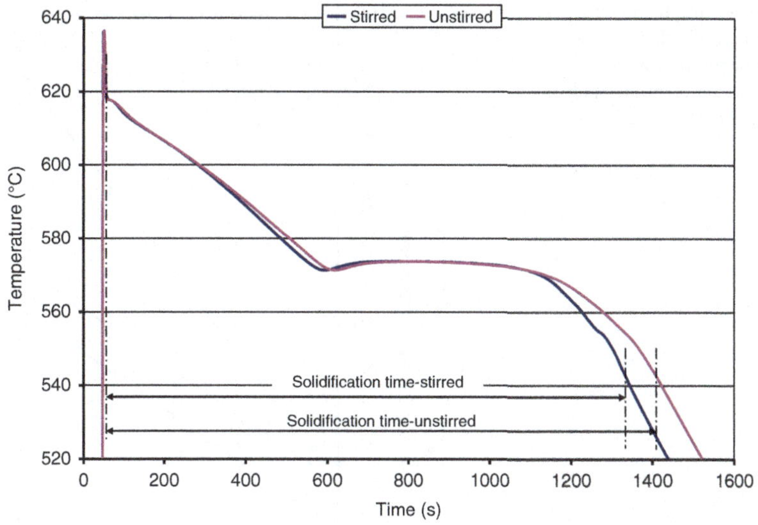

Fig. 3.13 Cooling curves comparison of unstirred and stirred samples

Fig. 3.14 Schematic of different thickness of boundary layer: (**a**) unstirred, (**b**) stirred melts [25]

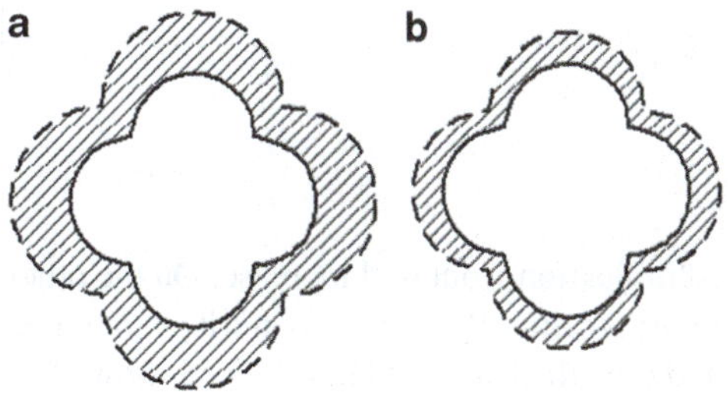

melt. In the presence of forced convection, solute distribution is more uniform. By stirring, the thickness of the stagnant boundary layer which exists around the primary particles is decreased due to thermal–solutal homogenization (Fig. 3.14) and directional heat flow is no longer dominant. The columnar growth of the grains gradually fades away and the growth morphology becomes smoother and more equiaxed. Beyond the boundary layer, the solute concentration is only governed by the convection field. Furthermore, vigorous agitation causes warmer liquid to shave off the tip of columnar dendrites to promote the formation of equiaxed structure in the stir cast parts. The so-called globule particles will eventually form with further stirring. Furthermore, due to vigorous agitation, no stable diffusion field can be formed and therefore growth would be sluggish in all directions, i.e., rounded refined structure.

According to Griffiths et al. [26], electromagnetic stirring promotes the columnar to equiaxed transition by increasing bulk liquid velocity as well as by increasing the silicon content of the Al–Si alloy. With increasing the bulk liquid velocity, the number of fragments produced by the interaction of the bulk liquid with the

Fig. 3.15 Graph showing the variation in the extent of the equiaxed region of different compositions and EMS currents, Al8.5Si [26]

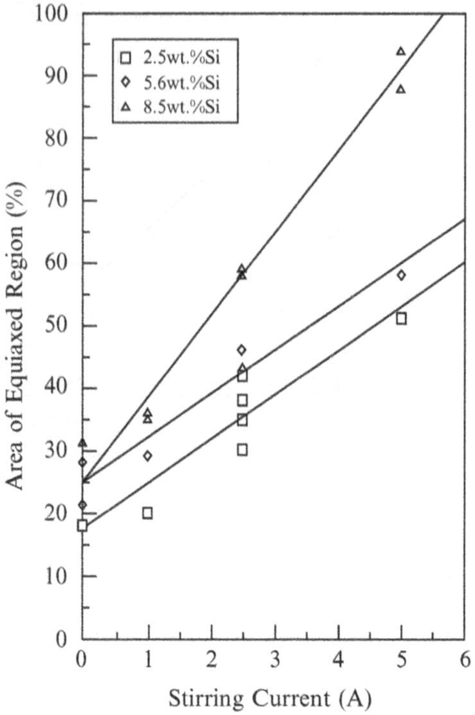

solidification front will increase. On the other hand, by rapid superheat removal due to the stirring, the broken dendrites are more stable with greater chance of survival and growth. The graph in Fig. 3.15 shows the percentage of the equiaxed regions for all the ingots cast with different currents. For example, in a series of experiments on Al8.5%Si, it has been shown that in the solidified ingot without stirring, the equiaxed region covered 31 % of the area while with EMS application, the equiaxed region increased to 88 % with a maximum current 5A. Also a decrease in the size of equiaxed grains was reported [25].

3.3.2 *Alloy Distribution*

Figures 3.16a and 3.17a are secondary electron images of α-Al particles after the microprobe ablations. The arrows show the scan direction and the dots are the beam affected zones. The point analysis was taken at about 5–10 μm intervals depending on the line length. Figures 3.16b and 3.17b show typical line scan results for samples having different titanium levels, and without/with stirring, respectively (of note that an individual particle was selected for microprobe analysis based on the polarized light microscopy).

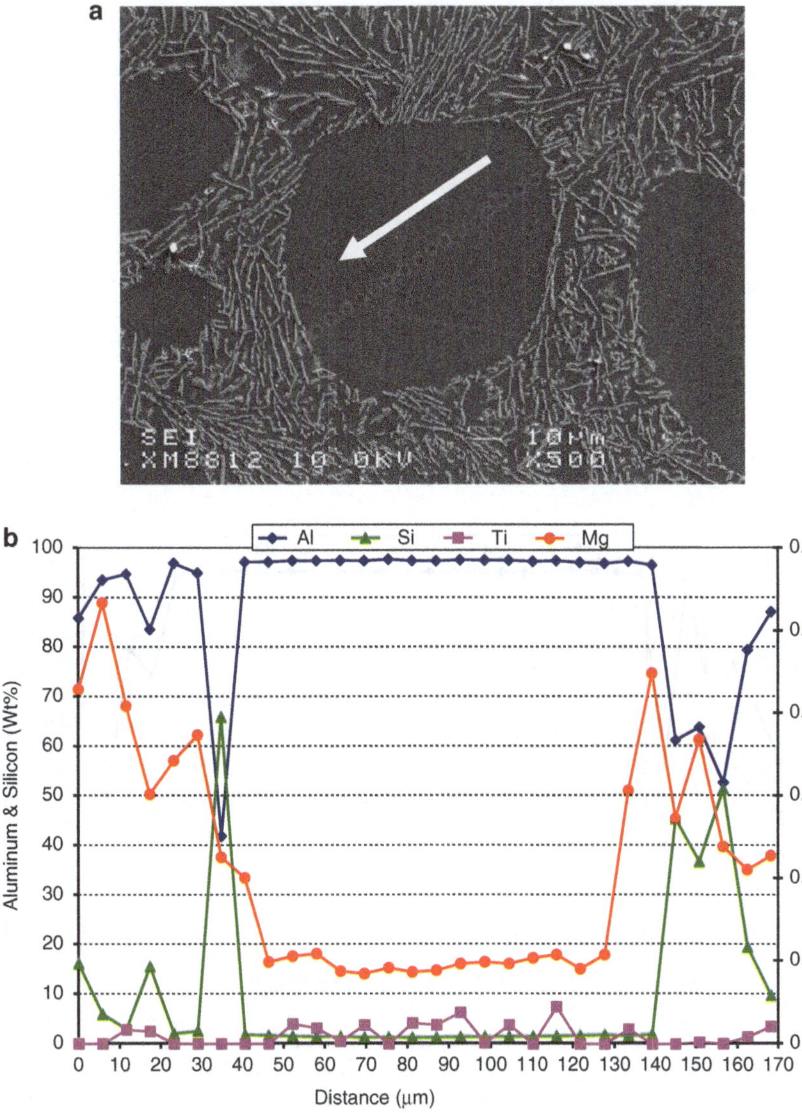

Fig. 3.16 A356 without Ti, nonstirred condition; (**a**) secondary electron image (SE), and (**b**) microprobe results for point scans in the direction shown by *arrow* (sample A1) [25]

The grain or globule boundaries could be easily detected by decreasing or increasing the alloying elements. The changes in aluminum concentration depend on the point where the analysis is taken and reduces whenever the probe passes across silicon flakes. The magnitude of reduction in aluminum concentration depends on whether the electron beam strikes exactly the silicon or a part of it. This is also the case for silicon percentage. For magnesium, the curve fluctuates

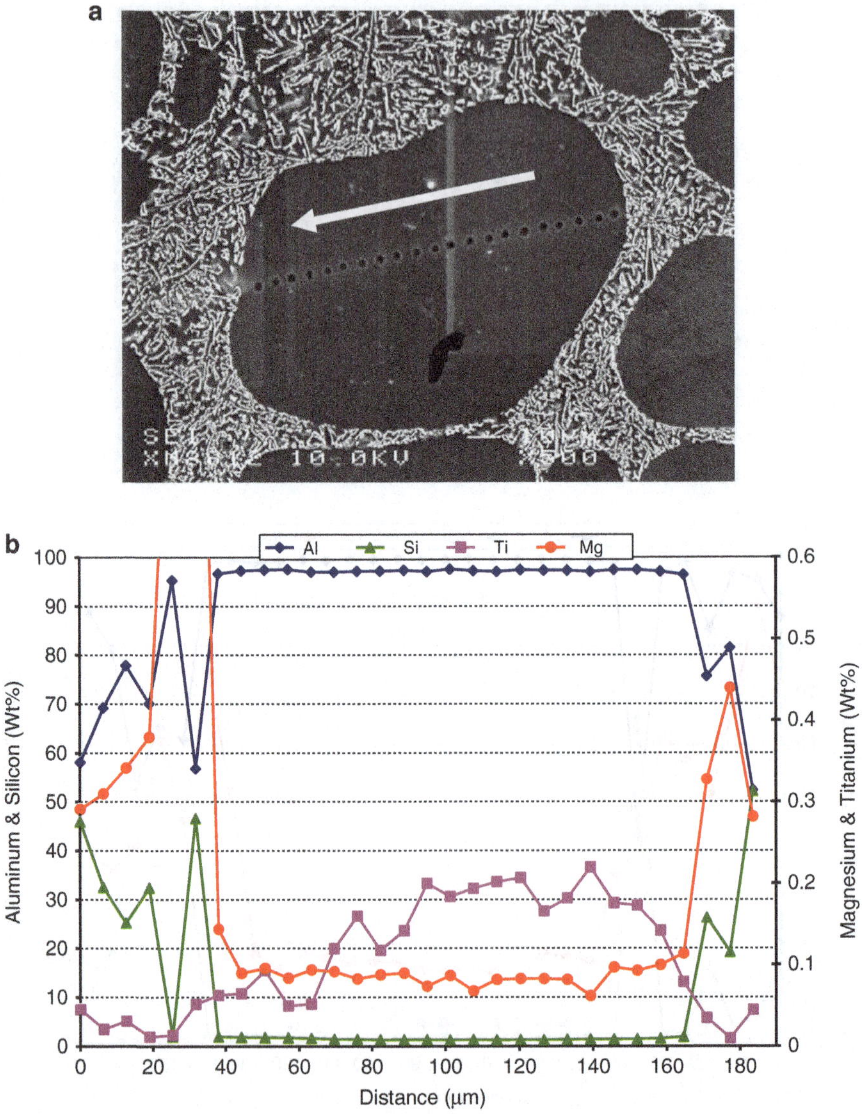

Fig. 3.17 Alloy 356 stirred; (**a**) secondary electron image (SE), and (**b**) microprobe results for point scans in the direction shown by *arrow* (sample B1) [25]

the same way as aluminum or silicon. The line scans for Mg are similar to the Si with higher values in the eutectic region. In the case of Ti, the reversed trend is observed. As depicted in Fig. 3.17b, a peak of Ti surrounded by areas of Ti depletion. The results of Ti are similar to those observed by Easton and Setiukov et al. [27, 28].

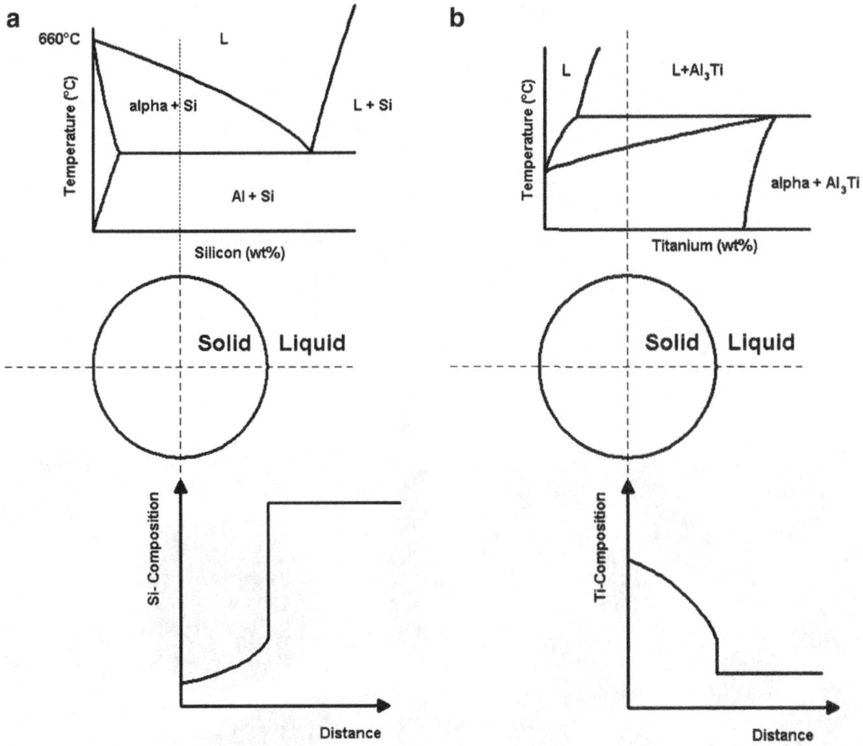

Fig. 3.18 Schematic representation of the binary diagrams for (**a**) Al–Si and (**b**) Al–Ti systems together with chemical compositional variation in a primary α-Al particle, assuming complete diffusion and mixing in the liquid [29]

The reason for decreasing of silicon, magnesium, or iron from the edge toward the center of α-Al cells is contributed to the solubility of these elements in α-Al as shown by the phase diagram of Al–Si alloy as a representative in Fig. 3.18a. According to the binary phase diagram, the first solid solution primary phase formed just below the liquidus is purer than the subsequent solid layers. In other words, the solubility of alloying elements in the primary α-Al particles increases with lowering the temperature as for instance for silicon, it reaches 1.65 wt% at eutectic temperature, Fig. 3.18a. For Ti, the case is different, where the first solid, center of the α-Al cell, is richer than the subsequent solid which is attributed to the following reasons:

- the peritectic reaction having partition coefficient $k > 1$, which means lower titanium content with decreasing the temperature and thus more titanium in the center of the particle as the first formed solid.
- Ti capability in generating nucleation sites for the primary α-Al particles due to the formation of Al_3Ti or TiB_2 compounds as described elsewhere (Fig. 3.18b), see Chap. 6.

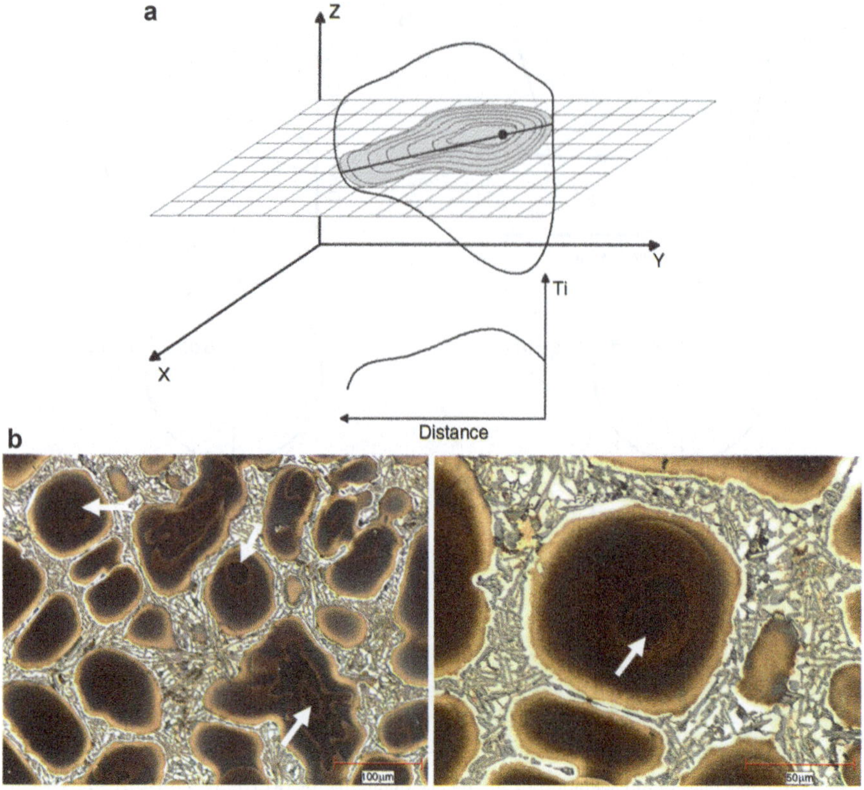

Fig. 3.19 (**a**) Schematic of an α-Al particle to show the nucleus and the variation of Ti in the polished surface, (**b**) typical optical micrographs to show nuclei locations (*arrows* show the nucleation points) [25]

It is also noticeable that the maximum Ti concentration may not be exactly at the center of α-Al globules. Such shifting of concentration peak to the left or right is attributed to the angle of sectioning of the sample and therefore detecting a centrally located particle is typically a fortuitous incident. In order to identify segregation, samples were etched by a $KMnO_4$, and NaOH and color were enhanced by using polarized light plus a sensitive tint filter. Figure 3.19a shows the more likely expected shape of primary α-Al particle which indicating nucleus forming on the right side of the particle. Figure 3.19b shows micrographs of actual segregations in the as-cast SSM billet. As it is evident, solidified layers are around the nuclei which are located almost anywhere within the primary α-Al cells.

Figures 3.20 and 3.21 show the silicon distribution for two alloys with different silicon concentrations and stirring speeds where the stirred alloy is compared with the unstirred one (silicon was chosen due to its higher solubility limit and hence better distinction in the solubility lines). As a reminder, in binary Al–Si phase diagram, the maximum solubility limit of Si in the primary α-Al is 1.65 % at the

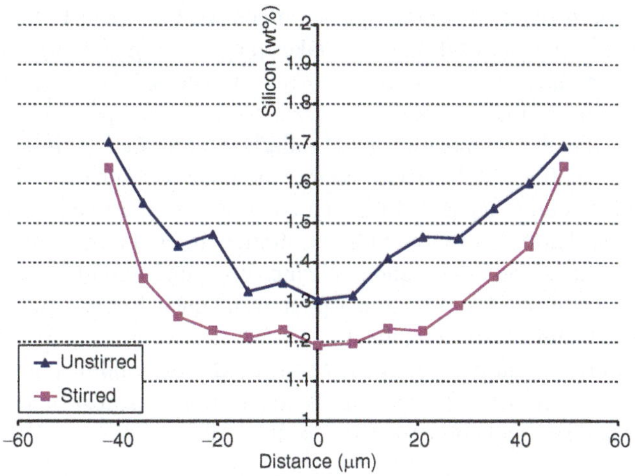

Fig. 3.20 Microprobe results showing the differences in Si distribution between stirred and unstirred conditions (samples A1 and A2) [25]

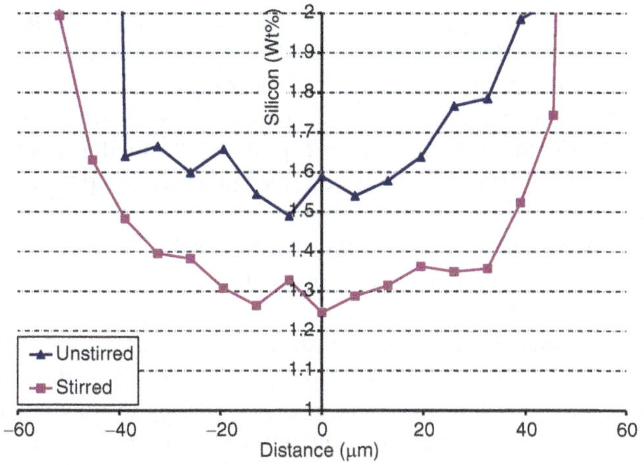

Fig. 3.21 Microprobe results show the differences in Si distribution between stirred and unstirred samples (samples B1 and B2) [25]

eutectic temperature. However, the maximum solubility limit may vary depending on the alloying elements and cooling rate, i.e., ternary or multi alloying systems may have lower solubility limit for Si.

Solute buildup is the nature of alloy systems. Atoms of solutes build up a boundary layer which has a different composition from both solid and liquid. According to the binary phase diagrams with partition coefficient $k < 1$, the first solid is always purer than the liquid and solute is rejected into the liquid. With decreasing the temperature, the next layer of solid will be slightly richer in solute than the previous layer. As this sequence proceeds, the liquid becomes progressively richer in solute and solidification takes place at progressively lower temperature. However, since there is almost no diffusion in the solid, the separate layers of solid retain their original compositions. Thus, the mean composition of the solid is always lower than the composition at the solid/liquid interface. This is what one can see from Figs. 3.20 and 3.21. These figures show the percentage of Si in the α-Al particles and it is quite clear that the first solid, center of Al cell, is always purer than the others. Furthermore, the Si concentration at any point across the primary α-Al particle varies with and without stirring. It is quite important to note that all line scans were collected for specimens poured and quenched at same temperatures and had almost similar silicon content.

In order to better clarify the effect of stirring during solidification, the issue is discussed under two separate headings of nucleation and growth which provide better understanding and comprehension of the topic.

3.3.2.1 Variation of Chemical Composition Due to Stirring

Nucleation in stir cast solidification starts at higher temperature compared to the conventional unstirred samples (Fig. 3.12). Thus, according to the phase diagram, the first solid to form is more depleted in solute, e.g., with less silicon concentration. This is obvious through Figs. 3.20 and 3.21 showing less silicon concentration in stirred samples. If the phase diagram does not shift the same way as the nucleation temperature, then such conclusion is correct. Since, the solidus and liquidus lines movement are not clear, the concept is described in term of fluid flow and boundary layer concentration. As discussed, the applied convection removes solute concentration at the solidification front and thus the next solid to form is not as rich as expected from phase diagram or the unstirred alloy.

From the growth standpoint, these nuclei grow in spherical manners due to unstable thinner boundary layers and also effect of "shaving-off" action of the flow. This thinner boundary layer therefore provides less growth (primary phase can grow within the extent of boundary layer) comparing to conventional method. Figure 3.22 shows a schematic representation of solute distribution in boundary layer due to the stirring. The boundary layer thickness decreases due to the stirring and thus the degree of constitutional undercooling is also decreased. Also with purer nucleation sites due to the stirring, the bulk liquid becomes more and more saturated in alloying elements and thus a new composition forms for bulk liquid, named "C'." Martorano et al. [30] suggested a new mechanism for the columnar to equiaxed transition and they have shown that the columnar as well as the equiaxed dendritic growth will stop if the solute rejection by the equiaxed grains ahead of the columnar front is sufficiently large. As a result, more accumulation of solute in the remaining liquid could be considered as the reason for the obstruction of further growth according to the solutal blocking theory [30]. On the other hand, the liquid has more alloying elements and as a result the remaining liquid has lower liquidus temperature. This lower temperature plus thinner boundary layer causes smaller

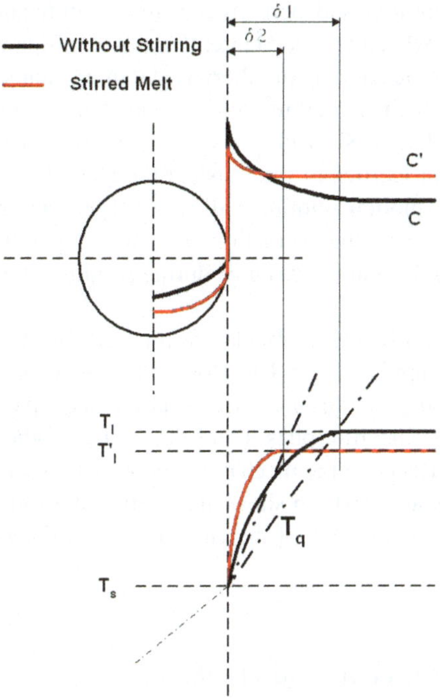

C_0: Initial solute concentration

C: Average solute concentration in liquid

C′: Average solute concentration in liquid after stirring

δ: Thickness of boundary layer, $\delta_1$before and δ_2 after stirring

T_l: Liquidus temperature

T'_l: Liquidus temperature after stirring

T_s: Non equilibrium solidus temperature

T_q: Temperature imposed by temperature gradient arising from heat flow in the casting

Fig. 3.22 Schematic of solute distribution in solid/liquid interface [25]

constitutional undercooling and therefore the growth is hindered due to lack of undercooling.

In another study by Fan and coworkers [15], it was reported that the growth concept in the stirred alloys is not the same as those mentioned in the past, for example, Vogel and Cantor [14]. Using Monte-Carlo based simulation, they showed that growth enhancement occurs under pure laminar flow, but when the flow becomes pure turbulent, there is progressive growth retardation occurring as compared to pure diffusive flow (Fig. 3.7). In contrast to laminar flow that has greater growth probability, for turbulent flow the growth rate is initially high due to the faster transport of solvent atoms but gradually retards with the progress of solidification [15].

3.3.3 Particle Size Alteration During Stirring

In semi-solid casting of hypoeutectic Al–Si alloys, the morphology and size of the eutectic silicon are important parameters affecting the mechanical properties of the as-cast products. In addition, the chemistry of the billet with respect to impurities and alloying elements such as Fe, Mn, Cu, Zn, Mg, and Ti may result in some complications in the microstructure due to the formation of intermetallic phases.

For instance, iron is one of the most common and perhaps the most significant alloying or impurity elements in Al–Si alloys; either added intentionally to mitigate certain characteristics such as soldering or die sticking or otherwise as an unwanted impurity. The Fe addition imparts two distinct features; on one hand, it forms different intermetallics together with Al and Si, with most of them regarded considerably harmful to the mechanical properties of the finished product and on the other hand, it reduces the interaction of molten aluminum alloys with permanent molds, i.e., soldering or die sticking, due to the formation of a thin layer of intermetallics at the interface between the die and the castings during solidification [31–33].

In the SSM processing, impurities and alloying elements are rejected into the liquid and solidify under shear forces and rapid cooling. Therefore, it is of technical and technological importance to study the possible microstructural changes, particularly evolution of silicon and iron-intermetallic phases and those of the other intermetallics phases, and their possible effects on the mechanical properties of the as-cast products. In this section, attempts are made to show that stirring not only produces a globular structure, but also alters the size of the eutectic silicon flakes and iron intermetallic needles.

3.3.3.1 Effects of Superheat and Stirring on the Eutectic Silicon and Iron Intermetallic Phases, EMS

Sand Mold

The selected optical micrographs of the as-cast conventional and EMS processed alloys are shown in Fig. 3.23. The micrographs signify that: (a) the Fe intermetallics in the both cases are β-type with platelet morphology, and (b) the eutectic silicon has coarse flake morphology due to the slow solidification rate of the alloy in the

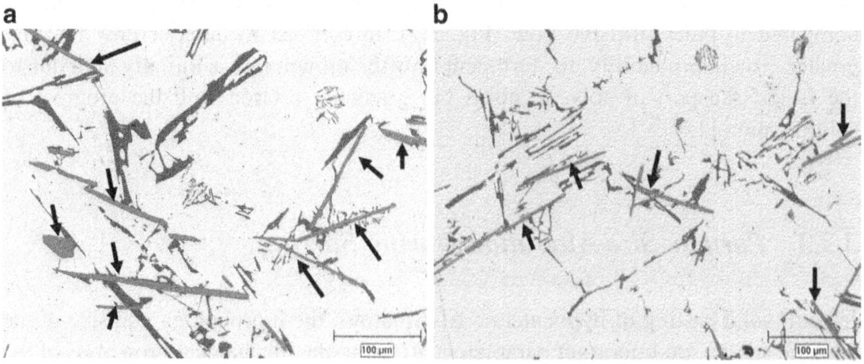

Fig. 3.23 Optical micrographs showing the effect of; (**a**) conventional process and (**b**) EMS process on the formation of β-Fe compound for Al7Si0.8Fe (Sand mold, 690 °C) (*arrows* show couple of iron intermetallics)

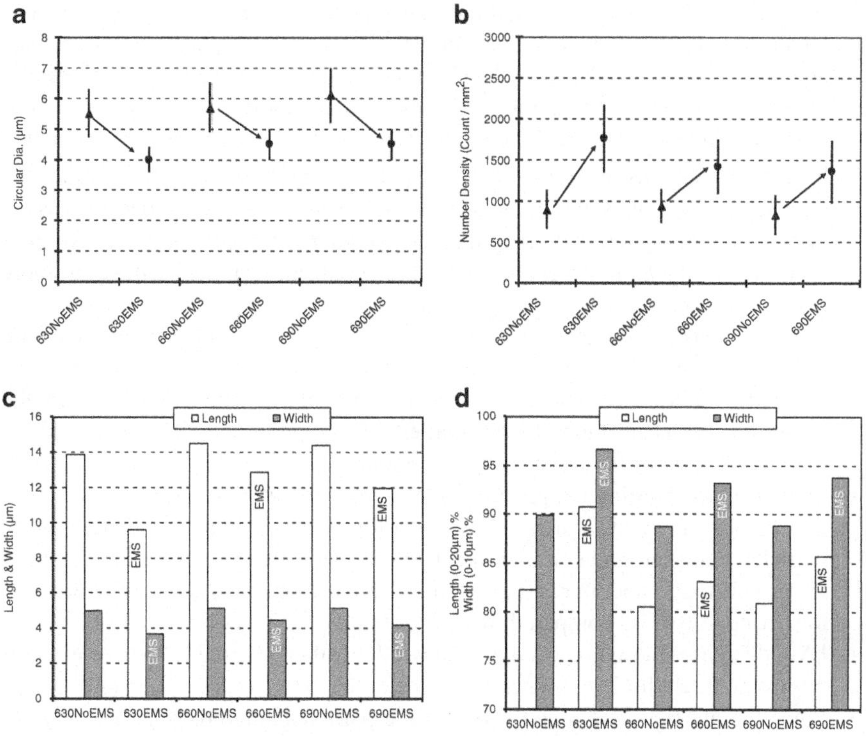

Fig. 3.24 Image analysis parameter for sand casting, eutectic silicon: (**a**) average circular diameter, (**b**) number density, (**c**) average length and width, and (**d**) percentage of particles having length and width less than 20 and 10 μm, respectively [34]

sand mold. The electromagnetic stirring not only leads to the formation of globular primary α-Al phase but also refines the iron and the eutectic silicon phases.[2]

Quantitative metallographic results for eutectic silicon are shown in Fig. 3.24 and summarized below (quantitative metallography will be explained in Sect. 4.3.2) [34];

- Average silicon equivalent circular area diameter increases with increasing the pouring temperature (Fig. 3.24a). This is more evident in the conventional castings. Higher pouring temperatures deactivate (remelt) most of the α-Al nucleation sites and thus promoting the formation of larger columnar dendrites. The mass of the remaining liquid to undergo eutectic transformation is distributed among smaller number of primary dendrites giving larger individual liquid

[2]One of the most complicated issues in image analysis is thresholding between iron intermetallics and other phases especially eutectic silicon since both have approximately the same gray level. For this case, a special routine was written for exactly finding the color difference, eliminating small debris remaining from polishing, and finally doing the analysis.

pools across the billet. The larger individual pools coupled with lower number of localized heat sinks due to lower number of primary α-Al phase should encourage slow and prolonged growth of eutectic constituents such as Si to render coarse flakes. Generally, the average silicon diameter decreases by stirring. Of note that diameter of a circle having the same area as the Si particle is calculated by the image analysis software, i.e., $2\sqrt{\text{area}/\pi}$, with the assumption that the examined object shape is close to a circle. As a result, the error is increased as the particle is more lamellar/rectangular. The smaller error bars for stirred samples may suggest the particles geometry is changing from rectangular flakes to more equiaxed flakes.

- Number density (number of silicon particles per unit area) increases with reducing the pouring temperature in conventional casting (Fig. 3.24b). The vigorous agitation also increases the number of the eutectic silicon in the structure. This is attributed to the smaller eutectic pools and thermal and solutal homogenization of the remaining bulk liquid.
- Increasing the pouring temperatures both with/without EMS increases the average length of silicon particles (Fig. 3.24c). The difference in length of Si particles between conventional and EMS treated alloys is more significant at lower pouring temperatures. For example, the difference is 31 % reduction due to stirring at 630 °C compared to 16 % at 690 °C.
- EMS decreases the average length and width of silicon particles, but this effect is more significant for length than width due to the nature of the flake morphology.
- Since the average length and width of the particles may not be so sensitive to show the changes due to stirring and pouring temperature, the percentage of particles with length less than 20 μm and width less than 10 μm are presented in Fig. 3.24d. The percentage of smaller particles increased for the stirred alloy and such rise is greater at lower pouring temperature.

The modification of Si particles achieved by stirring is of great importance as porosity is not an issue in contrast to the modification of Si by Sr addition, which is usually accompanied by increasing in porosity content (detrimental to mechanical properties). The refinement of eutectic silicon flakes and Fe-intermetallics by EMS could lead to the utilization of the presently undesirable high Fe content aluminum alloys in producing high integrity cast components.

The results for iron intermetallics are shown in Fig. 3.25 and summarized later [34]; (in the present work, the casting conditions such as superheat, cooling rate, iron level, and alloying elements were in the range that prevent the formation of α-iron or the Chinese script morphology [35]);

- In conventional casting, the average number density of iron particles is almost the same for various pouring temperatures but increases due to the stirring in all the pouring conditions. The greatest rise however belongs to the lowest pouring temperature (Fig. 3.25a).
- The percentage of the Fe-based particles having aspect ratio greater than 10 increased with increasing the pouring temperature (Fig. 3.25b). This is due to the larger α-Al grains and thus greater space for growing β-platelets. The entire

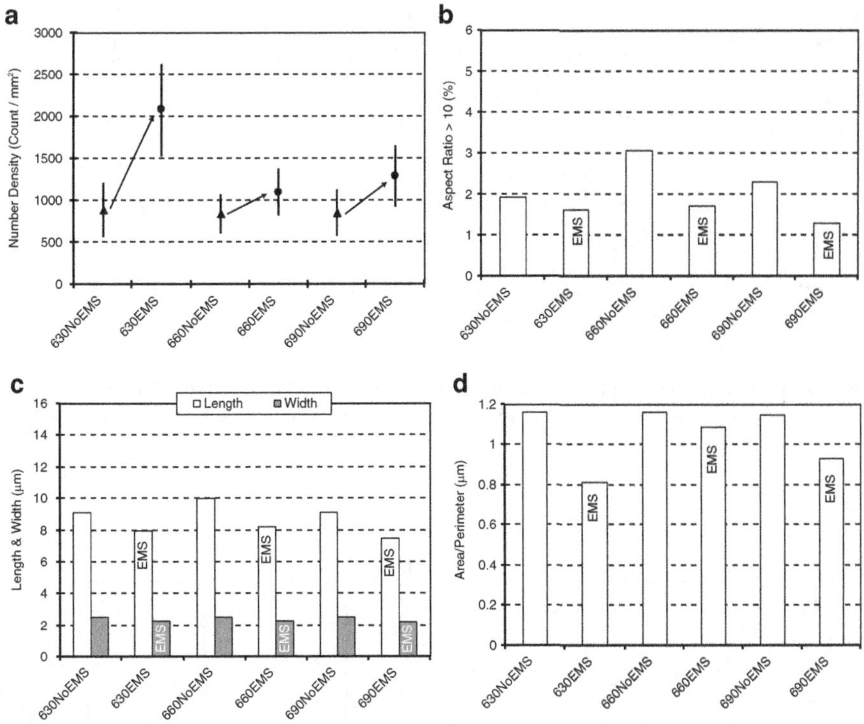

Fig. 3.25 Image analysis parameter for sand casting, iron intermetallics: (**a**) number density, (**b**) percentage of particles having an aspect ratio greater than 10, (**c**) average length and width, and (**d**) A/P ratio [34]

stirring results show approximately the same values since stirring diminishes the outcomes from superheat. This graph confirms the results of number density.

- With stirring, the average particle size of iron intermetallics decreased (Fig. 3.25c). The average size mainly referred to the length of the needles since the width is so small and any changes in the width are insignificant to overall particle size. As it was mentioned before, thin platelet morphology of the particles causes less difference in the width values. As the last parameter, A/P decreased with stirring and hence particles become smaller.

Copper Mold

The selected micrographs of the samples cast in copper mold are presented in Fig. 3.26. Compared to sand casting (Fig. 3.23), the β-iron intermetallics and silicon flakes have been refined and are thinner and smaller. This makes image analysis of the resulting structure more complicated. Therefore, all the analyses for iron and silicon flakes were performed at ~×500 magnification. For eutectic silicon, the results are summarized later (Fig. 3.27) [34];

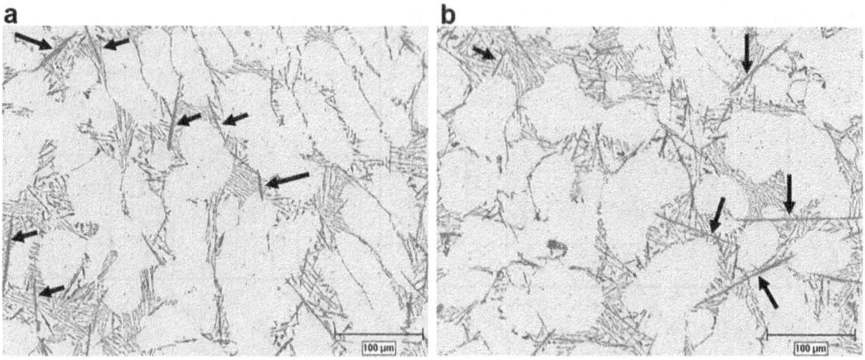

Fig. 3.26 Optical micrographs showing the effect of: (**a**) conventional and (**b**) EMS casting on the formation of β-Fe compound for Al–7%Si–0.8%Fe (copper mold, 690 °C) (*arrows* show some of the iron intermetallics)

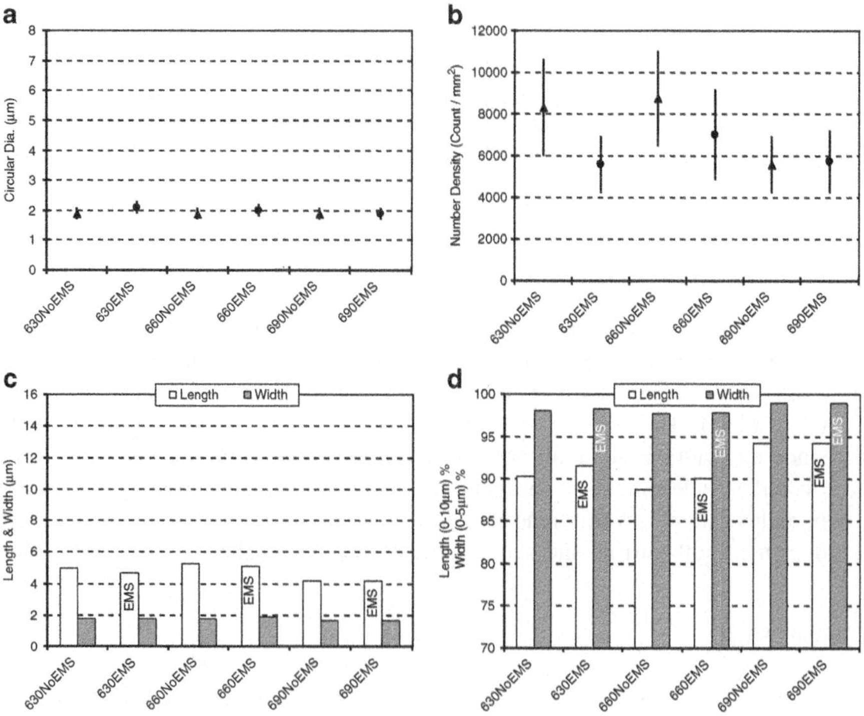

Fig. 3.27 Image analysis of the eutectic silicon for the copper mold casting: (**a**) average circular diameter, (**b**) number density, (**c**) average length and width, and (**d**) percentage of length and width less than 10 and 5 μm, respectively [34]

- Average circular diameters in all cases were almost the same. This is coming from the fact that cooling rate is the dominant parameter affecting the size.
- Number density for alloys poured at lower superheat exhibits a contradictory trend to that of sand casting—Fig. 3.24, where stirring reduces the number of silicon particles and at best, 690 °C, there is no difference between stirred and unstirred conditions. However, it is quite important to note that number density of silicon particles increased remarkably comparing to sand casting. Such finding may be superficial due to the limited resolving power of light microscopy where agglomerated silicon particles, due to stirring, have been counted as one. Such hypothesis may be corroborated by close examination of the optical micrographs in Fig. 3.23, where there is a distinct reduction in the silicon size. As mentioned before, it is the authors' belief that in contrast to sand mold billets, the influence of low pouring temperature in copper mold is overshadowed by the faster cooling rate.
- Results from length and width also have the same trend and high cooling rate overshadows any effect of electromagnetic stirring. The final conclusion is that the difference between the stirred and unstirred conditions is not significant and cooling rate diminishes the effect of stirring.

The results of the iron addition to the alloy with/without EMS are shown in Fig. 3.28. Cooling rate plays a critical role on the nucleation and formation of different Fe intermetallic phases and hence different cooling rates stabilize different morphologies. In the absence of manganese, the iron compounds crystallize only as β-phase as the stable phase at conventional cooling rates.

It is well proven that with increasing the cooling rate, the average length of β-AlFeSi platelets and also its crystallization temperature decrease exponentially with cooling rate and at the critical cooling rate of ~20 $°Cs^{-1}$, the morphology also changes to the Chinese script or α-AlFeSi phase [36]. The effect of decreasing the size of β-AlFeSi intermetallics is obvious in Fig. 3.26. The image analysis results also confirm and support this concept. The general trend is the same as the sand mold billets with small differences due to the higher cooling rate. In fact, the effect of EMS on iron morphology was masked by faster cooling of the water-cooled copper mold.

In conclusion, stirring leads to earlier nucleation of primary particles and reduces the entire solidification time. It further causes the disintegration of dendrites branches. The agitation prevents the formation of stable diffusion fields around solid particles and without a stable diffusion field, growth is interrupted and the fragmented dendrite branches transform to rounded globules without much growth. It was also shown stirring changes the alloy content of the primary α-Al particles and especially reducing the silicon concentration of the primary particles and resulting in the formation of a more enriched eutectic. The forced convection due to stirring is beneficial by inducing homogenous composition, refining primary particles phase, refining silicon and intermetallic particles and therefore improving mechanical properties of the alloy.

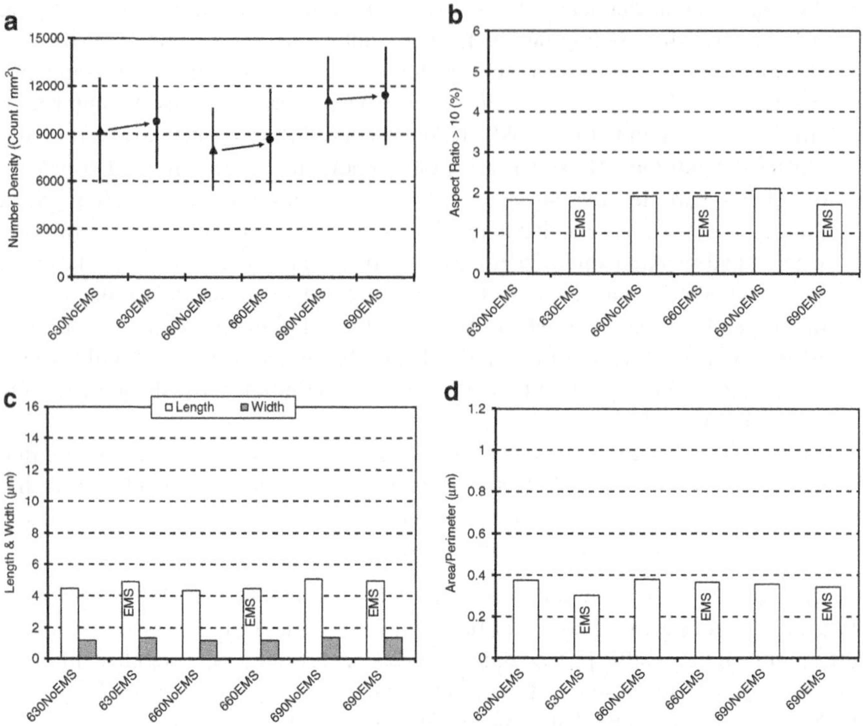

Fig. 3.28 Image analysis parameter for copper mold casting, iron intermetallics: (**a**) number density, (**b**) percentage of particles with aspect ratio >10, (**c**) average length and width, and (**d**) A/P ratio [34]

References

1. M.C. Flemings, *Solidification Processing* (McGraw-Hill, New York, 1974)
2. F.A. Crossley, R.D. Fisher, A.G. Metcalfe, Viscous shear as an agent for grain refinement in cast metals. Trans. Metal. Soc. AIME **221**, 419–420 (1961)
3. F.A. Crossley, Magnetic stirring: a new way to refine metal structure. Iron Age **186**, 102–104 (1960)
4. V. Kondik, Microstructure of cast metal. Acta Met. **6**, 660 (1958). Letter to editor
5. D.R. Uhlmann, T.P. Seward, B. Chalmers, The effect of magnetic fields on the structure of metal alloy castings. Trans. Metal. Soc. AIME **236**, 527–531 (1966)
6. B. Chalmers, *Principles of Solidification* (Wiley, New York, 1964)
7. W.J. Boettinger, F.S. Biancaniello, S.R. Coriell, Solutal convection induced macrosegregation and the dendrite to composite transition in off-eutectic alloys. Metal. Trans. A **12A**, 321–327 (1981)
8. M.H. Johnston, R.A. Parr, The influence of acceleration forces on dendritic growth and grain structure. Metal. Trans. B **13B**, 85–90 (1982)
9. S. Wojeiechowski, B. Chalmers, The influence of mechanical stirring on the columnar to equiaxed transition in aluminum copper alloys. TMS-AIME **242**, 690–698 (1968)

10. R.D. Doherty, H.I. Lee, E.A. Feest, Microstructure of stir-cast metals. Mater. Sci. Eng. **A65**, 181–189 (1984)
11. M.C. Flemings, Behavior of metal alloys in the semi-solid state. Metal. Trans. A **22A**, 952–981 (1991)
12. A.Hellawell, Grain evolution in conventional and Rheo-castings. in *4th International Conference on Semi-Solid Processing of Alloys and Composites* (Sheffield, England, 1996), 60–65
13. Z. Fan, Semisolid metal processing. Inter. Mater. Rev. **47**, 49–85 (2002)
14. A. Vogel, B. Cantor, Stability of a spherical particle growing from a stirred melt. J. Cryst. Growth **37**, 309–316 (1977)
15. A. Das, S. Ji, Z. Fan, Morphological development of solidification structures under forced fluid flow: a Monte-Carlo simulation. Acta Mater. **50**, 4571–4585 (2002)
16. D.B. Spencer, R. Mehrabian, M.C. Flemings, Rheological behavior of Sn-15 Pct Pb in the crystallization range. Metal. Trans. **3**, 1925–1932 (1972)
17. P.A. Joly, R. Mehrabian, The rheology of a partially solid alloy. J. Mater. Sci. **11**, 1393–1418 (1976)
18. A. Vogel, R.D. Doherty, B. Cantor, Stir-cast microstructure and slow crack growth. in *Proceedings of the Solidification and Casting of Metals* (The Metals Society, London, 1979), pp. 518–525
19. K.A. Jackson, J.D. Hunt, D.R. Uhlmann, T.P. Seward, On the origin of the equiaxed zone in castings. Trans. Metallur. Soc. AIME **236**, 149–157 (1966)
20. A. Mortensen, On the influence of coarsening on microsegregation. Metal. Trans. A **20A**, 247–253 (1989)
21. S. Nafisi, J. Szpunar, H. Vali, R. Ghomashchi, Grain misorientation in Thixo-billets prepared by melt stirring. Mater. Charact. **60**, 938–945 (2009)
22. T. Campanella, C. Charbon, M. Rappaz, Grain refinement induced by electromagnetic stirring: a dendrite fragmentation criterion. Metal. Trans. A **35A**, 3201–3210 (2004)
23. N. Limodin, L. Salvo, E. Boller, M. Suery, M. Felberbaum, S. Gailliegue, K. Madi, In situ and real-time 3-D microtomography investigation of dendritic solidification in an Al-10wt% Cu Alloy. Acta Mater. **57**, 2300–2310 (2009)
24. M. Chen, TZ. Kattamis, Dendrite coarsening during directional solidification of Al-Cu-Mn alloys. Mater. Sci. Eng. **A247**, 239–247 (1998)
25. S. Nafisi, R. Ghomashchi, Effect of stirring on solidification pattern and alloy distribution during semi-solid-metal casting. Mater. Sci. Eng. **A437**, 388–395 (2006)
26. W.D. Griffiths, D.G. McCartney, The effect of electromagnetic stirring during solidification on the structure of Al-Si alloys. Mater. Sci. Eng. **A216**, 47–60 (1996)
27. M.A. Easton, H. StJohn, Partitioning of titanium during solidification of aluminium alloys. Mater. Sci. Technol. **16**, 993–1000 (2000)
28. O.A. Setiukov, I.N. Fridlyander, Peculiarities of Ti dendritic segregation in aluminum alloys. Mater. Sci. Forum **217–222**, 195–200 (1996)
29. S. Nafisi, R. Ghomashchi, The effect of dissolved Ti on the primary α-Al grain and globule size in the conventional and semi-solid casting of 356 Al-Si alloy. J. Mater. Sci. **41**, 7954–7963 (2006)
30. M.A. Martorano, C. Beckermann, C. Gandin, A solutal interaction mechanism for the columnar-to-equiaxed transition in alloy solidification. Metal. Trans. A **34**, 1657–1674 (2003)
31. A. Couture, Iron in aluminum casting alloy—a literature survey. AFS Int. Cast Metals J. **6**, 9–17 (1981)
32. P.N. Crepeau, Effect of iron in Al-Si casting alloys: a critical review. AFS Trans. (1995), 361–366
33. H.R. Shahverdi, M.R. Ghomashchi, H. Shabestari, J. Hedjazi, Microstructural evolution due to interfacial reaction between liquid aluminum and solid iron. J. Mater. Proc. Tech. **124**, 345–352 (2002)

34. S. Nafisi, D. Emadi, M.T. Shehata, R. Ghomashchi, Effects of electro-magnetic stirring and superheat on the microstructural characteristics of Al-Si-Fe alloy. Mater. Sci. Eng. **A432**, 71–83 (2006)
35. N.A. Belov, A.A. Aksenov, D.G. Eskin, *Iron in Aluminum Alloys* (Taylor & Francis, New York, 2002)
36. L. Backerud, G. Chai, J. Tamminen, *Solidification Characteristics of Aluminum Alloys, Volume 2, Foundry Alloys* (American Foundry Society, Des Plaines, 1990)

Chapter 4
Methodology of SSM Characterization

Abstract The microstructure of the SSM alloy billets plays an important role during fabrication of finished engineering component and its performance in service. As a result, it is critically important to characterize the microstructure to ensure high-quality feed stock. This chapter provides a detailed account of the characterization techniques available to SSM researcher to confidently examine the quality of as-cast billets through thermal analysis, rheological characterization, and quantitative metallography.

Conventional solidification of hypoeutectic Al–Si foundry alloys takes place with dendritic formation of primary α-Al phase within the liquid. The alloy composition, temperature gradient within the melt, thermo-fluid convection, and rate of heat extraction and the resulting constitutional supercooling are the most effective parameters on the morphology of the primary α-Al phase. Variation of any of these factors during solidification would alter the as-cast structure. For instance, introduction of agitation (forced convection) into the solidifying melt changes the distribution of alloying elements and localized chemical composition, could remove constitutional supercooling, and promote dendrite-to-equiaxed transition, i.e., break down and globularization of the α-Al phase. Degeneration of the α-Al phase results in some opportunities which are of interest from commercial viewpoints.

The advantages of the SSM processing along with different available technologies as discussed in Chap. 2 are enormous. However, the lack of industrial interest in the 80s and 90s stemmed mainly from the high cost of billet preparation, the issue of recycling the returned and scraped parts, and to some extent the lack of proper characterization of the both semifinished billets and finished engineering components. Recently, the cost issue is resolved by introducing novel cost-effective rheocasting techniques and development of new alloying systems.

In order to generate a semi-solid structure, the alloy system plays the key role where the coexistence of liquid and solid within a temperature range is the prerequisite for the slurry preparation. The mechanics and mechanisms of the primary particles' evolution, dendrite to equiaxed transformation, is the next concern since the formation of globular morphology is expected to enhance die filling and improve mechanical properties of as-cast parts. The ideal microstructure for SSM slurry is fine spherical solid particles uniformly distributed within a liquid matrix.

S. Nafisi, R. Ghomashchi, *Semi-Solid Processing of Aluminum Alloys*,
DOI 10.1007/978-3-319-40335-9_4

The solid fraction should be considered carefully, since low fraction solid may lead to SSM slurry handling and mold filling difficulties due to insufficient viscosity and turbulence while high fraction solid adversely affects the die filling and requires more powerful machinery and thus increase the cost of manufacturing.

Based on the aforesaid requirements, characterization of semi-solid material is a necessity to confirm, modify, and obtain optimum structure for SSM-shaped components. This knowledge not only provides an idea about the material, but also leads to better understanding of rheological behavior and eventually improve the mechanical properties of cast pieces.

The term "*characterization*" covers a wide range of thermal, mechanical, and microstructural analytical techniques used to evaluate physical, mechanical, and metallurgical parameters of SSM billets and finished products. The SSM billets are often required to be examined for their solidification pattern, rheological behavior, and microstructure. The outcome of such studies helps in understanding / predicting the fluid flow and die-filling behavior of the SSM billets along with the possible mechanical and load-bearing characteristics of finished product.

The following sections provide an account of thermal (solidification), rheological, and structural methods available to characterize SSM products. In addition, the experimental procedures employed to prepare samples for characterization of SSM billets are also explained.

4.1 Solidification Characterization

The quality of finished product is tied up with the quality of feedstock (billets). The integrity of the billets, however, is dependent on the SSM processing parameters and these parameters are interpreted in terms of solidification condition. The cooling curve is the fingerprint of an alloy solidification behavior as it provides information on nucleation and growth of phases form during liquid to solid transformation. The following explains how the solidification behavior of the billets should be monitored and characterized.

Following the preparation of molten alloy in a resistance heating furnace, melting of Al–Si ingots and degassing using argon, a portion of the molten charge was poured into graphite cups of 25 mm inner diameter and 5 mm wall thickness (Fig. 4.1). Cups were held inside the crucible for approximately 1 min prior to tests to reach equilibrium condition ensuring uniform temperature distribution across the sample at the beginning of solidification. Each cup with ~50 g of alloy was transferred to the testing platform and two K-type thermocouples (0.8 mm diameter) were quickly immersed into the melt near the center and wall of the mold with their tips at 10 mm from the mold bottom. Temperature readings were collected by a high-speed high-resolution data acquisition system (National Instrument SCXI-1102) at ten readings per second sampling rate. To ensure the radial heat flow, insulating plates were placed above and below the sample cup. To improve data consistency and reproducibility, the same thermocouples were used for all tests by

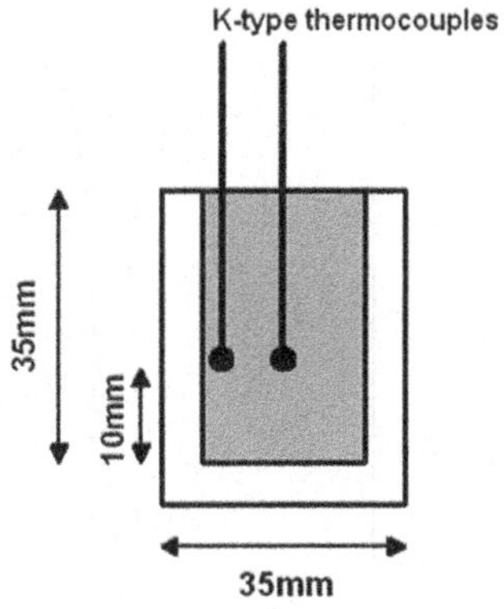

Fig. 4.1 Graphite cup geometry for thermal analysis purpose

placing the thermocouples in a 1 mm (inner diameter) stainless steel sheath. The protective sheath saved thermocouples after solidification where they could be easily removed from sample and reused.

As for metallography, transverse section at the tip of the thermocouples was cut and for consistency of results, the area between the center and the wall of the entire samples was examined (quarter area).

According to the literature, the cooling rate can either be calculated above liquidus temperature or in the mushy zone. However, the cooling rate is often registered high right after mold filling due to initial rapid heat dissipation. The cooling rate gradually slows down in the mushy zone. As an example, the cooling rates for the graphite molds used were between 1.5–2 °Cs^{-1} and 0.5–0.6 °Cs^{-1} above the liquidus and in the mushy zone, respectively. The analysis of the thermal data was carried out following Backerud et al. and Tuttle [1, 2]. Since there are some dissimilarities in the definition of the critical parameters in the literature, the following nomenclatures are defined in this book. The points are also identified on an actual cooling curve in Fig. 4.2 (only for graphite cup samples):

- $T_{\mathrm{nuc}_{\mathrm{Al}}}$: Start of primary α-Al dendrites nucleation
- $T_{\mathrm{min}_{\mathrm{Al}}}$: Unsteady state growth temperature, the temperature beyond which the newly nucleated crystals grow to such extent that the latent heat liberated surpasses the heat extracted from the sample
- $T_{\mathrm{g}_{\mathrm{Al}}}$: Recalescence of steady-state growth temperature due to release of latent heat of primary α-Al dendrites
- ΔT_{Rec}: Temperature difference between unsteady ($T_{\mathrm{min}_{\mathrm{Al}}}$) and steady ($T_{\mathrm{g}_{\mathrm{Al}}}$) state growth temperatures of primary α-Al particles (recalescence)
- t_{Rec}: Recalescence time, time difference between $T_{\mathrm{min}_{\mathrm{Al}}}$ and $T_{\mathrm{g}_{\mathrm{Al}}}$, the times associated with $T_{\mathrm{min}_{\mathrm{Al}}}$ and $T_{\mathrm{g}_{\mathrm{Al}}}$ (in the literature, it was labeled as liquidus undercooling time [2, 3])
- $T_{\mathrm{nuc}_{\mathrm{eut}}}$: Start of eutectic nucleation
- $T_{\mathrm{min}_{\mathrm{eut}}}$ and $T_{\mathrm{max}_{\mathrm{eut}}}$: Minimum and maximum of eutectic temperatures
- $\Delta\theta$: Variation of eutectic recalescence ($T_{\mathrm{max}_{\mathrm{eut}}} - T_{\mathrm{min}_{\mathrm{eut}}}$)

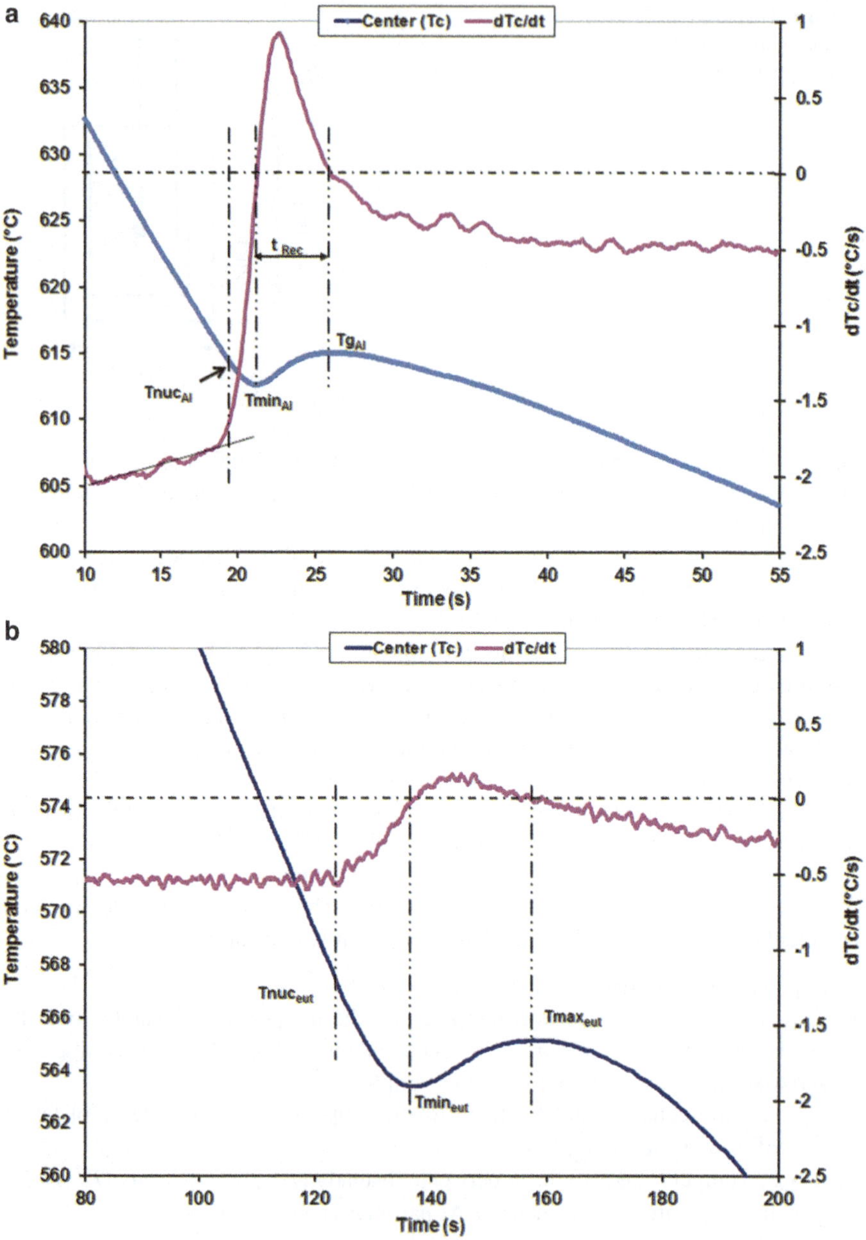

Fig. 4.2 Cooling curve and first derivative of 356 alloy: (**a**) α-Al formation region, (**b**) eutectic region

- T_{end}: Solidification termination
- ΔT_{α}: α-Al solidification range ($T_{nuc_{Al}} - T_{nuc_{eut}}$)
- ΔT_{eut}: Eutectic solidification range ($T_{nuc_{eut}} - T_{end}$)

Undercooling was defined in many academic literatures (e.g., [4]) as the difference between $T_{min_{Al}}$ and $T_{g_{Al}}$ with $T_{min_{Al}}$ defined as the starting point of solidification. The authors would like to reemphasize on the finding of Backerud et al. [1] where the actual solidification starts above the $T_{min_{Al}}$ and it is detectable by the first derivative ($\frac{\partial T}{\partial t}$) curve as shown in Fig. 4.2. The change in the slope of $\frac{\partial T}{\partial t}$ is an indication of energy change in the system and the only energy change during solidification is the formation of solid phase, start of nucleation. So if the solidification start is the point shown in Fig. 4.2, then how undercooling can be defined as undercooling is an integral requirement to trigger nucleation and to onset solidification. The actual undercooling may be defined as the difference between the equilibrium melting temperature (discernable from equilibrium phase diagram) and the $T_{nuc_{Al}}$. Therefore, the value of ($T_{min_{Al}} - T_{g_{Al}}$) is neither undercooling nor nucleation range and can only be defined as the recalescence range (of note that this topic is not intended to be opened in this book).

Chemical composition of the melt plays an important role in the thermal analysis results. As for instance, the liquidus temperature of the Al–Si melts varies with Si% according to the following equation (4.1) [5]:

$$T_l(^\circ C) = 662.2 - 6.913 * [\%Si] \tag{4.1}$$

So, by a simple calculation, it is evident that a compositional difference of 0.1% Si between two samples means a difference of about 0.7 °C. Hence the chemistry variation is a key in the cooling curve analysis.

4.1.1 Various SSM Processing

The solidification characterization is explored in the context of four SSM processing routes as detailed below.

4.1.1.1 Rheocasting, Low Pouring Temperature Technique

As indicated in Chap. 2, among different techniques available for SSM processing, casting at low superheat is regarded as the least expensive alternative to produce thixo/rheo billets. For these series of tests, the molten metal was poured into a coated cylindrical steel mold of 75 mm diameter and 250 mm long and the bottom of the mold was sealed by a refractory material. After melt preparation, the alloy was poured at different superheats, in the range of ~ 0–80 °C. Prior to pouring, the mold was tilted to reduce the turbulence. Figure 4.3 shows schematically the actual

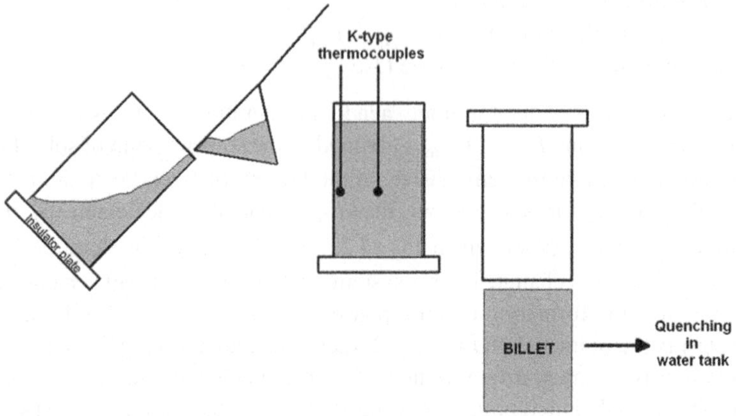

Fig. 4.3 Schematic of low pouring temperature process

experimental setup. In all cases, two K-type thermocouples were installed at the mold center and near the wall with their tips at 80 mm from the bottom of the mold to monitor temperature distribution of the bulk liquid during solidification. Solidification of the alloy continued up to the point where the melt temperature at the center of the billet reached 593 ± 2 °C. It is expected to have a solid fraction of 0.3–0.35 at this temperature according to equilibrium lever rule and Scheil's equation [4]. The wall temperature was registered at 591 ± 1 °C. After preparation, the billets were taken out of the mold, still in the mushy zone, and quenched quickly in cold water. For metallography, transverse section at the tip of the thermocouples was used and for consistency, the area between the center and the wall (quarter area) was examined.

4.1.1.2 Rheocasting, SEED Technology

In addition to low pouring process, the solidification behavior of billets produced via SEED technology was examined in detail. For these series of tests, about 2 kg of the alloy was poured into the same mold as explained in Sect. 4.1.1.1. For turbulence reduction during pouring, the mold was tilted during initial stage of pouring and was brought back to the vertical position (Fig. 4.4). Immediately after pouring the mold and its content were swirled at 150 RPM or 2.5 Hz with an eccentricity of 12 mm. The swirling motion is expected to not only distribute the solid particles formed on the mold wall, but also to break the dendrite arms and rendering the formation and uniform distribution of spherical particles. The duration of this stage depends on the mold dimension and the charge mass and for these series of experiments it was set to 60 s.

In the next step and after stopping the swirling motion and pausing for a short time (5–10 s), the bottom plug was opened to allow a portion of the residual liquid

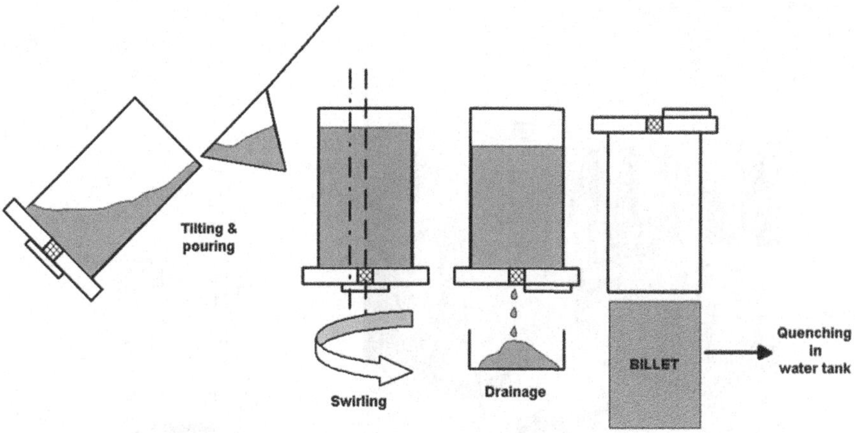

Fig. 4.4 Experimental procedure for the SEED process

to drain out. As a result, the fraction solid in the billet (slug) increases and a free-standing billet is produced. After a specific time of 20 s, the prepared billet was unloaded and transferred into the water tank to be quenched to room temperature at about 20–25 s. In all experiments, two 0.8 mm diameter K-type thermocouples were inserted near the mold wall and also the center with their tips at a distance 80 mm from the bottom of the mold to collect thermal data. The quenching temperatures for all the tests were 598.5 ± 2.5 °C. For metallography, transverse section at the tip of the thermocouples was used and for consistency, the area between the center and the wall (quarter area) was examined.

4.1.1.3 Electromagnetic Stirring (EMS)

For better understanding of the effect of stirring on the solidification behavior and resulted microstructure of SSM billets, electromagnetic stirring method was used. Binary Al–7%Si alloys were prepared by melting 99.7 % commercially pure aluminum in a SiC crucible in an electric resistance furnace. Addition of silicon and iron was carried out at 720 ± 5 °C using pure silicon and Al–25%Fe master alloy. The chemistry is given in Table 4.1.

For attaining different cooling rates, two dissimilar mold materials were used. For higher cooling rate, a copper mold with a water cooling jacket (Fig. 4.5b) and for the lower cooling rate, CO_2-bonded silica sand mold was used to produce ingots of 76 mm in diameter and 300 mm long. The entire configuration was placed in EMS machine, Fig. 4.5. For these series of experiments, the frequency was set to 50 Hz and the current was 100 and 30A for copper and sand molds, respectively (stirring was stopped around 400 °C in EMS samples).

For the superheat variation, pouring temperature was changed between 630 and 690 °C. The cooling rate above liquidus in the copper and sand molds for the

Table 4.1 Chemical analysis of the melt (wt. %)

Si	Fe	Al
6.7–6.9	0.8–0.81	Bal.

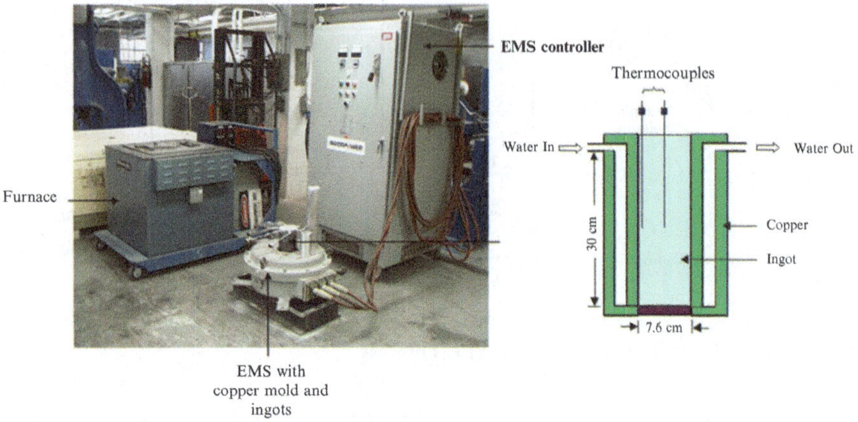

Fig. 4.5 (**a**) Setup for EMS experiments, (**b**) schematic details of the water-cooled copper mold

conventional ingot (with no stirring) was about 4.8 °Cs^{-1} and 3.3 °Cs^{-1}, respectively (thermocouples were inserted 200 mm from the bottom). The cooling rate in the sand mold is relatively high at the beginning of the pouring which is due to the large volume of sand compared to the liquid metal. However after the initial rapid heat dissipation, the bulk liquid temperature decreases slowly due to low heat diffusivity of sand mold. As a result, the sand could absorb significant amount of heat after filling the mold. The cooling rate is then reduced in the mushy zone. For the experiments with no stirring, the liquid was poured into the same molds and allowed to air cool. The samples with no stirring will be referred as "*conventional.*" For structural study, samples were sectioned 200 mm from the bottom and the area between the center and the wall (quarter area) was examined.

4.1.1.4 Thixocasting of Refined/EMS Specimens

In addition, a limited number of experiments were carried out to study the solidification behavior of thixocast billets. For this purpose, graphite cup samples with ~25 mm diameter and samples from transverse section of EMS billets (200 mm from the bottom, in areas between center and wall) were reheated in a single coil 5 kW induction furnace operating at 80 kHz, Fig. 4.6 [6].

Samples were placed vertically on an insulator plate. Temperature variation during the tests was monitored by putting two thermocouples in the same manner as described in Fig. 4.1. Induction furnace was controlled by central thermocouple and the wall thermocouple was connected to the data logger system.

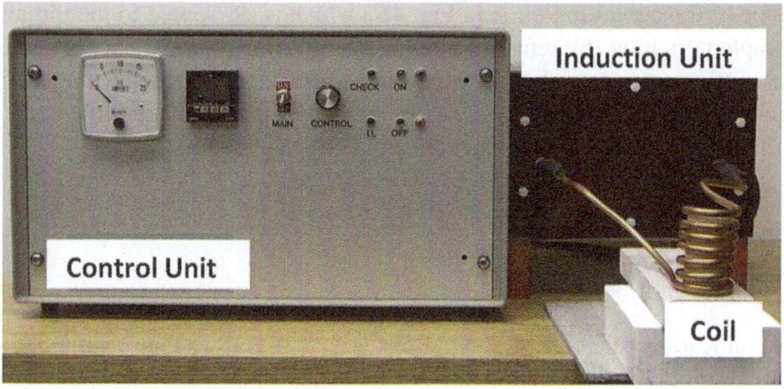

Fig. 4.6 Experimental setup for reheating of samples

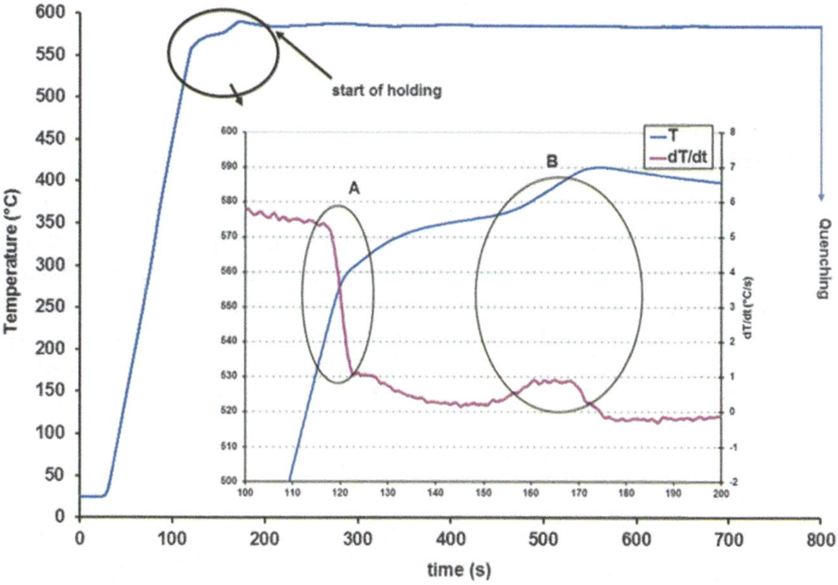

Fig. 4.7 Typical reheating profile of wall thermocouple

A typical reheating curve during partial remelting is presented in Fig. 4.7. During the tests, it was ensured that the temperature difference between center and the wall is less than 2 °C. Heating rate was set to 5–6 °Cs^{-1} and reheating range was 583 ± 3 °C (about 38–40 % fraction solid according to ThermoCalc[1] calculations). Initial heating rate was rapid (5–6 °Cs^{-1}) while there is a sudden change in

[1]ThermoCalc is a commercial thermodynamic-based software package (www.thermocalc.com)

the slope of the curve (region A in Fig. 4.7) which is associated to the initiation of the sample remelting (by remelting of Mg_2Si constituent and ternary eutectic [1]). This variation is followed by another transformation due to the main eutectic reaction in region B of Fig. 4.7. After predetermined holding time (e.g., 5 or 10 min), samples were rapidly quenched in cold water.

4.1.2 *Melt Treatment*

The solidification behavior of SSM alloys is expected to change with the application of grain refining and modification treatments. The followings explain sample preparation and test procedures to characterize solidification behavior of melt-treated SSM billets.

In this study, three refiners and one Sr-based modifier with the following specifications were used:

- Al5%Ti1%B, rod form
- Al5%B, rod form
- Al4%B, waffle form
- Al10%Sr, rod form

To achieve different degrees of refinement, modification, and combined effects, the above-mentioned master alloys were added to the melt to raise the concentration of specific elements, i.e., Ti, B, and/or Sr levels in the alloy. Specified amount of these master alloys, wrapped in aluminum foil, were added to the melt between 720 and 730 °C using a preheated graphite bell. The entire melts were degassed with argon after master alloys addition and the contact time for the first sampling was set to 20–30 min depending on the treatment (~30 min for modification due to incubation period and ~20 min for sole grain refiner). Prior to taking any samples and after each step, each melt was cleaned by skimmer.

4.1.3 *Chemical Analysis*

Samples for chemical analysis were cast into a standard scissor mold. Disk shape samples having a 56 mm diameter and 10 mm thickness were prepared for chemical analysis by machining of the surface. Finally, 6–8 point OES ablations (Optical Emission Spectroscopy) were performed on the machined surface using ThermoARL-4460 and the results are presented by averaging all ablations. The measurement of the grain refiner and modifier elements concentrations is essential and thus calibration of the OES detector regarding these elements especially boron is vital. Two standards especially produced for analysis of 356 Al–Si alloys were used with a 1 % accuracy of the results. Table 4.2 shows the chemistries used in this book.

Table 4.2 Chemical composition of the base alloys (wt. %)

	Si	Mg	Fe	Mn	Cu	Ti	B	Sr	Al
Binary alloy	7.0–7.3	Nil	Max 0.09	Nil	Nil	Nil	Nil	Nil	Bal.
	6.7–6.9	Nil	0.8–0.81	Nil	Nil	Nil	Nil	Nil	Bal.
A356	6.4–6.7	0.36–0.4	Max 0.08	Max 0.003	Max 0.001	Max 0.0058	Nil	Nil	Bal.
356	6.8–7.0	0.33–0.36	Max 0.09	Max 0.003	Max 0.002	0.1–0.13	Nil	Nil	Bal.

4.2 Rheological Characterization

The flow of SSM billets within the die is an important issue to control the integrity of the finished cast product. Therefore, it is essential to characterize SSM billets flow when injected into die cavity. The following explains briefly the fundamentals of rheology and gives some equipment and procedures applied to complement the theoretical principles.

4.2.1 Principles of Rheology

4.2.1.1 Introduction

Since the inception of Semi-Solid Metal (SSM) processing concept in early 1970 [7, 8], and the fact that the deformability of SSM billets is shear rate and time dependent, the issue of rheology has been studied to some extent. The "S2P" biannual international conferences over the last two decades, (1990–2016) have also dedicated specific chapters covering rheological behavior of SSM alloys. Rheology is an integral part of SSM research efforts, even though the concept may not be as clear and familiar for metallurgists.

Rheology is the science of deformation and flow of materials. It is a branch of physics concerned with the mechanics of deformable bodies. It also deals with the simultaneous deformation and flow of materials. In this context shear flow is an important type of deformation in rheology and may be visualized as a process in which infinitely thin, parallel planes slide over each other as in a pack of rigid cards. The breakup or rupture of solids and liquids into smaller segments or droplets and the rejoining and sticking together, "*cohesion*," of particles or droplets to make a continuum body and mass are often included in rheology [9].

The rheological behavior and properties of a substance may sometimes exhibit considerable changes with time or with continuing deformation [10]. These changes occur either reversibly or irreversibly with reversible deformation called elasticity and irreversible deformation is known as flow. The work used in deforming a perfectly elastic body is recovered when the body is restored to its original undeformed shape, whereas the work used in sustaining flow is dissipated as heat and is not mechanically recoverable. Elasticity relates to mechanically recoverable energy and viscous flow to the conversion of mechanical energy into heat. Elastic deformation is always a function of stress, whereas the rate of deformation for flow is a function of shear stress.

From the rheological point of view, the mechanical properties of material can be described in terms of elastic, viscous, and inertial contributions. For elastic deformation, the material deforms reversibly and is able to recover instantly to its original shape and size when the applied stress is removed. For viscous body however, stress and resulted strain cannot be sustained for long and will be relieved by flow with rate of flow being a function of stress. Of course, extremely viscous materials may exhibit elastic strain for considerable period of time. Periods which are short with respect to the time needed for appreciable flow. This means that a given material may be considered as an ideal elastic body for relatively short times and as an ideal viscous body for long time periods. Regardless of geometry of the body and deformation, the flow will always be in the form of laminar shear [11, 12].

With such simple definition, the interrelationship of rheology and mechanical properties of materials are closely tied up with materials' viscosity and deformation behavior within two phase region, liquid and solid, the so-called *mushy state*.

4.2.1.2 Viscosity

Viscosity is the main parameter for rheology of semi-solid metallic alloys and plays an important role equivalent to that of "fluidity" concept in liquid metals (e.g., [13–15]) and modulus of elasticity for solids [11]. Viscosity is an indication of SSM capability in filling the mold and determines the required force for deformation and flow of materials [16]. According to several review articles, viscometry is identified as an appropriate route for the rheological studies of materials [17–19].

Viscosity of fluids is expressed in terms of the coefficient of viscosity, η. Based on Newton's law of viscosity, the ratio of shear stress to the gradient of velocity is a constant showing the capability of momentum diffusion through the body of material as expressed mathematically in the following equation (4.2):

$$\tau_{yx} = -\eta \frac{\mathrm{d}v_x}{\mathrm{d}y} \tag{4.2}$$

Where $\frac{\mathrm{d}v_x}{\mathrm{d}y}$ is velocity gradient, τ_{yx} is the shear stress, and η is viscosity. Equation (4.2) may be rearranged and expressed as equation (4.3):

$$\tau_{yx} = -\eta \frac{-\frac{\mathrm{d}x}{\mathrm{d}t}}{\mathrm{d}y} = \eta \frac{\mathrm{d}x}{\mathrm{d}y} \times \frac{1}{\mathrm{d}t} = \eta \frac{\mathrm{d}\gamma}{\mathrm{d}t}$$

$$\tau_{yx} = \eta \dot{\gamma} \tag{4.3}$$

Where γ and $\dot{\gamma}$ are shear strain and shear strain rate, respectively. In contrast to Newton's law of viscosity where the viscosity (η) is constant, the term "apparent viscosity" is used when viscosity is influenced by applied shear rate.

The ratio of the viscosity, η, to the fluid density (4.4) is called *kinematic viscosity*, ν, which is a measure of momentum diffusivity, analogous to thermal and mass diffusivity [10]:

$$\nu = \frac{\eta}{\rho} \tag{4.4}$$

The value of η is interpreted in terms of a power law, relating shear stress (τ) to average shear rate, ($\tau = m\dot{\gamma}^n$). The apparent viscosity is then calculated as the ratio of shear stress to shear rate, [$\eta = m(\dot{\gamma})^{(n-1)}$], where m and n are material constants of consistency and power law indices respectively [10]. Except for Newtonian fluids where η is constant and the viscosity is independent of shear strain rate, the viscosity is found to vary by several orders of magnitude with changing shear rate (or shear stress) in non-Newtonian fluids, i.e., $\eta = \frac{\tau_{yx}}{\dot{\gamma}}$ is a function of $\dot{\gamma}$. Typical flow curves for non-Newtonian fluids are presented for time-independent fluids in Fig. 4.8 [10, 20]. Line A in this figure represents the Newtonian body for which η is constant. When the shear rate increases more than in proportion to shear stress, curve B, the material is called "*Pseudoplastic*" or "shear thinning" liquid. To a first approximation, the flow curve of a Pseudoplastic liquid can be represented by the power law, the rate of shear is roughly proportional to the power of the shearing stress or vice versa, (4.5 and 4.6):

$$\tau_{yx} = \eta(\dot{\gamma})^n \tag{4.5}$$

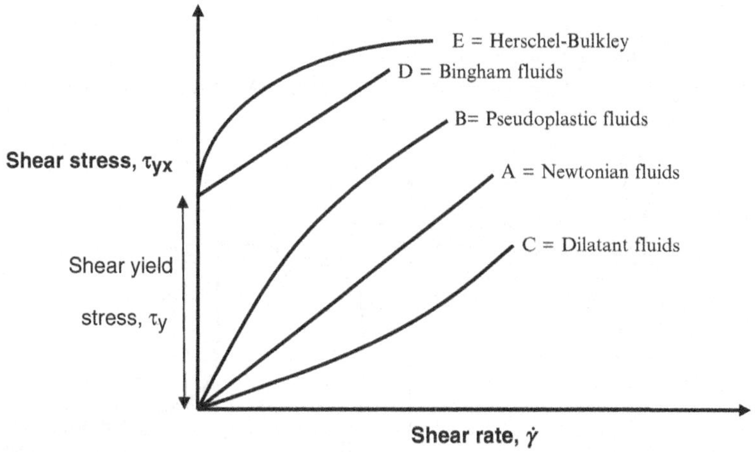

Fig. 4.8 Stress–strain rate curves for time-independent fluids (reproduced from [10] and [20])

$$\eta = m(\dot{\gamma})^{(n-1)} \tag{4.6}$$

The materials constants m and n (the consistency and power law indices, respectively [21]) are viscosity at shear rate of 1 s^{-1} ($m = \eta$) and a measure of fluid departure from Newtonian fluid ($n = 1$), respectively, i.e., $n < 1$ for pseudoplastic fluid (it is usually lies between 1/3 and 1/2, shear thinning) and $n > 1$ for dilatant fluid [10] or shear thickening.

It has to be mentioned that some of the non-Newtonian fluids may exhibit dual behavior, where both, shear thinning and shear thickening (pseudoplastic and dilatant, respectively) may be observed at different loading conditions. It has been reported that aluminum and tin–lead alloys have shown both the pseudoplasticity and dilatant behaviors depending on the applied shear rate when tested with high pressure die casting setup and rotational viscometer. Such alloys behave like pseudoplastic material at low shear rates, 2×10^3–2×10^4 s^{-1}, but show dilatant character when deformed at high shear rates, 10^6 s^{-1} [22, 23].

In addition, there are viscoelastic fluids which exhibit time-dependent recovery from deformation, i.e., recoil. This is in contrast with the behavior of Newtonian fluids which do not recoil. Such behavior may be compared with the behavior of thermoplastic polymers under loading with exception of elastic deformation which is due to stretching (straightening) of the polymer chains, instantly happens as load applied and is recovered instantly when load removed, ε_i in Fig. 4.9. After chain straightening, the chains start to move relatively with respect to each other and this is time dependent as is the case for viscoelastic fluids with layers of fluid moving with respect to each other. Such movement of chains or layers of fluid with respect to each other is known as "*viscous flow*" and therefore, viscoelastic deformation (ε_f–ε_i) is time dependent. Viscous flow, however, is dependent upon the viscosity. Such characteristic brings the time dependency of viscosity into the actual fluids.

The semi-solid materials with their thixotropic characteristics, as will be discussed later, do not store elastic energy and therefore do not recover with time when the applied shear stress is removed. The viscosity of thixotropic fluids decreases with time and approaches an asymptotic value at a constant shear strain rate and also the structure breaks down with time. As pointed out by Poirier and

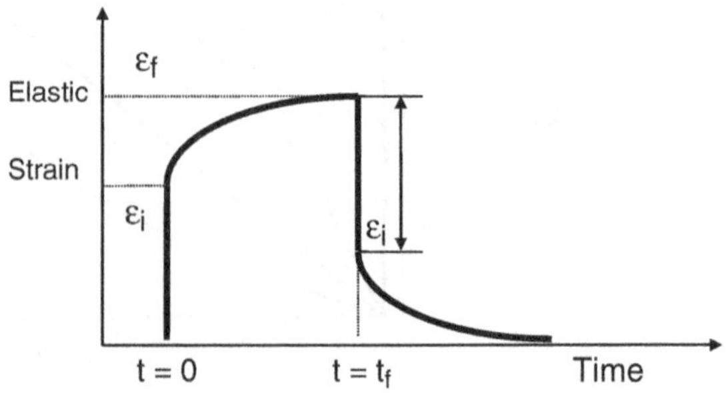

Fig. 4.9 Strain-time variation for viscoelastic deformation of polymers. For fluids, the recovery of instant elastic deformation is not applicable [24]

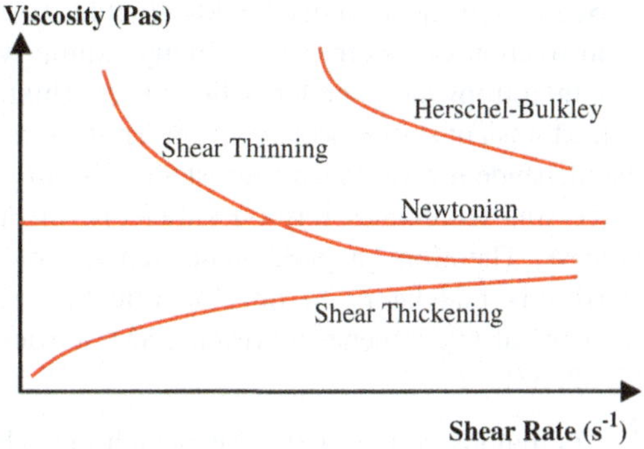

Fig. 4.10 Effect of shear rate on viscosity of different fluids [20]

Geiger [10], thixotropic fluids may be treated as general Newtonian fluids under steady-state conditions when the asymptotic value of viscosity is maintained.

The Bingham fluids exhibit a yield stress and then gives a linear relationship between shear stress and shear rate as expressed in (4.7)[20]:

$$\tau = \tau_y + k\dot{\gamma} \tag{4.7}$$

where τ_y is yield stress and k is a constant related to the viscosity. However, if the fluid behavior after yielding is not linear, it is referred as Herschel–Bulkley fluids where (4.8):

$$\tau = \tau_y + k\dot{\gamma}^n \tag{4.8}$$

Figure 4.10 represents viscosity variations against shear rate for different fluids [20]. In the eyes of mass producers of metallic artifacts, knowledge of viscosity is equivalent to die-filling characteristics, since lower viscosity causes better movement of material through the die [22, 25, 26]. The viscosity is always used as an input parameter for prediction of flowability in the simulation software [27–29]. Lower viscosity results in intricate thin wall component production with lower machine pressure and reduced rejects and scraps [30, 31].

4.2.1.3 Flow Behavior in the Semi-Solid Metals

The semi-solid metals are two-phase system containing a suspension of primary phase particles within the liquid phase matrix. The dynamic presence of two phases, "*the mush*", is dependent upon the alloy solidification range where a wider solidification range is translated into an easier and more controllable "*mush*". The growth mode of the solid phase is dendritic. Such system is shown to behave like Newtonian or non-Newtonian fluids depending upon the process parameters [17–19].

Semi-solid metal slurries with globular morphology of the solid phase and a solid fraction of less than 0.6 usually exhibit two distinct rheological characteristics: thixotropy and pseudoplasticity [19]. Thixotropy is about the time dependency of the transient state viscosity at a given shear rate, while pseudoplasticity discusses the dependency of the steady-state viscosity on the shear rate. All the SSM processing techniques rely on either one of these properties or both in a single process. Therefore, a good understanding of the rheological behavior of SSM slurries is mandatory to develop effective SSM processing technologies. The rheological phenomena in stirred SSM slurries can be divided into three groups [7, 32, 33]:

1. Continuous cooling behavior, which describes the changes in the viscosity of molten alloy during continuous cooling (at constant cooling rate and shear rate).
2. Pseudoplastic behavior, which describes the dependency of steady-state viscosity on the shear rate, or shear thinning behavior.
3. Thixotropic behavior, to describe the dependency of transient state viscosity on time.

There is important information that could be obtained on the rheological behavior of SSM slurries during continuous cooling with respect to the effects of solid fraction, shear rate, and cooling rate. In particular, it is more relevant to the practical conditions set in SSM processing techniques such as rheocasting and rheomolding. However as pointed out by Suéry et al. [16], such experiments are more relevant to exploiting the solidification behavior rather than studying the rheology of SSM slurries.

In contrast to continuous cooling tests, isothermal steady-state trials are more capable of characterizing the rheological behavior of SSM slurries and are a first step toward the determination of a constitutive equation. It is now generally accepted that the steady-state viscosity at a given shear rate depends on the degree of agglomeration between the solid particles, which in turn is the result of a dynamic equilibrium between agglomeration and deagglomeration processes [34].

The thixotropic behavior of SSM slurries was demonstrated by measuring the hysteresis loops during a cyclic shear deformation [35]. However, such procedure is not sufficient to quantify the kinetics of agglomeration and deagglomeration processes. As pointed out by Chen and Fan [34], in order to overcome this shortcoming, special experimental procedures involving an abrupt shear rate jump or a shear rate drop were developed to characterize the kinetics of structural evolution. It was found that the agglomeration process dominates after a shear rate drop, whereas a deagglomeration process dominates after a shear rate jump [34].

For the semi-solid billets with high fraction solid, the so-called self-standing slugs, there are other type of rheological studies which are based on keeping the shear rate constant within the bulk of a preheated sample and measuring the velocity of deformation under a constant dead weight, parallel plate compression viscometry, or deals with constant stress applied on the semi-solid slugs to develop the data of strain rate, extrusion methods. Such data can provide useful information about the viscosity of SSM materials having high percentage of fraction solid [36–38].

4.2.2 Rheological Behavior of SSM alloys

The flow and deformation of metallic alloys "rheological behavior" is mainly dependent on the SSM alloy viscosity (η), which itself varies with both metallurgical and process parameters. It may be represented by the following (4.9) [33, 34, 39, 40]:

$$\eta = f(\dot{\gamma}, t_s, T_a, \dot{T}, C_o, f_s, F, h_{\mathrm{istory}}) \quad (4.9)$$

Where η represents the viscosity, $\dot{\gamma}$ is the shear rate, t_s is shearing time, T_a is the temperature of semi-solid alloy, $\dot{T}$ is the cooling rate from liquid to T_a within the mushy zone, C_o is the alloy composition, f_s is fraction solid at T_a, F is the shape factor, and h_{istory} is history effect. Generally higher solute contents and cooling rates have been found to cause more dendritic morphology and hence higher viscosities for a given fraction solid. The apparent viscosity of the slurry at a particular point in time depends on its previous internal state (the history effect, h_{istory}). The internal state, which is continually changing, is described in some way by microstructural parameters such as size, size distribution, morphology, and dispersion of solid phase within the liquid matrix. Developing an all-encompassing model for semi-solid metals is therefore considered to be a difficult task and requires detail knowledge of the effect of the above-mentioned parameters on the viscosity. In addition, the process time before shaping is also important since it could change particle size, morphology, and distribution due to spheroidization and coarsening [41, 42]. The followings give a brief account of the effects of parameters on the viscosity of SSM slurries.

4.2.2.1 Metallurgical Parameters

The complex rheological properties of SSM alloys are strongly dependent on the local solid fraction, particle shape, and size and the state of agglomeration. The metallurgical parameters may be controlled through manipulation of solidification and alloy phase diagram.

Fraction Solid

One of the most important parameters affecting viscosity of the mush is the fraction solid of the primary phase, e.g., α-Al dendrites in case of Al–Si alloys [38, 43–45]. Fraction solid ($f_s = 1 - f_l$) at a temperature T within mushy zone may be calculated by Scheil equation, (4.10) and (4.11) [4].

The fraction solid (f_s) at a given location within the solid–liquid zone is predicted by a mass balance wherein a simplest case neglects ripening and diffusion in the solid. For this case the relation between weight fraction liquid in mushy zone,

f_l, and liquid composition at a given location, C_l, is given by the Scheil equation which may be written in the following form for the constant partition ratio, k:

$$f_l = \left(\frac{C_l}{C_o}\right)^{-1/1-k} \tag{4.10}$$

Or for fraction solid and solid composition (C_s) is given by:

$$C_s = kC_0(1 - f_s)^{k-1} \tag{4.11}$$

C_0 is the composition of the alloy.

Different methods are reported in the literature for measuring the fraction solid, among which the most commonly used are quantitative metallography, thermal analysis, and application of thermodynamic data based on equilibrium phase diagram [46]. Each route has its advantages and disadvantages and could be used as per requirements of the research program.

Chen and Fan [34] developed a microstructural model to describe the relationship between viscosity and effective solid fraction, rheological behavior, of liquid-like SSM slurries under simple shear flow. In their model, liquid-like SSM slurry is considered as a suspension in which interacting spherical solid particles of low cohesion are dispersed in a liquid matrix. In a simple shear flow field, a structural parameter (S) defined as the average number of particles in each agglomerated solid particle was used to describe the state of agglomeration and deagglomeration. Through effective solid fraction, f_s^{eff}, a parameter which is slightly different to that of fraction solid as a result of agglomeration/deagglomeration and possible entrapment of liquid, (4.13), viscosity can be expressed as a function of the structural parameter S (4.12).

$$\eta = \eta_o\left(1 - f_s^{\text{eff}}\right)^{-\frac{5}{2}} \tag{4.12}$$

The effective fraction solid is given as:

$$f_s^{\text{eff}} = \left(1 + \frac{S-1}{S}A\right)f_s \tag{4.13}$$

where η is instantaneous viscosity, η_o is the viscosity of liquid matrix (when effective fraction solid = 0), A is a model parameter related to packing mode which decreases with increasing the packing density (the fraction of space filled by solid particles). The parameter A was expressed by a linear equation of ($A = 3.395–4.96\ f_s$, where f_s = solid fraction) for Sn15%Pb alloy [47], and the effective fraction solid $\left(f_s^{\text{eff}}\right)$ was regarded as the sum of the actual solid fraction and entrapped liquid fraction.

Equation (4.13) indicates that the effective solid fraction is influenced by the actual solid fraction, agglomerate size, and the packing mode in the agglomerated chunks. It is interesting to note from (4.12) that the viscosity of semi-solid slurry is a direct function of the viscosity of the liquid matrix and the effective solid fraction. The flow conditions affect viscosity only indirectly through changing the effective solid fraction.

Viscosity steadily increases with increasing of solid fraction till the point where solid particles can no longer move easily and the already solidified segments have developed strength, a 3D solid skeleton is formed. This is specified as the dendrite coherency point (DCP). The dendrite coherency point marks the transition from mass feeding to interdendritic feeding during solidification. After DCP, viscosity increases abruptly [48].

With stirring in SSM processing, dendrite coherency point is postponed due to the shallow temperature gradient initiated from the forced convection within the melt. The breakdown of dendrites due to stirring coupled with multidirectional growth of fragmented dendrites due to more uniform temperature distribution within the mold, shallow temperature gradient, encourage the formation of equiaxed grains, thus postpone the rapid rise of viscosity to higher fraction of solids.

The effect of fraction solid was studied on the viscosity of A356 SSM billets cast conventionally [38]. Application of low superheat in conventional casting resulted in the formation of the desired primary Al morphologies for a series of SSM billets. It was confirmed that at low shear rate values (less than 0.01 s^{-1}), the SSM billets can be treated as the Newtonian fluid even though the two-phase fluids are basically non-Newtonian with engineering strain–time behavior comprising non-steady-state or transient state with thixotropic characteristics and quasi-steady-state from which the viscosity values calculated [38]. The SSM billets showed pseudoplastic behavior where the viscosity numbers decreased with increasing applied pressure (shear rate) as shown in Fig. 4.11. The SSM billets were also treated as the non-Newtonian fluids and the materials constants m and n were calculated and presented as an empirical relationship (4.14) for different fractions solid [38].

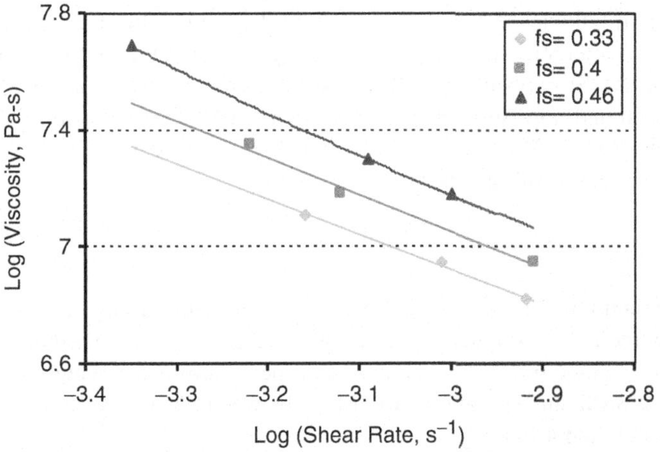

Fig. 4.11 The viscosity numbers are plotted against average shear rates for SSM billets having different fractions solid [38]

$$\log\eta = 5.56 - 1.39f_s - (1.56f_s + 0.14)\log\dot{\gamma} \quad (0.33 < f_s < 0.46) \tag{4.14}$$

The above-mentioned empirical relationship expresses the direct effect of fraction solid on the viscosity of semi-solid billets and if the fraction solid kept constant, the viscosity varies inversely with shear rate. The validity of equation (4.14) was further confirmed with experimental results reported in the literature and it was stated that the equation above is valid at very low shear rates, less than 0.01 s^{-1} [38].

Primary Phase Morphology

The morphology of primary phase has a pronounced effect on the flow behavior of semi-solid metal slurries [17–19]. It is found that dendritic structures at the same solid fraction exhibit approximately several orders of magnitude greater flow resistance than the equiaxed structures [49]. In fact the globular particles move easier over one another than dendritic ones which tend to interlock during the application of external force, resistance against flow [7, 33, 50, 51]. In addition, since the beginning of SSM processing research, it was the nondendritic structure which imparted the interesting and useful rheological characteristics, such as pseudoplasticity and thixotropy. Therefore, a good understanding of the effect of particle morphology on the rheological behavior is not only of scientific merits but also from technological point of view it has great significance on the development of new SSM processes.

Although several attempts were made to experimentally correlate the morphology of the primary solid phase to rheological characteristics (e.g., [37, 52]), there is not any theoretical model to account for the effect of particle morphology on the flow behavior of metal slurries. However, there are simple models and defined parameters to describe rheological characteristics of slurries with different solid particles morphologies [50]. Lashkari et al. [37] specified the effect of morphology on flow characteristics of SSM billets using aspect ratio of the primary α-Al particles in Al–Si hypoeutectic alloys. Non-Newtonian behavior for SSM billets was assumed and non-Newtonian power law model was employed to study rheological behavior of SSM slurries. The model expresses viscosity changes in terms of applied stress and resulting shear rates as (4.15):

$$\eta = m(\dot{\gamma})^{n-1} \tag{4.15}$$

where, as mentioned before, m and n are the material constants (consistency and power law indices), respectively [21]. Figure 4.12 shows the relationship between the consistency and power law indices and aspect ratio which eventually related to viscosity of SSM slurries. The relationships between n and m with aspect ratio are given as (4.16 and 4.17):

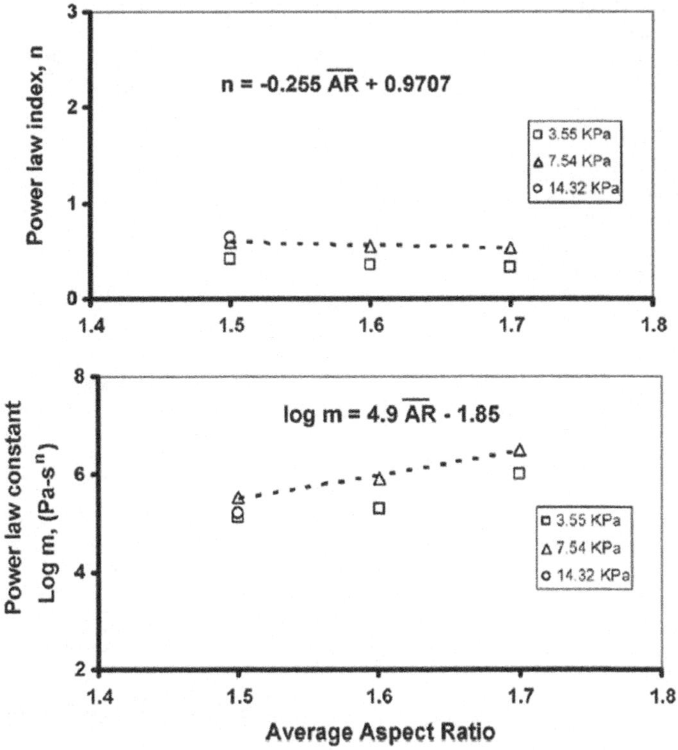

Fig. 4.12 Effect of primary α-Al particle aspect ratio on the power law (n) and consistency (m) indices [37]

$$n = 0.97 - 0.255\left(\overline{AR}\right) \quad \text{for } 1.5 < \overline{AR} < 1.7 \tag{4.16}$$

$$m = 10^{\left(1.85 - 4.9\overline{AR}\right)} \quad \text{for } 1.5 < \overline{AR} < 1.7 \tag{4.17}$$

In another study [19, 50], Fan and his coworkers studied the effect of morphology on the apparent viscosity (the viscosity of steady-state flow segment) of SSM slurry using the concept of fractal dimension, D_f. Fractal dimension defined as the parameter to characterize the morphology of solid particles using the concept of Hausdroff dimension. Hausdorff dimension of a single point is zero, of a line is 1, of a square is 2, and of a cube is 3. For dendritic morphology, the value of $D_f = 2.5$ and $D_f = 3$ illustrates the fully globular particles in SSM slurries.[2] They calculated D_f from experimental viscosity data of continuous cooling experiments on Sn–15 % Pb [35]

[2]For more information on fractal dimension, interested readers are encouraged to consult the following article;

R.S. Qin and Z. Fan, "Fractal theory study on morphological dependence of viscosity of semisolid slurries", Mater. Sci. Tech., 2001, 17, 1149–1152.

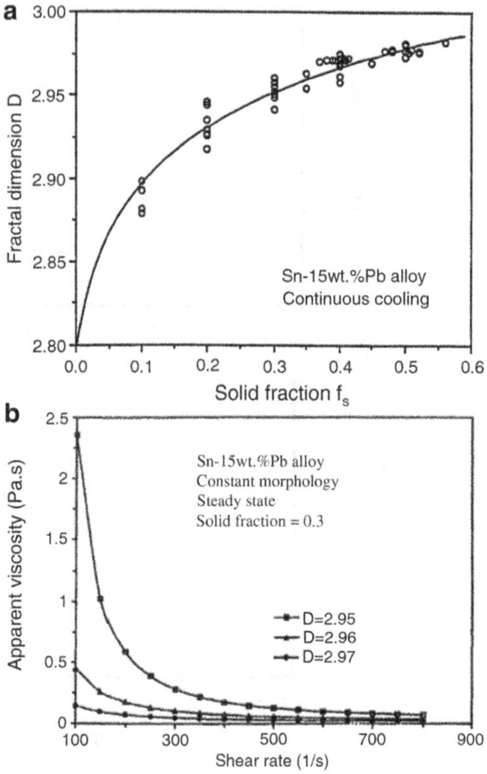

Fig. 4.13 The effect of primary particle morphology on the apparent viscosity as a function of shear rate; (**a**) Fractal dimension D deduced from experimental viscosity data from continuous cooling experiments reported by Joly et al. [35] as function of solid fraction [19], (**b**) calculated steady-state apparent viscosity at three different fractal dimension values for Sn15%Pb SSM alloy [50] (reprinted by permission of Taylor & Francis Ltd)

as shown in Fig. 4.13a and substituting the resulted values in (4.13) and (4.12) to calculate the effective solid fraction and apparent viscosity, respectively, Fig. 4.13b. As shown, for D_f value closer to 3, the apparent viscosity is lower [50].

Zoqui et al. [53] also reported the effect of morphology on the rheology of SSM alloys, using rheocast quality index. Rheocast quality index, RQI, defined as RQI = globule size/(grain size × shape factor), indicates quasi rosette shape feature with RQI = 0.1 and a perfect globular shape with RQI = 0.9.

Particle Size and Distribution

One of the main parameters controlling the complex rheological properties of semi-solid metal alloys is the particle size and distribution including the state of agglomeration. It is expected to have better flow with finer microstructure as there is easier movement and less collision among particles with lower viscosity [17–19].

It has to be pointed out that finer particles are prone to agglomeration due to larger surface area and therefore there may be changes in viscosity with time, dynamic state. Generally, size and distribution of primary particles are time-dependent variables based on two mechanisms of breakdown (deagglomeration), and coagulation (agglomeration) [54, 55]. For the breakdown mechanism (deagglomeration process), the bonds between particles break down due to shear forces during the early stages of the shearing and is responsible for the rapid decrease in the effective viscosity of the material [56, 57]. The second mechanism is the buildup of the solid structure where metallurgical bonds (necks) form between particles when the sample is at rest [57, 58]. It is understandable that the first mechanism is much faster than the second since the agglomeration mechanism relies on diffusion which is not as effective and fast as stirring and generation of convective currents during deagglomeration. The tendencies for the suspended particles within the liquid matrix to agglomerate could be decreased or increased with the application of external forces on the semi-solid mush.

The particle size is therefore dependent on the applied shear rate as it can promote agglomeration or deagglomeration due to particle–particle impact and bonding if the applied shear rate is low. It is also time dependent during isothermal processing of SSM billets where Ostwald ripening (dissolution of smaller particle and growth of larger ones) or particles coalescing (particles joined together by impact and due to high diffusion rate fully fuse together) could change particle size. Therefore, the dynamic interaction among solid particles may cause the formation of chunks, agglomerated particles, within semi-solid slurries and make the flow of mush harder. However, after a while under the influence of viscous forces, the equilibrium takes place between agglomeration and deagglomeration processes and the viscosity changes reach a steady state and uniform distribution of particles is observed [33, 47, 50, 51, 59, 60]. The uniform distribution of isolated particles within semi-solid slurries is always an important issue for researchers to lower viscosity and achieve sound final components.

Particle size analysis and distinction between grains and globules have paramount importance in SSM processing. Technically, there is a difference between globule and grain sizes. Globules are the primary particles which are apparently detached from each other; however, there is every possibility that the neighboring individual particles might be interconnected from underneath of the polished surface. It is worth noting that the concept of globule size measurement in conventional cast samples is not scientifically valid which is associated with the errors related to the sectioning of dendritic branches. This will be explained later in this chapter.

Alloy Chemistry and Pouring Temperature

Alloy Chemistry

The composition of slurry has a direct effect on the solidification morphology. Increasing solute concentration in the alloy causes enrichment of melt at the solidification front. The resulting constitutional supercooling breaks down planar interface to cellular morphology which eventually becomes dendritic with increasing constitutional supercooling at the solidification front. With the formation of dendrites, as stated earlier, the apparent viscosity of slurry increases and rheocasting in this condition becomes more difficult.

It is believed the higher concentration of solute in the alloy increases the volume fraction of liquid trapped between the primary crystals. Such phenomenon brings higher viscosity within the slurry. An empirical equation has also been proposed to relate the viscosity and compositional factors [61, 62].

$$\eta = \eta_o \left\{ 1 + \frac{\alpha \rho \, C^{1/3} \, \dot{\gamma}^{-4/3}}{2\left(\frac{1}{f_s} - \frac{1}{0.72 - \beta C^{1/3} \dot{\gamma}^{-1/3}}\right)} \right\} \tag{4.18}$$

where η_o is the apparent viscosity of liquid, ρ is the density of alloy, C is the solidification rate $\left(\frac{df_s}{dt}\right)$ (fraction solid forms per unit time, t), f_s is the fraction of solid, $\dot{\gamma}$ is the shear rate, and α and β are numerical values which are dependent on the chemical composition of alloy and become larger with increasing solute content. For Al–3.6 % Si, the value of $\alpha = 67.0$, $\beta = 6.27$ and density (ρ) are 2140 kg m^{-3} [61].

Furthermore, alloy composition could affect dendrite coherency point (DCP), as it is altered by solute concentration [48]. This is due to the effect of solute concentration on the formation of the primary solid phase as, for instance, for hypoeutectic Al–Si alloys, the percentage of primary α-Al decreases with increasing Si content and therefore the development of dendritic network is postponed, lower dendrite coherency temperature.

The effects of solute and trace elements on reducing grain size and improving mechanical properties of as-cast products are well-established facts (see Sect. 6.2.1.4). The alloy composition directly affects the percentage of primary phase solidifying within mushy zone. It is generally believed that small addition of alloying elements could interfere with grains nucleation and growth mechanisms and provide conditions required for new nuclei to form or prevent gain coarsening, to promote the formation of finer grains. With progress in solidification, the solutes form an enriched boundary layer ahead of the solidification front in which the actual temperature is lower than the equilibrium solidification temperature, constitutional undercooling zone [4]. Constitutional undercooling is responsible for dendritic growth. In other words, with controlling alloy composition, the type and percentage of solute elements, constitutional undercooling and thus the growth rate and morphology of primary phase, dendritic or equiaxed growth, may be controlled. The same concept could be extended for grain refiners which are specifically added to refine the as-cast structure. The effect of grain refiner and modifier addition during SSM processing of aluminum alloys is well discussed in this book (Sect. 6.2.1). The overall conclusions may be briefly summarized here as follows:

1. Grain refinement results in the formation of better and more uniform distribution of nearly spherical primary α-Al particles within the slurry.
2. The refinement of the grains has been identified as the main factor for better deformability of the billets. Modification also plays an important role in alloy deformability through reduction of the residual liquid surface tension, which reduces the apparent viscosity of the billets.

3. The microstructural study of the deformed area showed less segregation of liquid adjacent to the mold walls for the combined and overrefined semi-solid billets. This is attributed to the smaller particles size and lower viscosity of the billets.

Pouring Temperature

Pouring temperature or superheat is one of the important parameters to affect the evolution of primary phase during solidification. Several researchers have investigated the effect of pouring temperature on the microstructure of as-cast semi-solid metals (e.g., [63–69]). Low superheat is instrumental in establishing shallow temperature gradient within the slurry, thus encouraging equiaxed growth. Shallow temperature gradient removes directional heat extraction from the melt and prevents the formation of columnar dendrites within the mush [4]. This is an effective way to control the morphology of primary phase forms in the more recently introduced SSM processes, since agitation of the slurry is no longer the principal factor in promoting globular morphology [25, 68, 69]. Figure 4.14 shows the effect of pouring temperature on the microstructural evolution of 356 Al–Si alloy [67].

The importance of pouring temperature is even well pronounced in SSM processes where stirring with strong convective currents is applied as in case of electromagnetic stirring route which will be discussed in Sect. 5.4.

4.2.2.2 Process Parameters

Shear stress and shear rate ($\tau, \dot{\gamma}$)

One of the most important factors affecting the viscosity of SSM slurries is the applied shear force [39, 43]. It imposes laminar or turbulent flow within the slurry and induces disintegration of dendrites and the agglomeration or deagglomeration of the dendrite fragments, which is the main drive for fine distribution of primary phase particles. The applied shear force could eventually establish some sort of equilibrium between agglomeration and deagglomeration phenomena within SSM slurries, "steady state," and to prevent the formation of bulky particles; the main obstacle to SSM slurries flow within mold cavity. The great tendency for the formation of agglomerated chunky parts due to the presence of low energy grain boundaries within globular particles can be prevented by application of shear forces within the slurry [70]. The term "apparent viscosity" used for SSM slurries is to express the viscosity of steady-state flow and varies with shear rate and fraction of solid [7, 32]. As shown in Fig. 4.15, for a constant fraction solid, the viscosity decreases with increasing shear rate.

Shear rate is a material-related parameter which varies linearly with shear force in Newtonian fluids and nonlinearly in non-Newtonian fluids. Shear rate plays the same role as shear force where the increasing of shear rate decreases the viscosity within non-Newtonian fluids as shown in Fig. 4.11 and also Fig. 4.16. For ideal Newtonian fluids, the viscosity number is independent of shear rate [7, 32].

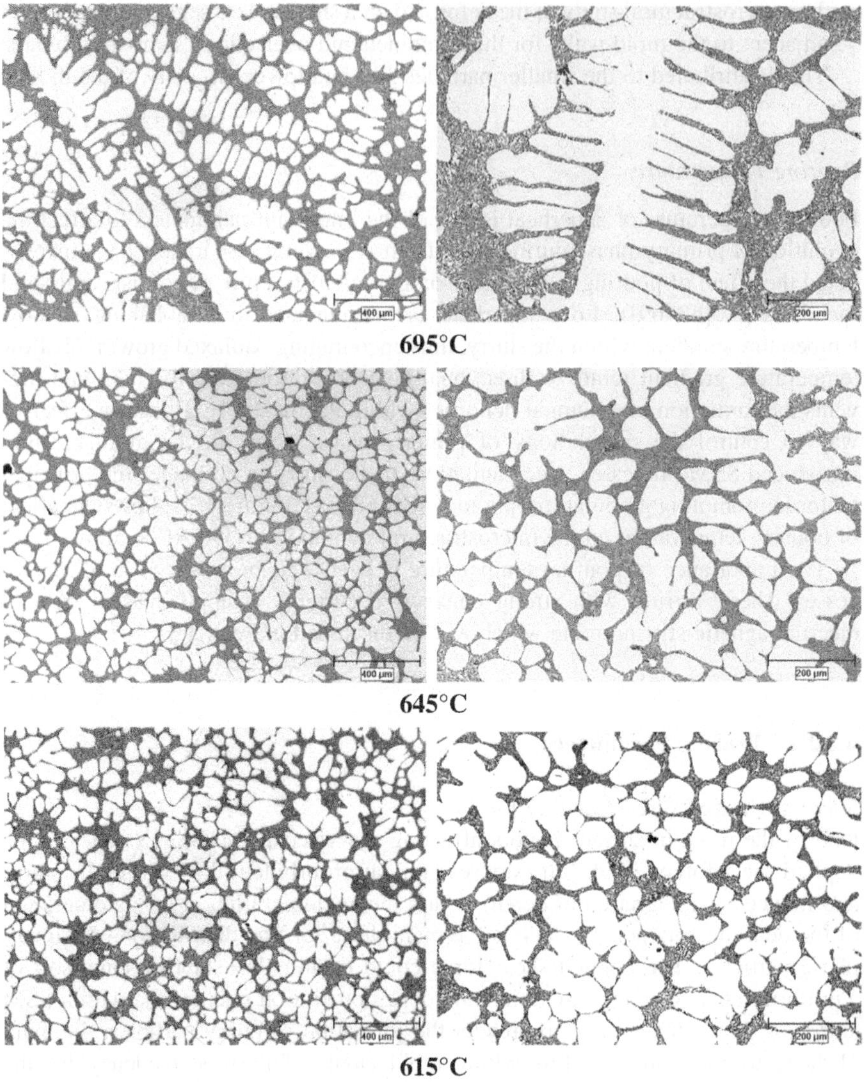

Fig. 4.14 Typical optical micrographs to show the effect of pouring temperature on the morphology of primary α-Al particles. Dendritic, rosette, globular (from 695 to 615 °C) [67]

The implementation of shear force and the resulting agitation of the slurry are brought about by different means which explained earlier in Chap. 2 for a range of SSM processes. Application of shear force also plays an important role during the course of preparing the primary feedstock for thixo- and rheo processes, where the mechanical rheocaster or agitator is the means by which the SSM billets are produced [71]. Swirling is another example of external force which not only induces shear stresses, but also brings thermal homogeneity within the SSM slurry

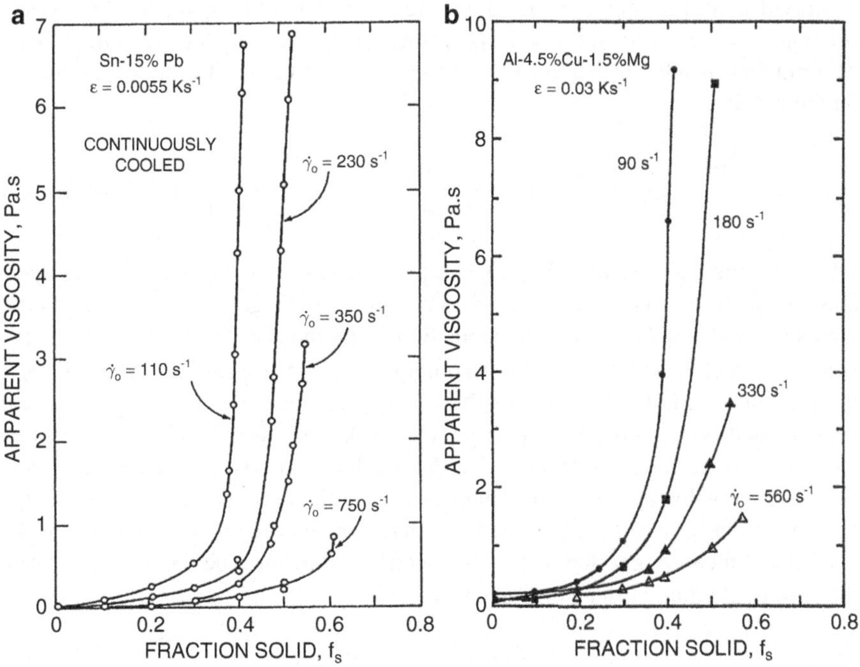

Fig. 4.15 Apparent viscosity versus fraction solid at different shear rates to confirm the viscosity decreasing with increasing shear rate, (**a**) Pb15%Sn [35] (**b**) Al4.5%Cu1.5%Mg [32]

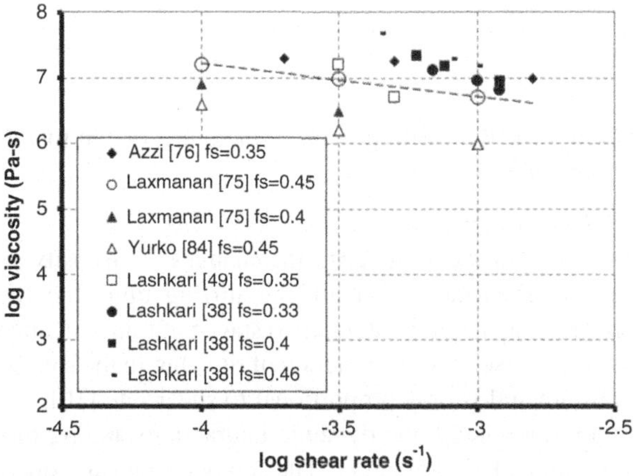

Fig. 4.16 The effect of shear rate on the viscosity of SSM slurries having similar fraction solid and globular morphology and tested within the same range of shear rates [38]

and establishes shallow temperature gradient to alleviate nucleation barrier within the bulk liquid. In addition, swirling may assist in disintegration of secondary and tertiary dendrites. Such phenomenon resulted in equiaxed grain growth in SSM cast billets with distinct deformation and flow characteristics (e.g., [72]).

Prasad et al. [73] have proposed a correlation (4.19) between particle size and fraction solid, cooling rate, and shear rate, which is in good agreement with experimental results. It is a useful tool to predict the effect of process parameters on the particle size.

$$d = \frac{\Phi D_l^{4/9}(T_l - T)^{1/3}}{\dot{T}^{1/3}\dot{\gamma}^{1/3}} \left(\frac{-f_s}{(1-f_s)\ln(1-f_s)}\right)^{1/3} \left(\frac{C_l - C_o}{C_o - C_s}\right)^{2/3} \quad (4.19)$$

where d is the particles diameter, D_l is liquid diffusivity, T_l is the liquidus temperature of the alloy, T is the temperature in the semi-solid region, C_o is the bulk liquid composition, C_l (4.21) is liquid composition at the interface, C_s (4.22) is the mean composition of the solid formed, $\dot{T}$ is cooling rate, $\dot{\gamma}$ is the shear rate, and f_s (4.20) is the fraction of solid. The value of Φ in (4.19) cannot be found from theoretical analysis as its value is dependent on the relationship between the fluid flow velocity and the velocity for particle–fluid slip. This relationship is not well defined, and therefore the value of Φ must be determined from the best fit of experimental data. However, Prasad et al. [73] pointed out that a detailed computer analysis of the experimental data suggested that for a Couette viscometer-type rheocasting system the empirical value of Φ can be taken as $\frac{0.119}{C_0^{2/3}}$.

$$f_s = 1 - \left[\frac{T_0 - T_l}{T_0 - T}\right]^{\frac{1}{1-k}} \quad (4.20)$$

$$C_l = \frac{C_0}{[1-(1-k)f_s]} \quad (4.21)$$

$$C_s = \frac{C_0[1-(1-f_s)/\{1-(1-k)f_s\}]}{f_s} \quad (4.22)$$

where T_o is the melting point of the pure solvent metal and k is the solute partition coefficient.

Shearing time (t_s)

For constant shear rate tests, the changes of viscosity are dependent on the duration time of shearing. Based on the stirring time, the rheological behavior of SSM slurries may be divided into two stages of transient and steady state. In the transient state, the viscosity is a function of time but in the steady-state condition, viscosity is constant and is only proportional to shear rate [10].

In more detail, the dynamic interactions among the solid particles result in the formation of agglomerates. The newly formed agglomerates may collide together to generate newer agglomerates with even larger size under viscous forces. Parallel to agglomeration, the already agglomerated particles may also break up, deagglomerate giving rise to agglomerates of a smaller size. At a particular time, the degree of agglomeration between solid particles depends on the nature of the system including particle size, its volume fraction, and the external flow conditions.

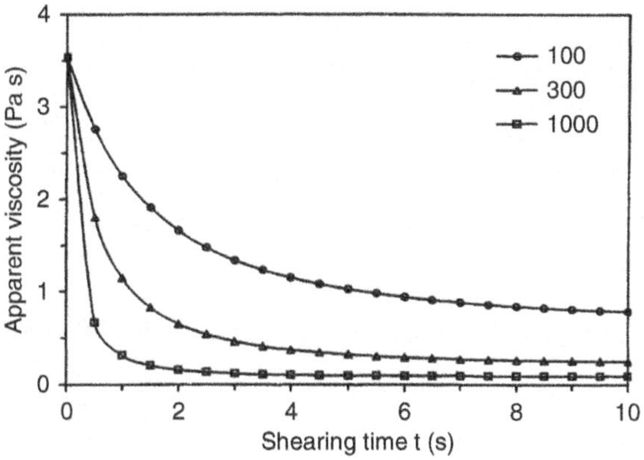

Fig. 4.17 Comparison of calculated transient state viscosity of Sn15%Pb alloy with solid fraction of 0.4 under different shear rates (s^{-1}) as function of shearing time [19] (reprinted by permission of Taylor & Francis Ltd)

At steady state, the degree of agglomeration is a direct result of the dynamic equilibrium between two opposing processes of structural buildup (agglomeration) and breakdown (deagglomeration) [59]. This condition is accessible after sufficient time from the beginning of shearing the slurry. But in transient state, which occurs at a time between the start of shearing and steady-state condition, there is not equilibrium between agglomeration and deagglomeration processes and the measured viscosity of slurry varies with time of stirring up to steady-state condition. The amount of transient time is dependent on the amount of induced shear force and decreases with increasing shear force. Other parameters such as shear rate can affect steady state and transient state conditions as seen in Fig. 4.17. It is clear that with increasing shear rate the slurry reaches steady-state condition at shorter time [19].

Cooling rate and holding temperature and time

Cooling rate, $\dot{T}$, shows the speed of heat extracted from slurry. It is well established that the magnitude of cooling rate has direct effect on the rate of solid formation or growth rate R. Increase of cooling rate causes increasing of growth rate and promotes dendritic solidification. The formation of dendrites and eventual dendritic skeleton increases the viscosity of slurry in comparison with the same fraction solid at lower cooling rate where the formation of a 3D skeleton is postponed. Figure 4.18 shows the effect of growth rate and temperature gradient on the solidification morphology of Sn–Pb alloy [4].

Holding temperature is the parameter to control the fraction solid in mushy zone. The relationship between the holding temperature and fraction solid may be determined by lever rule, Scheil's equation, thermal analysis, or microscopic examination of quenched specimens. It is clear that higher fractions solid cause greater viscosity values [40] as shown in Fig. 4.15. It is also apparent that higher shear rate values are needed to compensate for greater fractions solid if the viscosity remains the same [7, 32].

It is not easy to differentiate between the effects of temperature and fraction solid on the viscosity of slurries due to their intertwined close relationship in mushy

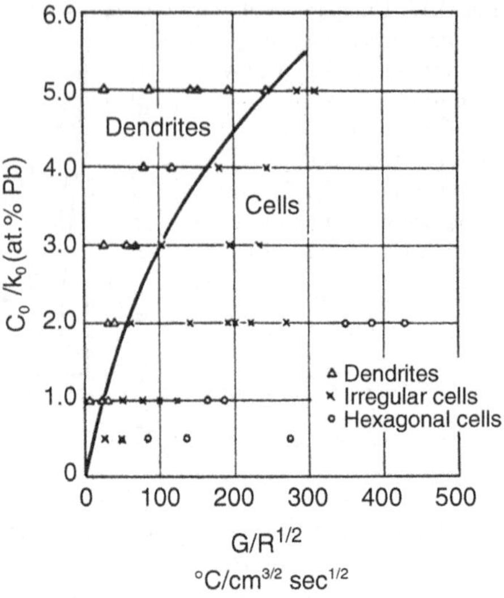

Fig. 4.18 Effect of temperature gradient and growth rate in different composition of Pb on the solidification mode of Sn–Pb alloys [4]

state. However, there is an inverse relationship for temperature and viscosity, (4.23), where higher temperatures impart lower viscosity values. This equation is only valid for systems where there is not any phase change due to temperature variation as, for instance, in polymeric materials.

$$\eta = \eta_o \exp\left(\frac{\Delta E}{RT}\right) \tag{4.23}$$

Where ΔE, η_o, R, and T are the activation energy, initial viscosity, the gas constant, and temperature, respectively.

Holding time (isothermal) is another parameter which affects viscosity due to particle growth. Particle growth in SSM slurries can take place by coalescence, Ostwald ripening, or a combination of both mechanisms. The growth rate of Al–Si alloys in the SSM condition is almost invariably controlled by cube law of the type $R^3 = k_c t$ [74], where the kinetics rate constant is given as (4.24):

$$k_c = \frac{8}{9} \frac{\Gamma D_L}{M_L (C_s - C_l)} f(f_\alpha) \tag{4.24}$$

Where $f(f_\alpha)$ is a function of the solid fraction (its value for globular α-Al particles in A356 is 3.17), Γ is the capillary constant (2×10^{-7} mK), D_L is the solute diffusivity in the liquid (3×10^{-9} m^2 s^{-1}), M_L is the liquidus line slope (6.8 K/at.%Si), C_s is the solute concentration in the solid phase (1.3 at.% Si), and C_l is the solute concentration in the liquid phase [74].

Sample size

Generally the sample size effect on viscosity is not a matter of concern if the homogeneity of temperature and shear rate distribution within the sample is maintained. Nonetheless it is preferred to perform tests on small size specimens to reduce the cost of testing, since for larger specimens more powerful machines are needed. In addition, smaller specimens may render less diverse results. In rotational viscometry for example, the distance of the annular gap between outer cylinder (cup) and inner cylinder (bob) [7, 32, 33, 51] is always preferred to be as close as possible to make the shear rate distribution more homogenized within the slurry during rotation.

Parallel plate compression viscometry method [75, 76] also uses small size samples. Such matter refers back to mathematical calculation of viscosity for cylindrical samples under parallel plate compression test. Such calculation has been done for Newtonian and non-Newtonian fluid assumption. The mathematical solution is always based on the samples with $h \ll R$ ($h =$ height and $R =$ radius) to reduce mathematical complexity. In fact the term v_z, the velocity along z axis, is neglected against radial velocity, v_r, and the continuity and momentum equations are solved. The effect of sample size was also studied at authors' laboratory using parallel plate viscometry. Lashkari et al. [36, 41] reported that one of the most important aspects of rheological testing is to confirm the irrelevance of sample size effect on the viscosity of SSM billets, at low shear rates, less than 10^{-2} s^{-1}. Two different sets of sample sizes with the aspect ratios of 0.4 (10 mm height 24 mm diameter), similar to those reported in the literature, and 1.8 (140 mm height 75 mm diameter) were prepared and compressed under the same initial pressures, holding temperature, 595 °C, and fraction solid, f_s of 0.33. Based on the assumption of "axial flow is negligible against the radial flow" during quasi-steady-state deformation, it was confirmed that large-scale samples could also be used during parallel plate compression viscometry to study rheological behavior of SSM billets through calculation of viscosity. It was also summarized that the sample size is not the important parameter as given below.

- Morphological evolution due to manipulation of pouring temperature has the greatest effect on the deformation behavior of the SSM A356 alloy where viscosity value decreases from dendritic to globular morphology.
- In the context of sample size, reheating of the thixocast samples may bring about minor morphological and size changes in the primary α-Al particles which could affect the viscosity value. This is particularly true for dendritic morphology while for rosette and globular morphologies the changes in viscosity value due to reheating are negligible.
- The viscosity numbers for the globular morphology prepared at low pouring temperature of 615–630 °C are the same in both rheo-billets (140 mm height 75 mm diameter) and thixo-disks (10 mm height 24 mm diameter) to confirm the irrelevance of $h \ll d$ criterion and the reliability of parallel plate compression test to determine viscosity of SSM slurries.
- The most important parameter to influence the flow characteristics of SSM slurries is the morphology of the primary phase, not the sample size.

4.2.3 Testing Methods

Viscosity is the main parameter to study the rheology of semi-solid metallic alloys. It is an indication of semi-solid metal capability in filling the mold and determines the required force for deformation and material flow. During conventional solidification viscosity rises up steadily with increasing solid fraction till the point where the solid can no longer move freely, dendrite coherency point (DCP), and the already solidified segment tends to develop strength.

In SSM processing, the DCP is postponed due to the forced convection or shallow temperature gradient within the melt through prevention of dendritic network development. The breakdown of dendrites due to stirring coupled with multi directional growth of fragmented dendrites due to more uniform temperature distribution within the mold, i.e., shallow temperature gradient, resulting from forced convection, encourages the formation of equiaxed grains, thus postponing the rapid rise of viscosity to higher fraction of solids as reported by Spencer et al. [7], in their pioneering work on the rheology of Sn15%Pb alloy during early 70s.

There are mainly three experimental routes to characterize the microstructure of SSM billets through analysis of their rheological behavior. They basically relate flow characteristics of SSM billets to its microstructure and viscosity. These are as follows:

1. Fluidity measurement
2. Viscometry
3. Cutting force measurement

4.2.3.1 Fluidity Measurement

For semi-solid metal slurries, the concept of mold filling, fluidity, is a critical issue, since as the concentration of solid fraction increases beyond 0.4–0.5, the viscosity increases so rapidly that makes die filling almost impossible. Furthermore, the behavior of the semi-solid alloy is more complex than one that is fully liquid due to its non-Newtonian nature. Therefore, it is not advisable to use the findings on the fluidity of molten alloys for semisolid metals, although the overall concepts of fluidity may apply for semi-solid metals as well.

The fluidity of molten metals has been widely studied (e.g., [4, 13, 14]) and it is well known that there are the molten metal factor (temperature, viscosity, and latent heat of fusion), the mold–metal interaction (heat flow, thermal conductivity, heat diffusivity, and mold temperature), and finally the test variables (metal head height, channel size, and pouring characteristics including pouring speed) that influence the fluidity. Despite the extensive work carried out on different areas of SSM processing, very few studies have been focused on the fluidity concept [77–83]. It is believed that the solidification mode of semi-solid slurries in a fluidity channel is similar to that of fully liquid alloys and that the slurry flow is stopped by a critical

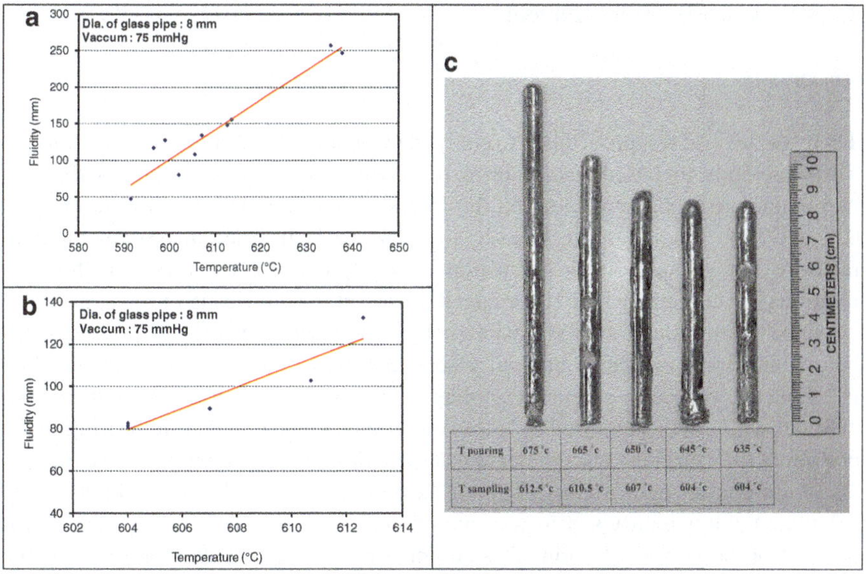

Fig. 4.19 Effect of sampling temperature on the fluidity distance for: (**a**) mechanical stirring, (**b**) SEED process, and (**c**) image of SEED fluidity test results, 356 Al–Si alloy ([78], reprinted with permission of The Minerals, Metals & Materials Society)

solid concentration near the tip of the flow where the solidification is most progressed [79]. Authors studied the fluidity of 356 Al–Si SSM slurries prepared by mechanical stirring and SEED process and showed that the fluidity is directly related to temperature as higher temperatures result in less solid fraction and thus higher fluidity, see Fig. 4.19 [78].

As expected, the distance travelled by the slurry increased with temperature. However, the nature of slurry flow within the glass tube is different for mechanically stirred and SEED processes. For mechanically stirred tests, it appears that as temperature is reduced, the initially formed solid particles acted as filter for the liquid since there is a tendency for agglomeration. Furthermore, the percentage of primary α-Al particles detected within the fluidity samples is only 20 % at its maximum. When compared with the actual solid fraction at temperatures for this alloy, it was clear that more than half of solid particles could not enter the tube used to take slurry sample at the partial pressure of 70 mmHg (used in this study). This was due to agglomeration for mechanically stirred samples and also the increase in particles size with reducing temperature. These particles are then appeared to have acted as filters and thus prevented solid particles entering the tube. Such behavior may suggest that fluidity test of this nature is not a true representative of the slurry characteristics.

4.2.3.2 Viscosity Measurement

Viscometry

There are several test procedures to study the viscoplastic behavior of SSM slurries. These methods are based on measuring the viscosity of slurries and are divided into two main categories depending on the fraction solid, i.e., low fraction solid up to 0.4, and high fraction solid, in excess of 0.4–0.5 [19]. The simplest methods to measure the viscosity of low fraction solid slurries are the direct methods of rotational viscometry where the induced torque in the slurries is measured.

Since the introduction of viscosity concept by Newton, nearly 200 years elapsed before the first practical rotational viscometer was devised by Couette in 1890 [9]. Couette concentric cylinder viscometer consisted of a rotating cup and an inner cylinder which was supported by a torsion wire and rested in a point bearing in the bottom of the cup. This viscometer was a large device with an inner radius, R_{bob}, of 14.39 cm and an outer radius, R_{cup}, of 14.63 cm. Couette design enabled him to calculate the apparent viscosities of non-Newtonian fluids with only a small error because of the very small ratio of gap to inner radius. In Couette-type viscometers, rotation of the cup while holding the bob stationary produces shear stress on the surface of the bob which are measured as torque. Figure 4.20a shows the schematic of the apparatus which was used by several researchers [7, 32, 74]. There is another type of cylindrical viscometers, Searle type viscometer, which is slightly different where the cup is stationary and the bob or inner cylinder rotates and induces shear into the melt or slurry. In both apparatus, the temperature of slurry during stirring process is maintained almost constant by using electric heating elements inside the body of the apparatus. Temperatures are controlled by using thermocouples which are embedded in different sections. Figure 4.20b shows schematic of Searle-type viscometer used by other researchers [23, 33, 51, 61, 62].

Mathematical Treatments

In order to arrive at the fundamental equations for coaxial rotational viscometers, the following assumptions are made [7, 33, 34];

- The liquid is incompressible
- The motion of liquid is laminar
- The streamlines of flow are circles on the horizontal plane perpendicular to the axis of rotation (the velocity is only a function of radius)
- The motion is steady
- There is no relative motion between the surface of the cylinders and the fluid in immediate contact with the cylinders, no slippage
- The motion is two-dimensional
- The system is isothermal

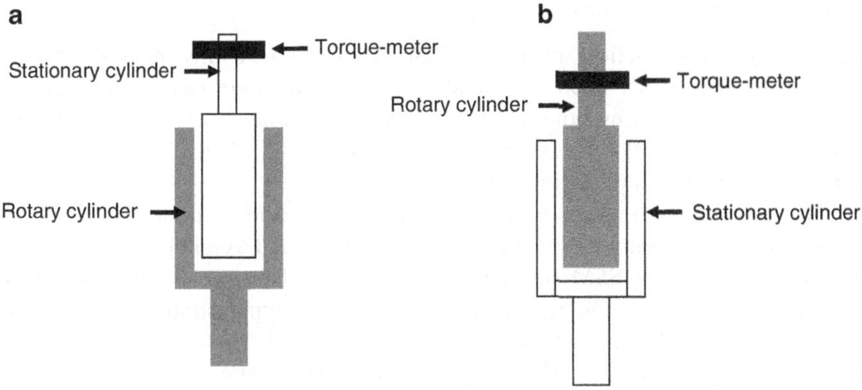

Fig. 4.20 Viscometers; (**a**) Couette type with rotary outer cylinder, (**b**) Searle type with rotary inner cylinder [40]

With such assumptions in both methods the apparent viscosity is calculated by a set of equations (4.25, 4.26 and 4.27) given below using torque data [7, 33, 34].

$$\tau = \frac{T}{2\pi r^2 L} \tag{4.25}$$

$$\dot{\gamma} = \frac{2\Omega}{r^2}\left(\frac{r_i^2 r_o^2}{r_o^2 - r_i^2}\right) \tag{4.26}$$

$$\eta = \frac{T}{4\pi L\Omega}\left(\frac{1}{r_i^2} - \frac{1}{r_o^2}\right) \tag{4.27}$$

Where T is the measured torque, L is the liquid height inside the cylinder, $\dot{\gamma}$ is shear rate, Ω is angular speed of rotor, η is apparent viscosity, r_i is inner cylinder radius, r_o is outer cylinder radius, and r is the actual annular gap radius. There are problems to use coaxial rotating viscometers which are for fluids whose viscosity is related to the geometry of instrument. This is because of the change of shear rate across the gap from the region of higher stress to that of lower stress. Use of the mean rate of shear is permissible only for Newtonian and plastic fluids. Because there is no set relationship between stress and shear rate for pseudoplastic and dilatant fluids, the actual shear rate, presented at (4.26), at any point in the gap cannot be estimated except for extremely small gaps.

Torque measurement is useful to investigate the dendritic coherency point during equiaxed/dendritic solidification. The coherency point is determined as the critical point where the torque increases sharply [48].

Parallel Plate Compression Test

For the high solid fraction slurries, the viscosity is not generally measured by the rotational viscometers. Such self-standing billets (slugs) are more rigid and can only be characterized by other methods including parallel plate compression test [36, 38, 75, 76, 82–89], direct or indirect extrusion [27, 85, 90, 91], indentation test, tensile test [92, 93], and cutting tests [94].

The most common way to examine rheological behavior of paste-like materials is by parallel plate compression test. In this method, a dead weight is simply applied on the top surface of SSM slug and its deformation behavior is investigated by analyzing strain variation versus time [95]. The resulting strain–time graph is further treated mathematically to calculate viscosity and characterize the rheological behavior of tested alloy. The interpretation of results obtained from such graphs can be treated differently depending on the assumption of the SSM slurries behaving as Newtonian or non-Newtonian fluids. In the case of low applied shear rates, less than 0.01 (s^{-1}), the resulting graphs could be treated similar to that of Newtonian fluids [75] with the following equations to calculate the viscosity of the semi-solid cylindrical billets.

The classical Newton's law of viscosity (4.2) was rewritten in terms of the applied force, F (4.28), for a cylindrical sample squeezed between two parallel plates, and with the assumption of the billet not filling the space between the two plates during the course of deformation [21]:

$$F = -\frac{3\eta V^2}{2\pi h^5}\left(\frac{\mathrm{d}h}{\mathrm{d}t}\right) \tag{4.28}$$

Integrating (4.28) for $h = h_0$ at $t = 0$ and $h = h$ at $t = t$, (4.29), and knowing the initial pressure, $P_o = \frac{Fh_o}{V}$, at the onset of deformation, the viscosity–time relationship is given in (4.30):

$$\frac{1}{h^4} - \frac{1}{h_o^4} = \frac{8\pi F t}{3\eta V^2} \tag{4.29}$$

$$\frac{3Vh_o}{8\pi P_o}\left(\frac{1}{h^4} - \frac{1}{h_o^4}\right) = \frac{t}{\eta} \tag{4.30}$$

The viscosity is calculated as the inverse slope of a graph where the left-hand side of (4.30), $\left[\left(\frac{3Vh_o}{8\pi P_o}\right)\left(\frac{1}{h^4} - \frac{1}{h_o^4}\right)\right]$, is plotted against time, (t). For Newtonian fluids, the average shear rate, $\dot{\gamma}_{av}$, at any instant during compression test is calculated as (4.31) [21]:

$$\dot{\gamma}_{av} = -\sqrt{\frac{V}{\pi}}\left(\frac{\mathrm{d}h/\mathrm{d}t}{2h^{2.5}}\right) \tag{4.31}$$

where v_x, η, V, h_o, h, F, and t are deformation speed (ms^{-1}), viscosity (Pa-s), volume of specimen (mm^3), initial height (mm), instantaneous height (mm), applied dead weight (N), and deformation time (s), respectively. The mathematical solutions employed to reach the above-mentioned equations, (4.28)–(4.31), could be found in Appendix A [38].

If the SSM billets are treated as non-Newtonian fluids, the solution to the flow equations for cylindrical sample squeezed between two parallel plates is as follows (see Appendix A):

$$\frac{h_o}{h} = \left\{1 + \left(\frac{3n+5}{2n}\right) k h_o^{\frac{n+1}{n}} t\right\}^{\frac{2n}{3n+5}} \tag{4.32}$$

where

$$k = \left\{\left(\frac{2n}{2n+1}\right)^n \left(\frac{4(n+3)}{\pi m d_o^{n+3}}\right) F\right\}^{\frac{1}{n}} \tag{4.33}$$

Equation (4.32) is only valid for deformation under steady-state condition where the engineering strain changes linearly with time. (4.32) could be further treated mathematically [38] to include engineering strain (e) as given in (4.34).

$$\log(1-e) = -\left(\frac{2n}{3n+5}\right)\log t - \left(\frac{2n}{3n+5}\right)\log\left(\frac{3n+5}{2n} k h_o^{\frac{n+1}{n}}\right) \tag{4.34}$$

In order to calculate the values of m and n, the logarithmic of engineering strain, log $(1-e)$, should be plotted against time, log t, and the slope of such graph and its intercept with strain axis should provide the necessary means to calculate m and n [21]. Figure 4.21 shows schematically a parallel plate compression test machine and the actual in-house designed and fabricated of a large-scale parallel plate compression machine to characterize the microstructural evolution of SSM prepared billets (slugs) [95].

For rheological tests, the as-poured billets were transferred (at temperature 598 ± 2 °C) and compressed uniaxially in the parallel plate compression test machine by applying 2.2 kg dead weight at 598 ± 2 °C. The applied force and resulting displacement were monitored using a load cell with 0.02 % precision and a displacement transducer with $\pm$(0.1–0.2)% full stroke precision. A cylindrical furnace was installed on the press bed to keep the billet temperature constant during the compression tests. The furnace was equipped with quartz heat-resistant windows to allow one to view the billet. Two K-type thermocouples were positioned within the furnace to control the chamber temperature with a precision of $\pm$2 °C. All samples were taken out of the furnace after 10 min of compression and quenched in water to room temperature.

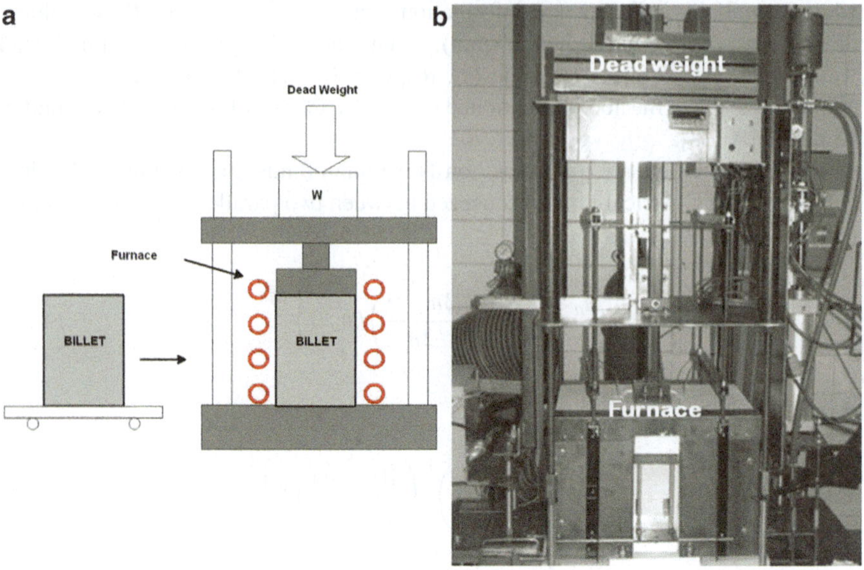

Fig. 4.21 Schematic (**a**) and a real overview of the press (**b**) [95]

Drop Forge Viscometry

Drop Forge Viscometer is a special case [84, 86] of parallel plate compression test where similar equations are used to calculate the viscosity. Such instrument is designed to provide a wide range of shear rates for viscometry tests, 10^{-5} to 10^4 (s^{-1}). Usually the high shear rate tests is performed within thousandth of second and can just investigate the transient part of the viscosity for thixotropic SSM materials. The developed equation used in such manner is given in (4.35) as a modification of (4.28):

$$m_p\left(\frac{\mathrm{d}^2h}{\mathrm{d}t^2}+g\right)=\frac{-3\eta V^2}{2\pi h^5}\left(\frac{\mathrm{d}h}{\mathrm{d}t}\right) \tag{4.35}$$

Where m_p is the mass of upper plate and g is gravitational acceleration. After calculating the derivatives and knowing the variables in (4.35), the viscosity can be calculated as a function of time.

Sherwood et al. [96, 97] have also developed mathematical equations to calculate viscosity of non-Newtonian materials squeezed plastically between two parallel plates, where the squeezed material fills the space between two platens. Assuming "σ_f" as the friction stress in plate/work piece interface and "σ_y" as the yield stress of the material, the following equation (4.36) is presented to specify the deformation behavior of the substance during compression viscometry.

$$F = \frac{2\sigma_f \sigma_y \pi r^3}{3h_o} + \frac{\sqrt{3}\sigma_y \pi r^2}{2}\left[\sqrt{\left(1 - {\sigma_f}^2\right)} + \frac{1}{\sigma_f}\sin^{-1}\sigma_f\right] \tag{4.36}$$

where F is total force required to compress the material, h_o is initial height, and r is the radius of circular platens, where the squeezed material fills the space between them. The force, F, is independent of the compression rate, $\left(\frac{dh}{dt}\right)$, as should be expected from an analysis based upon rate-independent plasticity theory. Kolenda et al. [83] have also used similar equation as (4.36) for evaluating the viscosity of two types of ceramic powders blended together as a ceramic paste. The resulted data indicated the ceramic paste behaving as non-Newtonian materials.

Direct and Indirect Extrusion

Extrusion tests may also be used to study the ability of the SSM billets to fill die cavity by providing information on the viscosity of semi-solid material. Figure 4.22 illustrates extrusion tests schematically along with related graphs [90, 93]. More experimental details and suitable theoretical treatments, mathematical equations, for such methods are accessible in the literature [74, 90, 93, 98].

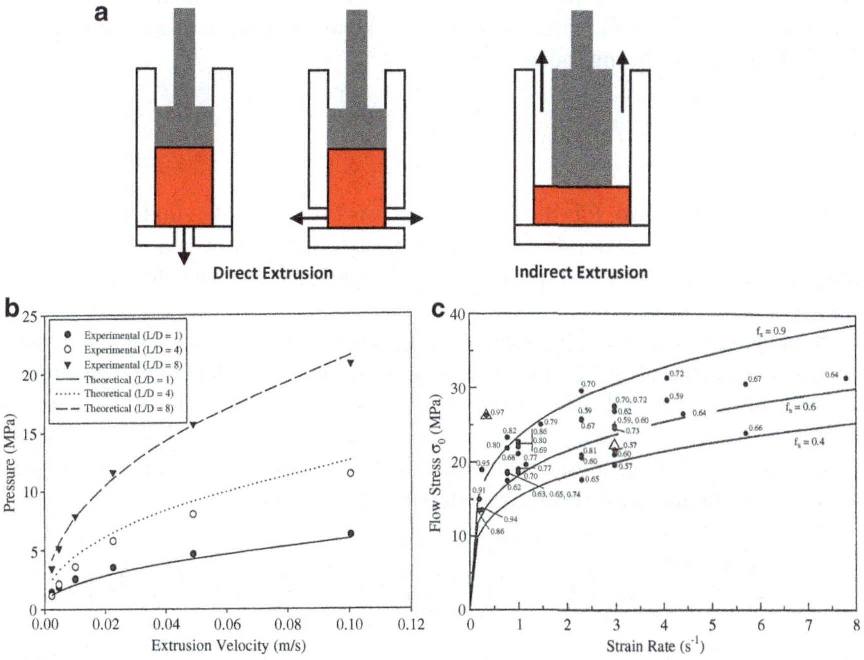

Fig. 4.22 (**a**) Different ways of extrusion viscometry, (**b**) Pressure versus Extrusion velocity graph, comparison of theory and experiments at different dimensions [90], (**c**) Flow stress versus strain rate graph at different fraction solid [91]

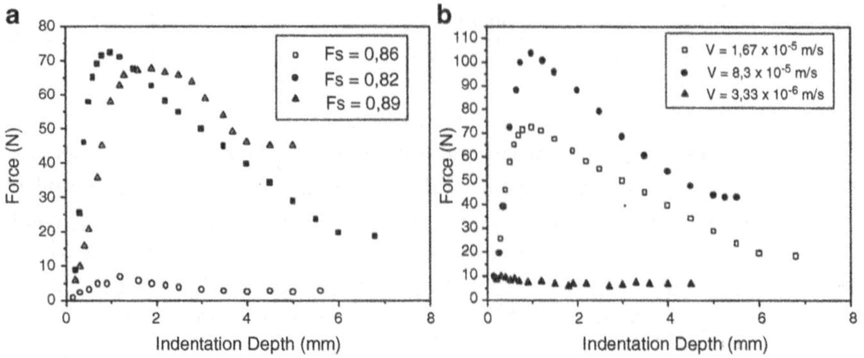

Fig. 4.23 Results of the indentation test for (**a**) constant penetration speeds and different fractions solid (**b**) different penetration speeds and constant 0.86 fraction solid [92]

Indentation Test

Indentation test is another simple way to study mechanical properties of SSM slugs. In this method, the depth of penetration of a steel cylinder under a constant pressure in the semi-solid billet is taken as an indication of viscosity of the alloy. This is a simple test and can be commercialized for online testing of SSM billets. Figure 4.23 shows the effect of testing parameter and fraction solid on the indenter/SSM billets interaction [92]. The following equation (4.37) was used to calculate the viscosity of SSM specimens during indentation test:

$$\eta = \frac{16\pi[(1-\upsilon^2)F]}{6r\dot{\varepsilon}} \tag{4.37}$$

where υ is Poisson ratio, F is the applied load, r is the radius of cylindrical indenter, and $\dot{\varepsilon}$ is the strain rate. It has been shown the pseudoplasticity behavior of Al–4%Cu alloy around 0.82–0.85 fractions solid and solid-like behavior for fractions in excess of 0.85 [92].

Some researchers have also applied the tensile test method for rheological study of semisolid materials [92]. However, both of tensile and indentation tests are not suitable for fraction solid less than about 0.85 [16, 92]. The indentation test is similar to compression tests with more constraints.

Lahaie et al. [99] proposed a physical model to show the relationship between tensile strength and semi-solid parameters, (4.38):

$$\sigma = \frac{\eta\dot{\varepsilon}}{9}\left(\frac{f_s^m}{1-f_s^m}\right)^3\left[\left(1-1/2\left(\frac{f_s^m}{1-f_s^m}\right)\varepsilon\right)^{-3}+2\left(1+\left(\frac{f_s^m}{1-f_s^m}\right)\varepsilon\right)^{-3}\right] \tag{4.38}$$

Thus, from (4.38), the mechanical response, i.e., tensile strength (σ), of the semi-solid body is dependent on the viscosity of intergranular liquid, η, the applied strain rate, $\dot{\varepsilon}$, the accumulated strain, ε, the solid fraction, f_s and a microstructural parameter, m, bounded by the values 1/2 and 1/3, corresponding to columnar and equiaxed microstructure, respectively. A similar equation has also been used by Wahlen [27] during compressive rheological test of aluminum alloys.

4.2.3.3 Cutting Force Measurement

The apparatus is basically designed as a quality control tool in production line. In this method, semi-solid slurry is cut by a steel blade or wire. The machine has a load cell to measure the force and sensors to determine the speed as well as the wire/blade tension. The speed of cutter to pass through the SSM billet and the resistance of slurry against the passage of cutting blade, the force to be exerted on the blade, is an indication of microstructural feature [52, 94]. Figure 4.24a shows the apparatus during a test. Figure 4.24b describes four indicators used to characterize the cutting curve:

(a) The maximum force
(b) The peak distance
(c) The area under the curve for the first half diameter of the slug corresponding to the cutting energy
(d) The central force relating to the average cutting force at the center of the slug (between 30 and 40 mm)

An example of the results is presented in Fig. 4.25a for three pouring temperatures of 675, 655, and 635 °C for A357 aluminum alloy (test performed at 590 °C billet temperature). A higher pouring temperature results in more dendritic microstructure which is expected to put more resistance against the blade to cut through the billets than globular microstructure. The average value for all four indicators increase with the pouring temperature and one example is presented in Fig. 4.25b.

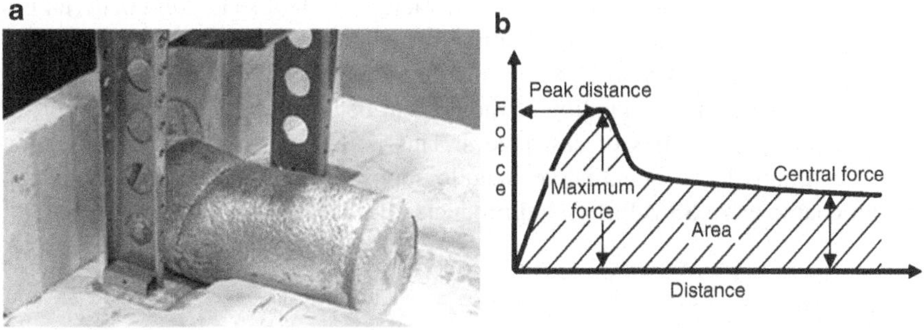

Fig. 4.24 (**a**) Cutting test, (**b**) schematic of a typical Force-cutting distance with four indicators [52]

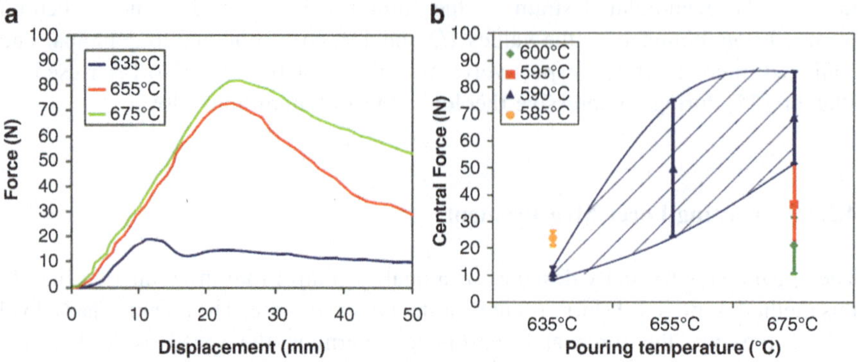

Fig. 4.25 (**a**) Force-displacement curves obtained at various pouring temperatures, (**b**) maximum force - pouring temp for different cutting temperatures [52]

4.3 Micro/Macro Structural Analysis

In addition to thermal characterization of the solidification process during SSM processing and rheological analysis of the as-cast SSM billets, the quality of semifinished SSM billets or the finished SSM products may be studied by macro and microstructural analysis of the constituent phases.

In order to generate a semi-solid structure, the alloy system plays the key role where the coexistence of liquid and solid within a temperature range is the prerequisite for the slurry preparation. The mechanics and mechanisms of the primary particles' evolution, dendrite-to-equiaxed transformation, is the next concern since the formation of globular morphology is expected to enhance die filling and improve mechanical properties of as-cast parts. The ideal microstructure for SSM slurry is fine, spherical solid particles uniformly distributed within a liquid matrix. The solid fraction should be considered carefully, since low fraction solid may lead to handling and mold filling difficulties due to insufficient viscosity and turbulence while high fraction solid may have die filling troubles or increase the cost of machinery.

Based on the aforesaid requirements, characterization of semi-solid material is a necessity to confirm, modify, and obtain an optimum structure for SSM component shaping process. This knowledge not only provides an idea about the material, but also leads to better understanding of rheological behavior and eventually improve the mechanical properties of cast pieces. This is the main theme of this section where qualitative and quantitative metallography principles are implemented to further characterize SSM billets and products [100].

4.3.1 *Qualitative Metallography*

In order to be able to accurately characterize SSM microstructure, it is necessary to understand the features and complexity of the resulting solidified structures and be able to differentiate between the observed two-dimensional (2D) structure and the actual three-dimensional (3D) morphology.

Normally after initial visual inspection, the microstructure of the solidified alloy is observed on the plane of polished surface using optical microscopy. The 2D analysis may not present a complete picture of the structure and sometimes could lead to invalid conclusions. This is particularly true for cases where in spite of a well distributed primary phase in 2D plane, the interconnectivity of isolated primary particles from underneath the plane of polish has resulted in misleading conclusions on the process efficiency.

In literature, there are different techniques employed to reveal the true morphological evolution of primary phase including reconstruction of 3D images by serial sectioning, X-ray micro-tomography, and finding the crystallographic orientation relationships by electron backscatter diffraction pattern (EBSD).[3] Serial sectioning is a destructive technique based on the successive grinding and polishing of the sample with capturing the consequent images. The major difficulties are calibration of the sectioning distance and also the working frame. These shortcomings are overcome by automatic polishing procedure and drilling guides and controlling holes perpendicular to the polished section. The morphology of primary particles together with their possible interconnectivity is characterized based on the position and shape of each feature along these serial sections. The final 3D image is constructed with the support of computer software [103–107]. An example is presented in Fig. 4.26.

For 3D visualization, X-ray microtomography was developed by Suery and his coworkers at Grenoble, France. The nondestructive technique is based on the X-ray beam passing across the sample with consequent capturing the transmitted image by a CCD or CMOS camera. The specimen is placed on a high precision rotating table in the trajectory of an X-ray beam. Part of the X-rays is absorbed whereas the transmitted X-rays are converted into visible light using a scintillator. The sample is then rotated over 180° or 360° and during this rotation, sufficient projections are recorded. The final stage is the same as serial sectioning method which is retrieving

---

[3]Electron Backscatter Diffraction "EBSD" is a technique which provides crystallographic information by analyzing crystalline samples in the scanning electron microscope "SEM." In EBSD, a stationary electron beam strikes a 70° tilted sample and the diffracted electrons form a pattern on a fluorescent screen. This pattern is unique to the crystal structure and orientation of the sample region from which it was generated. The diffraction pattern is used to measure the crystal orientation, grain boundary misorientations, discriminate between different materials, and provide useful information about local crystalline perfection. By scanning in a grid across a polycrystalline sample and measuring the crystal orientation at each point, the resulted map will reveal the constituent grain morphology, orientations, and boundaries. In addition, the data shows the preferred crystal orientations (texture) present in the material [101, 102].

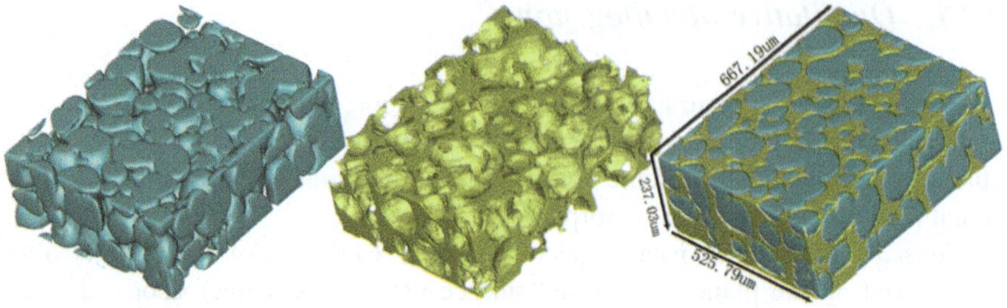

Fig. 4.26 Three dimensional morphology of A356 slurry reconstructed by serial sectioning of 60 consecutive images (Courtesy of Prof. Xiangjie Yang) [107]

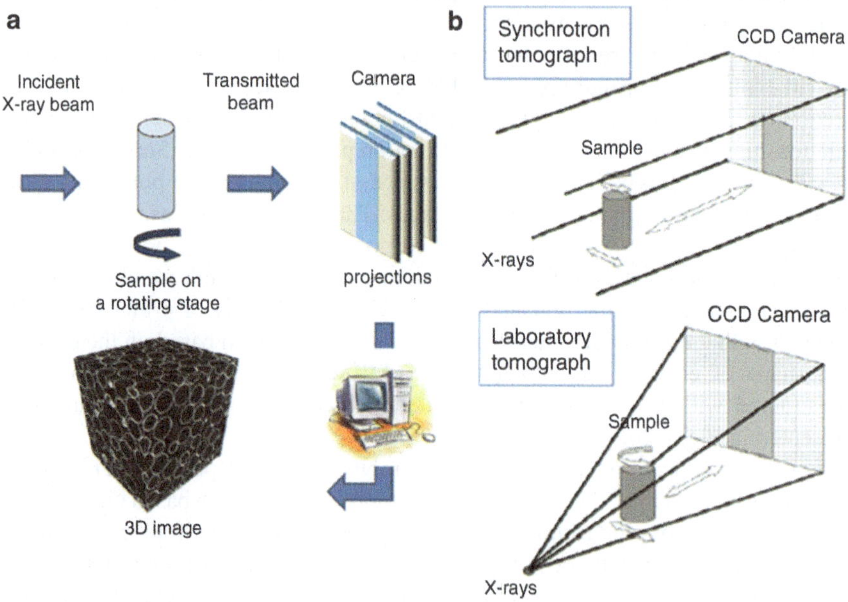

Fig. 4.27 (**a**) Principle of tomography, (**b**) synchrotron and laboratory tomography [108]

a 3D image of the sample [108]. In this method, the contrast between the phases is directly related to the atomic number difference between various phases and the fact that solid and liquid have good absorption contrast due to the large difference in chemical compositions. Previously, the technique was used at room temperature however by the development of the technique, in situ scanning in the semisolid state is viable [108–110].

According to Salvo et al. [108], tomographs are classified into two categories (Fig. 4.27): laboratory tomographs with divergent X-ray beam and polychromatic and the synchrotron sources (ESRF, SOLEIL in France, SLS in Switzerland, Spring 8 in Japan, etc.) where the X-ray beam is parallel, polychromatic, or monochromatic.

Kareh et al. [109] used synchrotron X-ray tomography to study the deformability of Al15Cu during indirect extrusion process in a small scale. The specimen originally heated to ~5 °C above the eutectic temperature for 200 h. A specimen from original material then heated to 10 °C above the eutectic temperature at around 558 °C to form 0.73 solid fraction. It was then held for 5 min before isothermal indirect extrusion at a ram speed of 0.01 mm s^{-1}. Figure 4.28 presents

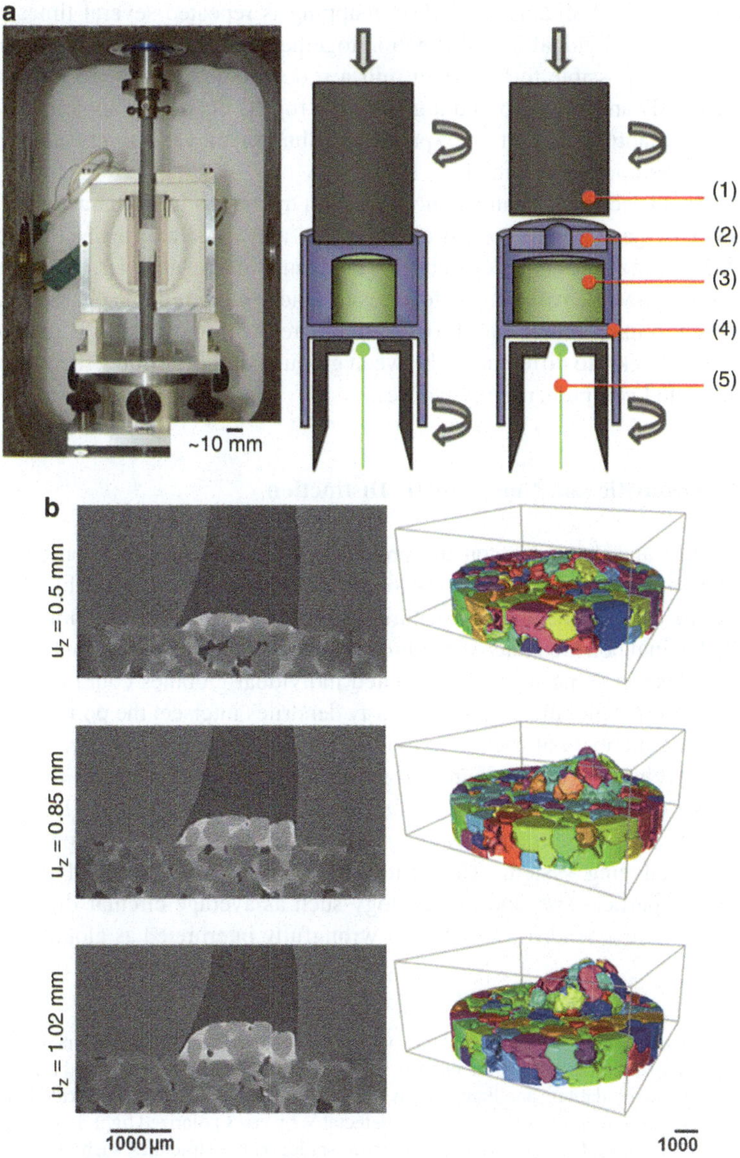

Fig. 4.28 (**a**) Experimental setup showing the actual machine and schematic of the compression setup "(1) ram, (2) boron nitride extrusion die, (3) specimen, (4) container, (5) thermocouple" (**b**) microstructure at three instances during semisolid extrusion (*left* is "xz" slices, *right* is 3D rendering of the solid) (α-Al gray, Cu-enriched liquid white, and pores dark) [109]

three snapshots in time showing filling process as well as a 3D rendering of the separated α-Al globules extruding upward only.

A modern technique is based on the automated serial sectioning using Focused Ion Beam (FIB) and generating crystallographic information with electron back-scatter diffraction (EBSD) which is installed in a combined FIB-SEM instrument. The process of FIB slicing and EBSD mapping is repeated several times before assembling the individual 2D information together into a 3D viewer. By using this technique, it is possible to Ion-beam mill a series of slices followed by mapping them by EBSD and using computing software for 3D crystallographic maps. This allows the quantification of the spatial distribution of various crystallographic features of the microstructure [111, 112].

The current section is mainly concerned with microstructural characterization of the SSM cast parts where certain complexities may generate misleading results. With the polarized light microscopy[4] [113] and image analysis, it is tried to differentiate SSM microstructural features to render more reliable characterization (in addition to microscopy-based routes, there are rheological tests, as production-line quality checks, to differentiate between globular and dendritic structures which was elaborated comprehensively before).

4.3.1.1 Dendritic and Nondendritic Distinction

Occasionally and depending on the process parameters, solidification conditions may lead to dendritic growth in all or portions of the semi-solid billet, as for example the most prone area for dendritic solidification is near the mold wall. As a result, the final polished microstructure may show dendrites' main trunks and their branches. Also in some segments, isolated individual globules could be observed which are not real globules. Hence, the way dendrites intersect the polished surface may generate numbers of pseudo individual and isolated particles (Fig. 4.29).

The conventional bright field and polarized light micrographs in Fig. 4.30 confirm the deficient nature of analysis carried out on 2D sections. The dendrite secondary branches are treated as individual and isolated particles if processed by an image analyzing system. This leads to erroneous calculation of parameters attributed to particle size and morphology such as average circular diameter and sphericity where a dendritic structure is wrongfully interpreted as globules.

---

[4]Metallographic etching is defined as the procedure required to be implemented to expose particular microstructural features of interest which normally are not evident in the as-polished condition. Anodic oxidation "anodizing" is an electrolytic etching depositing an oxide film on the surface and developing grain coloration that is detectable by cross polarized light. It was noted that in certain nonferrous alloys that have noncubic crystallographic structures such as magnesium, titanium, and zirconium, grain size can be identified efficiently in the polished condition while this is not possible for metals with cubic, isotropic crystal structure (using polarized light). Anodized surface films on cubic metals, such as aluminum, are usually examined under polarized light to reveal grain contrast [113].

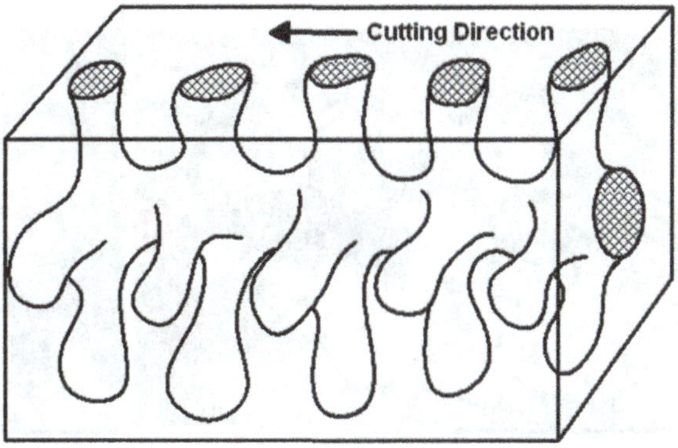

Fig. 4.29 3D view of dendrite-polished surface intersection (reprinted with permission of The Minerals, Metals and Materials Society) [100]

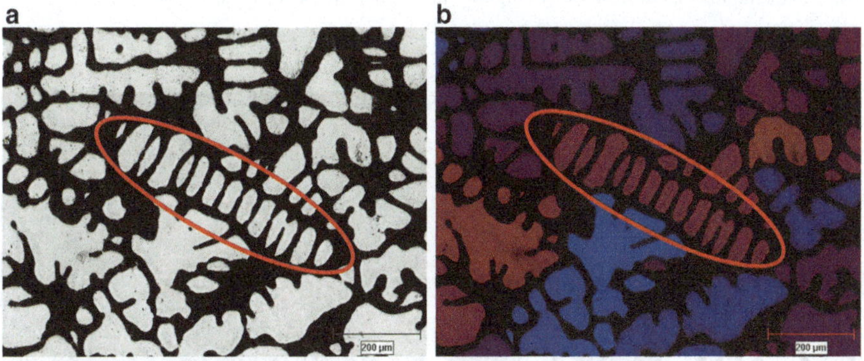

Fig. 4.30 Branches of dendrite with: (**a**) bright field, and (**b**) with polarized light, A356 alloy, quenched from 598 °C

Another case to explore the difference between dendritic and globular morphologies is using electron backscatter diffraction (EBSD) analysis as shown in Fig. 4.31. These grain maps belong to two different pouring temperatures (690 and 625 °C) for 356 commercial Al–7%Si alloy (melting temperature is around 615 °C). In EBSD, grains have different interpretation from that of the traditional metallography. In this method, two neighboring scan points belong to the same grain if the crystallographic misorientation between them is less than a defined value (normally 10 or 15°) [101, 102].

4.3.1.2 Pseudo-Globule

Primary particles in semi-solid microstructure have complex morphology. For instance, the particle that is supposed to be one globule may be interconnected to other globules and thus the proper nomenclature for such particle is "pseudo-

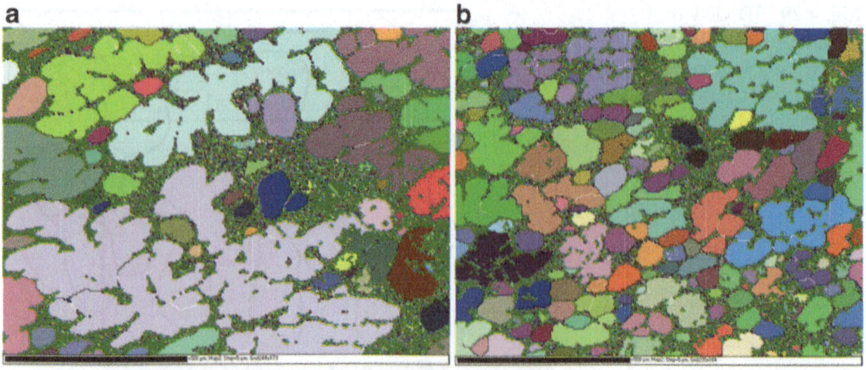

Fig. 4.31 Grain mapping of two different pouring temperatures for 356 commercial Al–7%Si alloy, (**a**) 690 °C, and (**b**) 625 °C (melting temperature is around 615 °C) (each grain is assigned a color to distinguish them from other neighboring grains) (black scale bar is 500 μm)

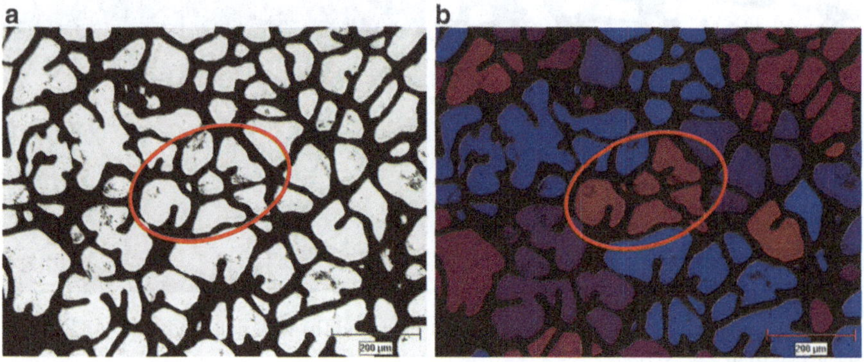

Fig. 4.32 Formation of cramped dendrite: (**a**) bright field illumination, (**b**) polarized light, A356 alloy quenched at 598 °C

globule." The concept of pseudo-globule could be easily understood by comparing different particles seen in the optical micrograph of Fig. 4.32. The typical encircled globules appear to be isolated. However, polarized light microscopy distinctly reveals the isolated particles have identical color contrast, and therefore originated from one particle.

This phenomenon is more prevalent in stir-based processes such as mechanical or electromagnetic stirring routes. By stirring, the resulting forced flow obliges the preformed dendrites to break up by mechanical fragmentation [4] or dendrite root remelting mechanisms [114]. However, if the applied shear force is not sufficient enough to disintegrate the branches, they may plastically bend to form the so-called cramped dendrites. Therefore, it is expected to have different colors for the isolated particles in the former case, true isolated particles, while similar color contrast should be expected in the latter case, cramped particle. The pseudo-globules seen in

Fig. 4.32 may have their origin in the cramped dendrites. These particles are not the result of agglomeration of isolated spherical particles since they have the same color contrast and therefore an interpretation based on the bending and intertwining of dendrites could be a valid conclusion. The only justification for such isolated particles is due to the metallographic 2D sectioning. Pseudo-globules have a key role on the rheological properties of SSM slurries and results in higher viscosity.

Breaking and fragmentation of primary dendrites by mechanical stirring is the original route for SSM processing. However in such structures, there are large primary agglomerations which have very complex shape. These groups of primary particles seem to have formed due to agglomeration of small and individual particles which is not an accurate assumption. Investigation on agglomerated particles goes back to the work of Ito et al. [106] on particles formed during mechanical stirring of Al–6.5%Si. The prepared SSM slurry was sheared at a rate of 900 s^{-1} for 2 h at 0.2 fraction solid and then quenched. The quenched sample was polished and examined in successive sections of 10–20 μm apart. By reconstruction of the 3D model from such serial sections, particles which appear to be isolated are actually interconnected to others. Figure 4.33 is a 3D reconstruction of this microstructure showing the bulk of the particles in the region studied are interconnected to form single agglomerate.

The idea was verified [58] by finding the crystallographic orientation relationship between agglomerate particles in Al–6.5%Si using electron backscatter diffraction technique, EBSD. In the same study, it has been stated that all particles belonging to the same agglomerate are related either by low angle grain boundaries

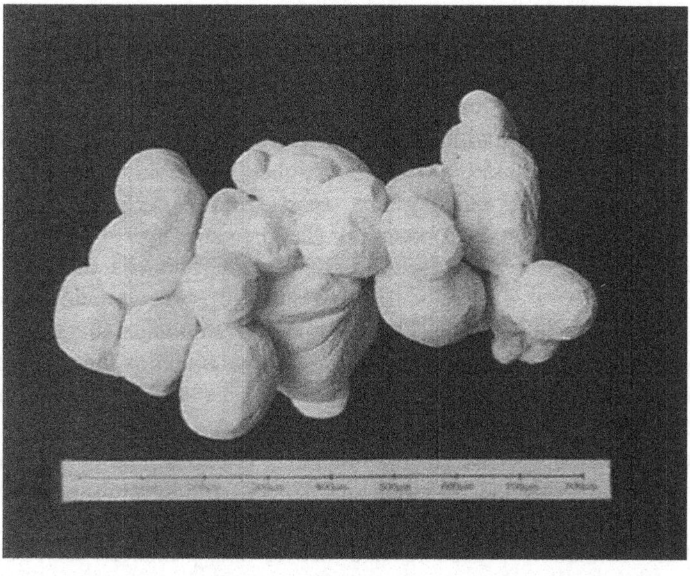

Fig. 4.33 3D reconstruction showing the interconnection of single agglomerate (reprinted with permission of The Minerals, Metals and Materials Society) [106]

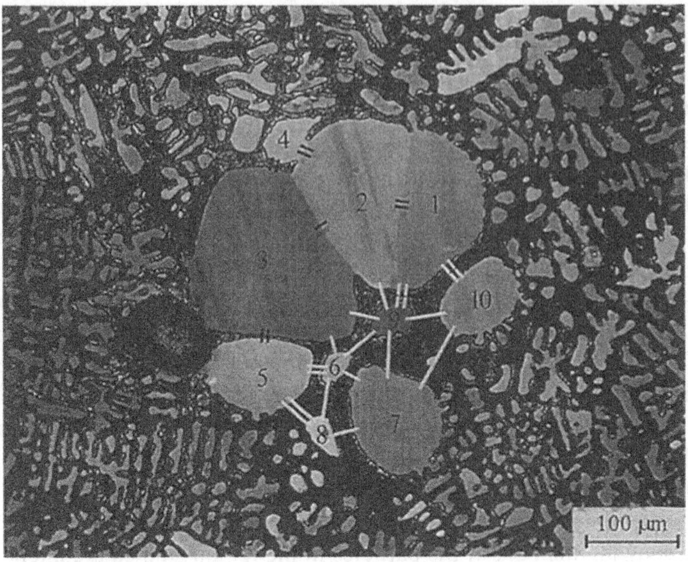

Fig. 4.34 Agglomerate of ten particles. The bonds marked (–) indicate a low angle grain boundary (0–10°), while bonds marked (=) indicate a misorientations of $60 \pm 5°$ with the rotation axis close to the [111] axis (reprinted with permission of The Minerals, Metals and Materials Society) [58]

with misorientations less than 10° or by misorientations about 60° around the [111] axis which corresponds to coherent twins. Figure 4.34 and Table 4.3 show an agglomerate of ten particles in which all particles have at least one neighboring particle with one of the two observed orientation relationship. As for instance, particle number 9 has an orientation relationship to particles 2, 3, 6, 7, and 10 corresponding to a low angle grain boundary, while it has an orientation relationship to particle 1 corresponding to a coherent twin boundary. This confirms that particle 9 is connected to the rest of the particles associated to the agglomerate through the third dimension [58].

In another study, Niroumand et al. [103] used Al–10.25%Cu and mechanical stirrer and by serial sectioning proved that the microstructure of slurry consists of pseudo-particles / clusters in polished surface. Many researchers have interpreted this phenomenon as a process of agglomeration of small globule particles and identified them as single primary grains, but in reality, the majority of these particles were interconnected from underneath (3D) three dimensionally. Figure 4.35 shows a pseudo-cluster and resulting 3D model of the sample indicating the structural complexity.

As described earlier, X-ray microtomography is a real-time tool which is quite capable of constructing the development of SSM structure during solidification process. In fact by analyzing the high resolution image, it is possible to visualize the solid phase formation procedure and see if the primaries are isolated or interconnected. Figure 4.36a shows a 3D view of A356 alloy isothermally held at 587 °C for 10 min. By removing the liquid from 3D image, the connection among solid particles become obvious, Fig. 4.36b [16].

Table 4.3 Misorientations data between particles in the agglomerate shown in Fig. 4.34 (relative misorientations θ, the rotation axis [h, k, l], and the deviation $\Delta\varphi$) (reprinted with permission of The Minerals, Metals and Materials Society) [58]

Particles	θ	[h, k, l]	$\Delta\phi$	Interface type
1–2	59.6	[1,1,1]	0.9	Coherent twin
2–3	3.7	[1,3,6]	2.8	Low angle
2–4	56.3	[1,1,1]	5.2	Coherent twin
3–4	55.8	[1,1,1]	8.8	Coherent twin
3–5	56.8	[1,1,1]	13.6	Coherent twin
3–6	3.4	[1,4,5]	12.4	Low angle
5–6	57.2	[1,1,1]	12.4	Coherent twin
5–8	55.6	[1,1,1]	7.9	Coherent twin
6–8	10.3	[0,2,3]	2.3	Low angle
6–7	4.1	[1,1,3]	2.5	Low angle
7–8	6.9	[1,3,3]	2.5	Low angle
7–10	6.7	[1,2,3]	8.4	Low angle
10–1	56.1	[1,1,1]	7.8	Coherent twin
9–1	59.1	[1,1,1]	3.3	Coherent twin
9–2	2.8	[2,3,3]	1.6	Low angle
9–3	6.4	[1,2,3]	0.7	Low angle
9–6	5.9	[1,6,8]	0.3	Low angle
9–7	4.9	[1,2,10]	2.1	Low angle
9–10	7.3	[2,3,6]	1.6	Low angle

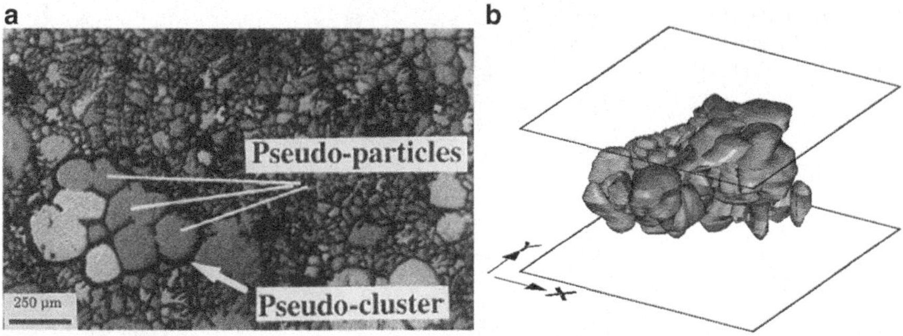

Fig. 4.35 (**a**) Definition of pseudo-particles and pseudo-clusters, (**b**) CAD generated view of a 3D rendered pseudo-cluster [103]

4.3.1.3 Sintering and Coalescence

Sintering and coalescence of primary particles are the other mechanisms active during SSM processes. It is envisaged during stirring the solid particles collide with each other to form welded joints which are eventually strengthened due to high temperature and easy diffusion.

Observation of unusual grain boundaries goes back to the work of Apaydin et al. [114] on the Al–10%Mg which was stirred with different stirring periods and shear rates at low cooling rate (1.5 K min^{-1}) (Fig. 4.37). It was claimed that these

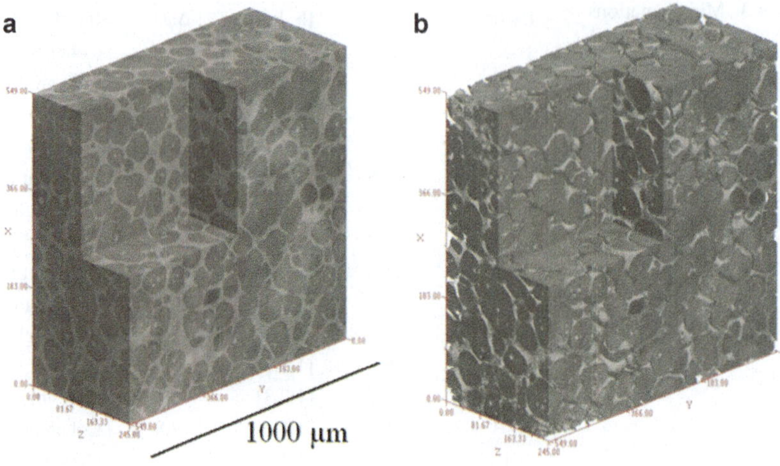

Fig. 4.36 3D image of A356 alloy isothermally held at 587 °C; (**a**) solid and liquid phases, (**b**) only solid phase [16]

Fig. 4.37 Optical micrograph of a particle showing four special boundaries in Al–10%Mg alloy stirred 7 min at 400 s^{-1}, ~×193 [114]

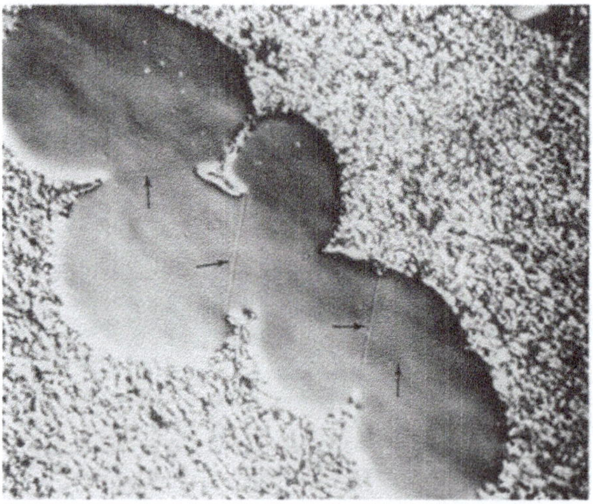

special boundaries are low energy boundaries since they did not show precipitation during solution and aging treatments. They postulated that these special boundaries are probably produced by sintering where the initial particle–particle contact resulted from particles collision due to stirring. The other possibility for special boundary formation is the growth twining mechanism. Generally in aluminum and its alloys, the (100) facets preferentially grow in the direction opposite to the heat flow [4]. However, if this growth preference is disturbed in any manner (such as by introduction of solute elements or by agitation), more of the (112) facets grow in

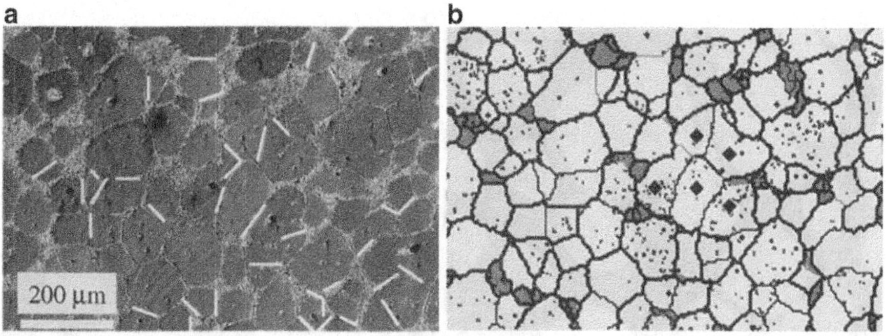

Fig. 4.38 Grain structure in EM stirred and grain refined Al-6082, poured at 700 °C, (**a**) secondary electron image, all special boundaries are marked with *white lines*, (**b**) grain boundary map. Low-angle and coherent twin boundaries are indicated with thin lines [60]

direction opposite to the heat flow and resulting in a nucleation for growth twining on the solid–liquid interface [114].

Figure 4.38a is a secondary electron image of an electromagnetically stirred "EMS" sample of Al-6082 alloy, isothermally held at 638 °C for 10 min [60]. By EBSD investigation, all special grain boundaries were characterized and shown by white lines in Fig. 4.38a. Figure 4.38b is a grain boundary map of the same area (the white grains represent unmelted material and the gray regions are quenched melt). High angle grain boundaries are marked with thick black lines whereas low-angle and twin boundaries are marked with thin lines. The diamonds highlight six grains separated by low angle and twin boundaries and accordingly belong to the same agglomerate. It was claimed that formation of a normal grain boundary would be energetically unfavorable as some solid–liquid interface with energy of 95 mJ m^{-2} would be replaced by a grain boundary with energy of 324 mJ m^{-2} [60]. Low angle grain boundaries and coherent twin boundaries, on the other hand, would form readily since they have lower interfacial energy than the solid–liquid interface having high angle boundaries.

An example of sintering in SSM materials is shown in Figs. 4.39 and 4.40. Two or more isolated solid particles joined and formed a pseudo cluster which is associated, for instance, to the stabilized contact during the stirring (arrows show solid necks). This effect could be seen in regions with low stirring or stagnant liquid and solid. Different globules present different primary α-Al particles with various orientations. Thus, sintering of individual primary particles could be easily detected by different color contrasts by color metallography or EBSD technique. In the semi-solid science, this is defined as "agglomeration" [17–19] where primary particles, basically consisted of simple isolated globular or rosette shape, come into contact and sinter together to form agglomeration. Particles become interconnected by solid necks and further spheroidized by material transport specifically through the neck area which has a negative curvature. The sintering mechanism has been assumed to occur under condition $\gamma_{gb} < 2\gamma_{sl}$ in which γ_{gb} and γ_{sl} are grain boundary and solid–liquid phase boundary energies, respectively. As mentioned by Doherty et al. [115],

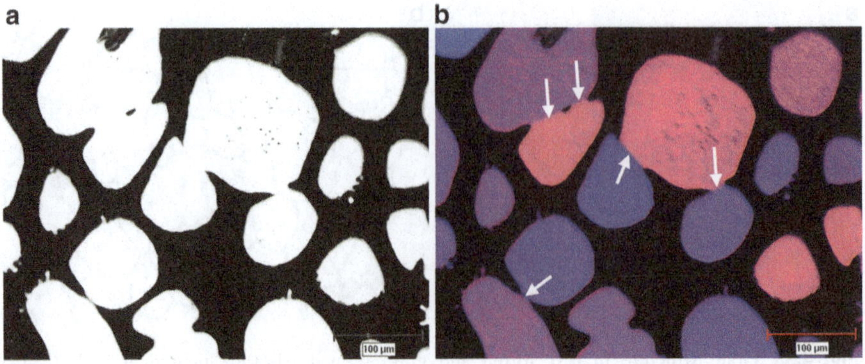

Fig. 4.39 Sintering effect of globules with bright field (**a**) and polarized light application (**b**), Electromagnetically Stirred A356 alloy, reheated at 593 °C

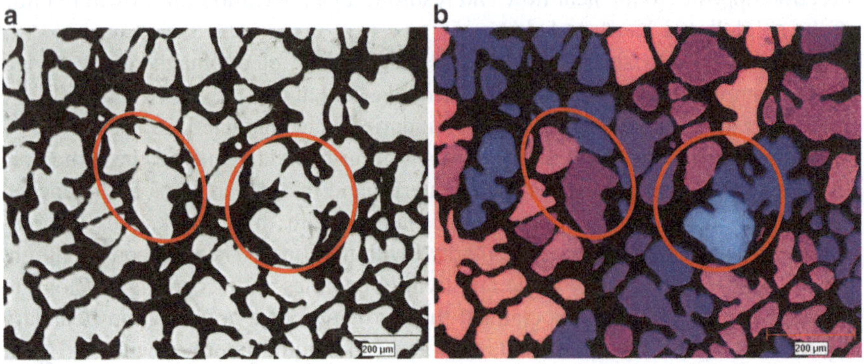

Fig. 4.40 Sintering effect of globules with bright field and with polarized light application, A356 alloy, quenched from 598 °C

small angle grain boundaries have lower energies and if formed with this condition, they could survive and remain in the microstructure.

An example of X-ray microtomography is shown in Fig. 4.41. Alloy Al–15.8% Cu was selected as solid and liquid phases in this alloy have good absorption contrast based on composition and the prepared specimen was held at 555 °C for 80 min while X-ray microtomography was carried out [116]. Two coarsening mechanisms were proposed. In one case, globules with very difference sizes were studied, Fig. 4.41a. The progressive dissolution of the small solid globule (1) is clearly seen (an Ostwald ripening like mechanism) however the coarsening of its neighbors is not large enough to be visible to the naked eye. The second coarsening mechanism is shown in Fig. 4.41b in case of equal size globules. While the neck size between particles 1 and 2 is increasing, necks between particles 2 and 3 or 2 and 4 remain stable. The neck diameter between particles 1 and 2 rapidly enlarges and the particles radii decrease slowly which is typical of coalescence and eventually leading into single particle [116].

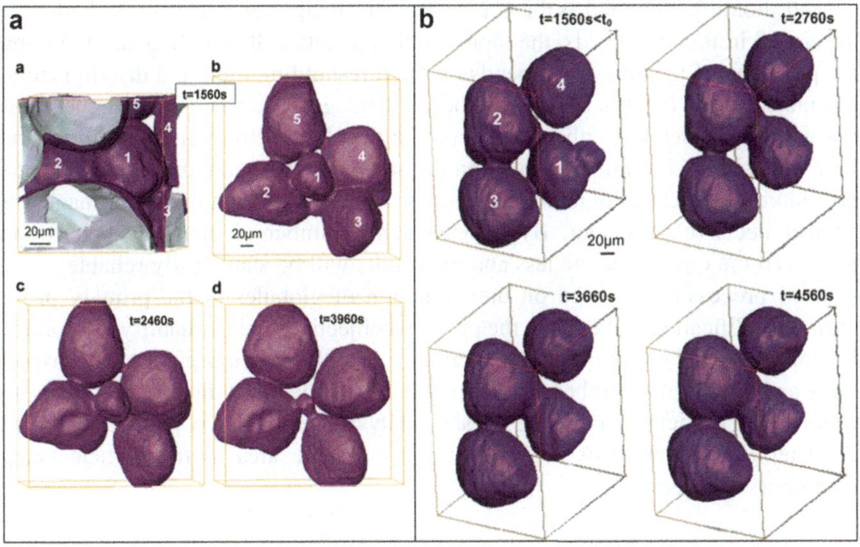

Fig. 4.41 X-ray microtomography of Al–15.8%Cu held at 555 °C for 80 min, (**a**) globules with various sizes, visible dissolution of globule 1 over time, (**b**) globule with equal size, coalescence toward formation of a single particle [116]

4.3.2 *Quantitative Metallography*

The application of polarized light microscopy provides a simple way to identify SSM structures as fully isolated globular or pseudo-globular morphologies including dendritic, degenerated dendritic, and rosette structures depending on the color contrast as mentioned and illustrated earlier. Once the distinction is made, image processing and analysis may be carried out to characterize the structure quantitatively [100].

Generally, it is assumed quantitative metallography is a simple technique; however, it should be remembered that there are a few but significant points where considering those should lead to acceptable results. According to the authors' experience, the major task during image analysis is the preparatory work done before processing and analyzing of the image. It means that by enhanced sample preparation including polishing and etching which is coupled with appropriate microscope selection and even working magnification, the resulting analyses would be reliable. For image analysis, the followings are of particular importance.

- Image acquisition
- Image enhancement
- Thresholding
- Image processing
- Data extraction

Thresholding is defined as the gray level of the image constituents. The better the difference in the gray levels, the more reliable results will be obtained. Therefore, the precision of the analysis depends on the thresholding limit and discriminating the unnecessary objects from the processed image. The main other issue is its repeatability which not only depends on the threshold setting, magnification, and light intensity and filtering but also on the sample preparation and using the same procedure in the metallographic preparation. Number of field of view (images) is another decisive parameter. By increasing the number of images captured and processed the errors become less and the results will be statistically reliable.

SSM processing is based on the formation of globules of the primary phase during solidification and as such the principal objectives are to quantify the volume fraction, size, and morphology of the primary particles. For morphology, the aspect ratio and sphericity number are proposed to be the most important factors to characterize particle shape while for size analysis average equivalent circular area diameter (the diameter of a circle having the same area as the particle being measured) is suggested.

4.3.2.1 Fraction Solid

Solid fraction is one of the main parameters in the rheological study and its variation alters the viscosity of the slurry. Solid fraction is measured in 2D polished surface (area fraction) and as mathematically shown (e.g., [117]), the volume fraction is equal to the mean value of area fraction. It is important to emphasize that the degree of the accuracy of the measurement is dependent on the number of field of view (images) analyzed.

Different methods are reported in the literature for measuring the fraction solid, among which the most commonly used ones are quantitative metallography, thermal analysis [118], lever rule (thermodynamic equilibrium condition), Scheil's equation [4], and application of software packages [119, 120]. The most common technique is quantitative metallography in which the sample that is rapidly quenched from SSM processing temperature to preserve the structure within the mushy zone is analyzed. The drawback is that the quenching is not rapid enough to prevent further growth of the solid particles and therefore the result is erroneous.

Thermal analysis is widely used to determine the solidification characteristics of metals and alloys. The technique could be divided into two groups: differential scanning calorimetry (DSC) and cooling curve analysis (CCA). Determination of latent heat is the typical application of DSC which is more accurate than the cooling curve analysis for many materials, including metals and alloys. DSC measures the energy (heat) evolved or absorbed by a sample as it is cooled, heated, or held at a temperature. However, DSC is limited to very small sample size, in the milligram range, and it is costly, requires technical expertise, and is not suitable for metallurgical or foundry shop floor operations. The alternative, cooling curve analysis (CCA), is simple, inexpensive, and most significantly suitable for commercial applications. Briefly, the amount of heat evolved from a solidifying test sample is calculated by measuring the area between the first derivative curve and base line. This amount is proportional to the fraction solid [118]. This method is used in the Chap. 6 to identify the critical temperatures during the course of solidification.

Computational thermodynamic packages such as CALPHAD (CALculation of PHAse Diagram) are supplementary tools that provide quantitative data for multicomponent alloying systems. In this method, the Gibbs free energy of individual phases is modeled as a function of composition, temperature, critical magnetic temperature, and sometimes pressure and the results are gathered in a thermodynamic database. It enables the calculation of multicomponent phase diagrams and tracking of individual alloys during heat treatment process or solidification by calculation of phase distribution and composition [119, 120].

It is important to comprehend the shortcomings of each route when comparing the results calculated/measured by any of these techniques. The lever rule, Scheil, and thermal analysis routes may provide results different from that of quantitative microscopy. As for example in the case of Al7Si and A356 alloys (Al7Si0.35 Mg), quenching from 593 °C leads to higher solid fraction of about ~60 % (Fig. 4.42) which has almost a difference of 30 % from thermal analysis, Scheil, and lever rule [46, 121].

The large discrepancy may have its origin in the following sources:

- Inefficient quenching method. The formation of the eutectic colonies confirms the ineffectiveness of water quenching in stopping further transformation from liquid to solid. This could be initiated from the vapor layer usually forms around the billet to reduce heat extraction from the sample. It was tried to change the quenching media. Tzimas [44] have changed the quenching medium to liquid Sn–Pb eutectic, which is supposed to offer higher thermal diffusivity and good wettability, but the method was still unsuccessful and a greater solid fraction was observed after the quenching process. Furthermore, the residual liquid may also segregate on the prequenched primary solid particles during transferring the billet to the quenching station.
- Growth of the primary particles is another possibility for inaccuracy in the measurement. During transferring and quenching, the growth is constrained by the temperature profile in the liquid phase and this residual liquid may precipitate on the prequenched primary solid phase. Martinez and Flemings [122] stated that the additional spherical growth occurs during the quenching process is responsible for this overestimation. They have shown that a primary spherical globule in Al4.5%Cu with 40 μm diameter grows almost 40 μm in diameter during the quenching period.
- The effect of higher cooling rate which was missed in all the previous reports. With higher cooling rate, not only the liquids line shifts up (start of the primary dendrites nucleation temperature), but also the eutectic line shifts down and a larger solidification range results [46]. As seen in Fig. 4.43, by increasing the cooling rate, the primary α-Al phase increases. This would result in the formation of new primary particles and coarsening of the former primaries.

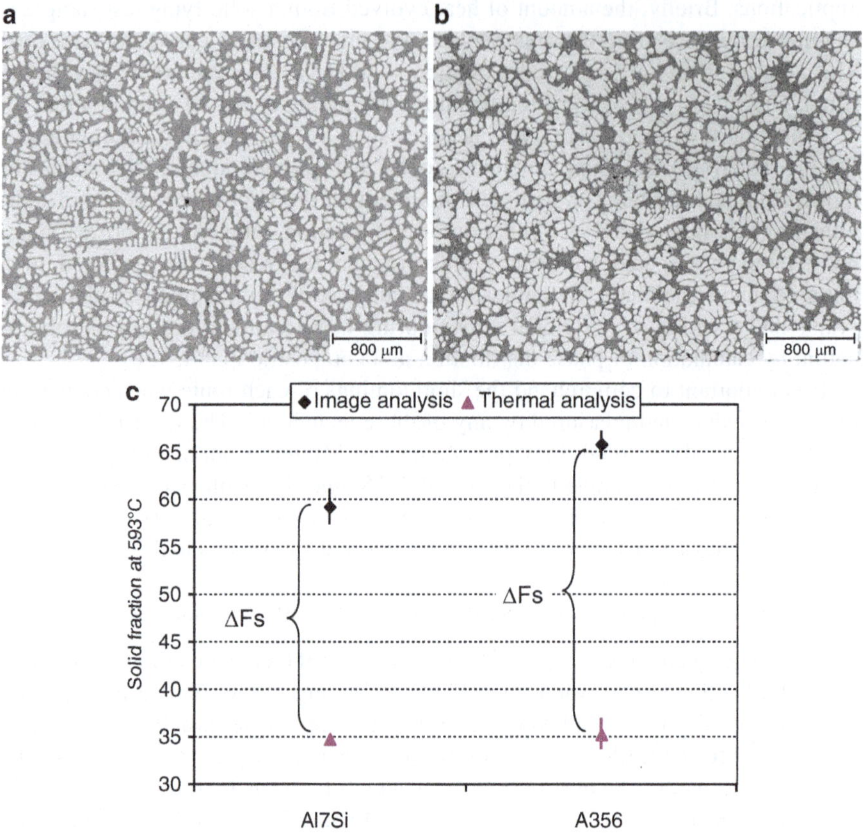

Fig. 4.42 Typical billet samples quenched from mushy zone at 593 ± 2 °C: (**a**) Al7Si, (**b**) A356, (**c**) solid fraction at ~593 °C, measured by thermal and image analyses [46] (ΔFs is the difference between two measurements)

4.3.2.2 Fraction of Entrapped Liquid

Liquid entrapment is a distinct feature of isothermal holding during semi-solid metal processing which normally is implemented during the thixocasting process. The first process occurring during isothermal holding of semi-solid structure is grain coarsening. In dendritic structure, the dendritic arms disappear by ripening and coalescence which has been already discussed in Chap. 3. The coalescence of isolated globules from same dendrite (same crystallographic orientation) or coalescence of neighboring arms may result in liquid metal entrapment (Fig. 4.44). The "entrapped liquid" adversely affects the deformability of material which is associated to a reduction in the interconnected liquid phase and therefore influences the rheological behavior of the slurry.

Volume of entrapped liquid depends on various parameters including semi-solid processing, cooling rate and morphology of as-cast structure, reheating time, and

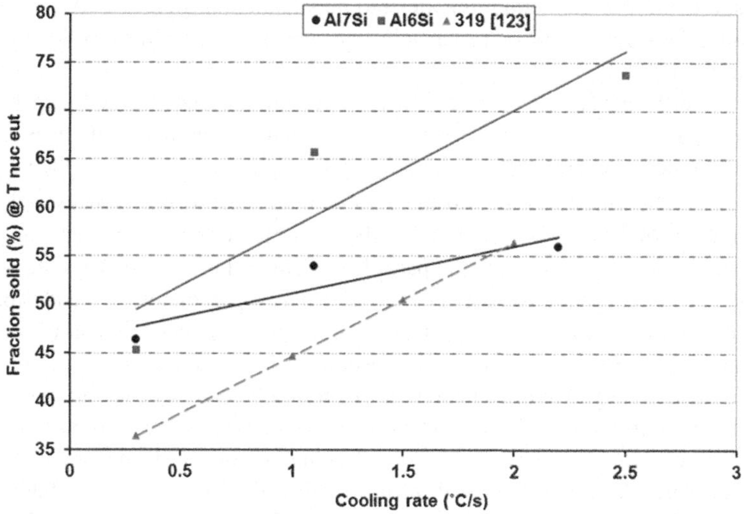

Fig. 4.43 Effect of various cooling rate on fraction solid, Al7%Si, Al6%Si, and 319 alloy (319 alloy data from [123])

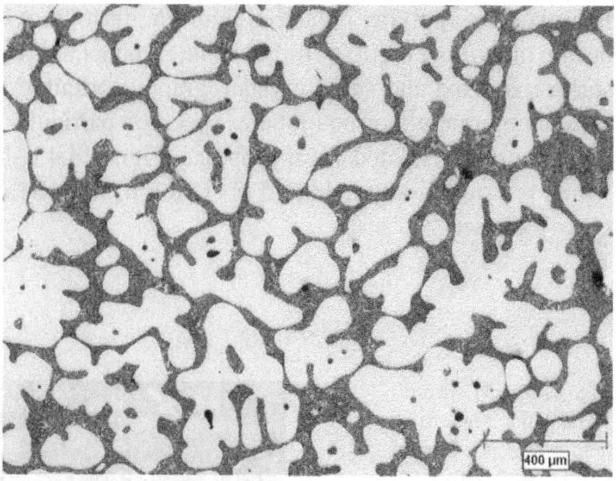

Fig. 4.44 Optical micrograph showing entrapped liquids, A356 cast in graphite mold, reheated and isothermally held for 10 min @583 °C

temperature. In fact contrary to its name, sometimes these pools which seem to be encapsulated in 2D surface may be interconnected in the volume.

4.3.2.3 Particle Size, Average Circular Diameter

The ideal microstructure for SSM processing is the one free from dendrites and having homogenous spherical particles. The globule size plays an important role on the castability and mechanical properties of the as-cast parts. However, it is important to factor in that the primary solid is a softer phase and minimizing its size is crucial for achieving the desired mechanical properties. As a general guideline, the optimum primary particle size reported for SSM alloys is less than 100 μm [124].

It is worth noting that image analysis technique has limited capabilities. Therefore, it should always be kept in mind that what image analysis cannot do, for instance, the definition of diameter in the software is based on the equation ($d = 2\sqrt{A/\pi}$), which A refers to the area of the object(s) being measured. Such measurement makes the assumption that the examined object has a shape close to a circle and for instance, the more the particle is rectangular, the greater is the error in the calculation (in this case, aspect ratio is a better parameter). Considering this point, it should be expected to detect small difference between the average circular diameter of a rosette and globule α-Al particle with the same area.

As indicated earlier in this chapter and with highlighting the disability of image analysis, it should always keep in mind that the analysis system could not distinguish between dendrite branches and individual globules. Therefore in case of dendritic structure, calculation will result in a lower value for circular diameter or rise in sphericity number which is an error source. In some cases, grain and globule sizes should be measured individually.

Technically, there are two different definitions for globule and grain size. Globules are primary particles which are apparently detached from one another. However, by applying polarized light microscopy, it becomes clear that the neighboring individual particles might be interconnected from underneath of the polished surface (having the same crystallographic orientation relationship). As a result, similar color neighboring globules specify a particular grain. By this method, grains could be differentiated from globules. Figure 4.45 shows an EMS stirred sample which is isothermally held at 583 °C for 10 min. Selected grains and globules are clearly shown.

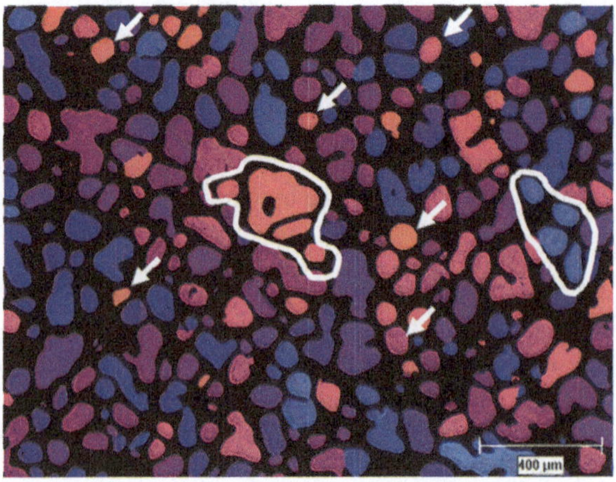

Fig. 4.45 EM stirred Al7Si0.8Fe isothermally held at 593 °C, Copper mold cast at 630 °C (*arrowed*: globules, *circled*: grains)

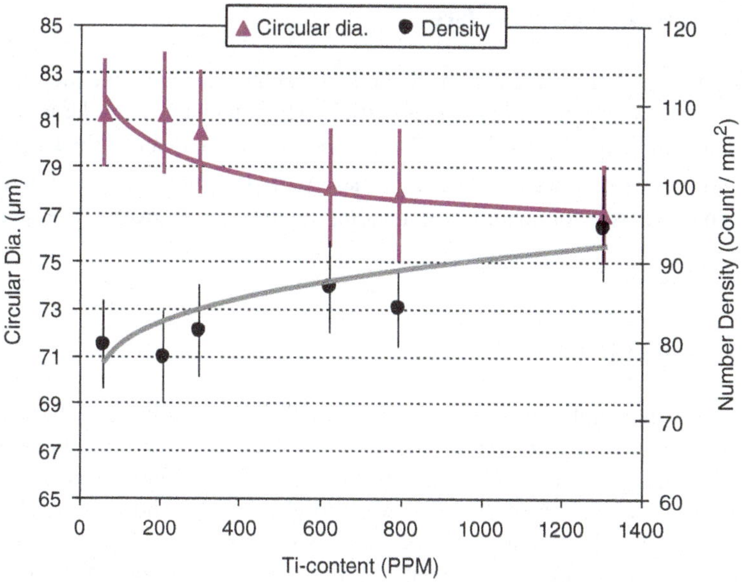

Fig. 4.46 Average circular diameter and number density for A356 with Ti and B addition (reproduced from [125])

4.3.2.4 Number Density

In some cases, it is necessary to have an estimation of the quantity of particles within a unit area, number density (mm^{-2}). Higher value of number density specifies smaller particles within the area. For instance, addition of grain refiner results in more effective nucleation sites which leads to limited growth and better distribution of primary particles. This is presented in Fig. 4.46 where with addition of AlTiB master alloy to A356 alloy, the density of primary particles increased. Figure 4.46 also shows number density increases when there are particles with smaller circular diameter [125].

4.3.2.5 Sphericity/Shape Factor

Primary particles' shape is a key factor in quantitative metallography affecting the viscosity of semi-solid slurry. Physically rounder particles have better flowability in comparison to rectangular ones. Sphericity is given by $\frac{4\pi A}{P^2}$ where A is the total area of primary particles and P represents perimeter of liquid–primary particles interface. The sphericity factor varies between 0 for objects having very elongated cross sections and 1 for those having circular cross sections. By using the average value of sphericity, it was concluded that this parameter is not responsive to small variation of the morphology and in most cases due to the averaging of numerous particle sphericity; the maximum difference is less than 0.1. Therefore, a more sensitive parameter was chosen which is the percentage of particles having sphericity greater than a certain value.

4.3.2.6 Specific Volume Surface

Also known as surface area per unit volume, it can be used to define the success of SSM process in producing individual particle which means having larger S_v. S_v is mathematically expressed as $S_v = \frac{4}{\pi}\frac{P}{A}$ where A is the total area of primary particles and P represents perimeter of liquid–primary particles interface. $\frac{A}{P}$ is a parameter which could be used for quantitative metallography which is inversely proportional to S_v.

4.3.2.7 Aspect Ratio

Aspect ratio is simply defined as the ratio of the longest over the shortest feret diameters, length per width. Feret diameter is defined as the distance between two parallel tangents on each side of an object (Fig. 4.47a). Values near one (1) are characteristics of spherical particles while higher values show more elongated (needle) like particles. Figure 4.47b provides aspect ratios of 356 Al–Si alloy slurry prepared at various pouring temperatures. For overstressing the effect of pouring temperature, the percentage of particles having aspect ratio greater than 2 is presented. With increasing the pouring temperature, the formation of elongated particles is promoted which is an indication of dendritic structure.

4.3.3 Microscopy/Image Analysis Setup

For characterizing the α-Al particles, 85 fields with the total area of 255 mm^2 were examined (Mag. ×50); while for evaluating of silicon morphology in graphite cups, 50 fields with 1.48 mm^2 were scanned (Mag. ×500). In the EMS samples and by

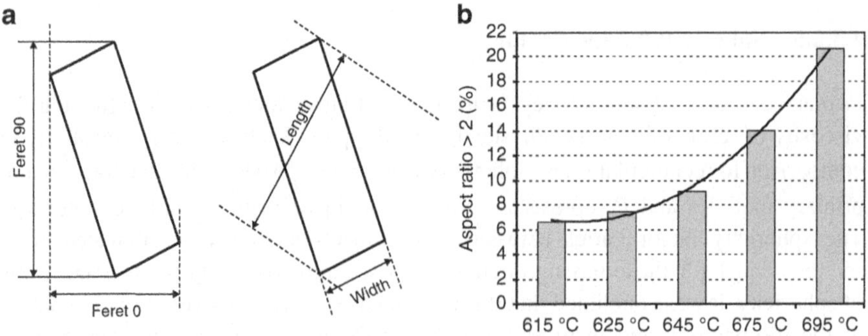

Fig. 4.47 (**a**) Definitions of feret, length, and width [100], (**b**) percentage of particles having aspect ratio greater than 2 [67]

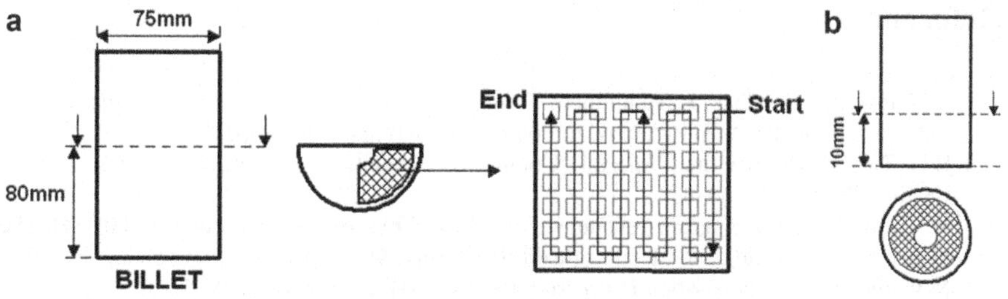

Fig. 4.48 (**a**) Area for quantitative analysis for SSM billets and applied scanning mode, (**b**) Area for quantitative analysis for conventional casting, graphite cup

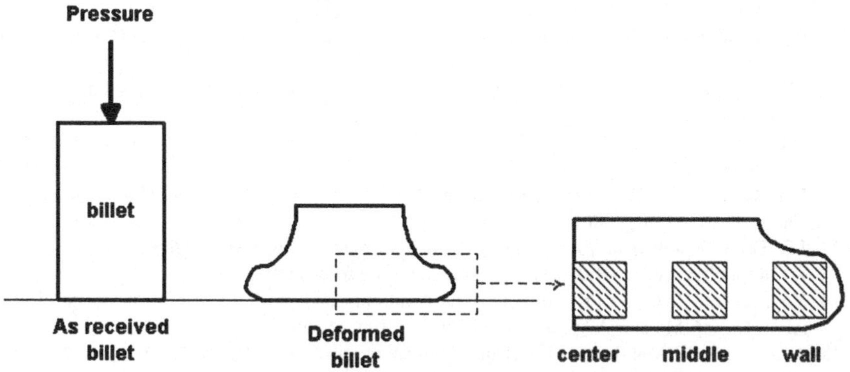

Fig. 4.49 Area for optical micrographs observations for a typical deformation

considering using different cooling rates, it was impossible to work at the same magnification and thus for sand and copper castings 50 and 80 fields at ~×200 and ~×500 magnifications with a total area of 9.3 and 2.3 mm^2 were used respectively. In aluminum alloys, the best resolution of individual grains could be achieved by examining electrolytically anodized specimens by polarized light microscopy. For grain size measurement, samples were polarized by fluoboric acid solution[5] and measurements were done between the wall and center of the graphite cup sample, Fig. 4.48.

In the case of rheological tests, the microstructure was studied on the longitudinal sections within the regions with the highest deformation rate as shown schematically in Fig. 4.49.

[5]In this method, a film of Al2O3 is deposited on the surface which its thickness depends mainly on the crystallographic orientation of the grain(s). When the sample is viewed by a polarized light which is passing through the analyzer, the film can rotate the plane of polarization regarding the orientation of the underlying grain and consequently producing various shades of colors by inserting a sensitive tint plate [113].

References

1. L. Backerud, G. Chai, J. Tamminen, *Solidification Characteristics of Aluminum Alloys, Volume 2, Foundry Alloys* (American Foundry Society, Des Plaines, 1990)
2. B.L. Tuttle, AFS Thermal Analysis Committee, Definitions in thermal analysis. Modern Casting (Nov 1985), 39–41
3. D. Gloria, Control of grain refinement of Al-Si alloys by thermal analysis. PhD Thesis, Department of Mining and Metallurgical Engineering, McGill University, Canada, June 1999
4. M.C. Flemings, *Solidification Processing* (McGraw-Hill, New York, 1974)
5. B.J. Yang, D. Stefanescu, J. Leon-Torres, Modeling of microstructural evolution with tracking of equiaxed grain movement for multicomponent Al-Si Alloy. Metal. Trans. A **32**, 3065–3076 (2001)
6. Norax, Canada http://www.noraxcanada.com
7. D.B. Spencer, R. Mehrabian, M.C. Flemings, Rheological behavior of Sn-15 Pct Pb in the crystallization range. Metal. Trans. **3**, 1925–1932 (1972)
8. R. Mehrabian, M.C. Flemings, Die castings of partially solidified alloys. AFS Trans. **80**, 173–182 (1972)
9. J.R. Van Wazer, R.E. Colwell, *Viscosity and Flow Measurement* (Wiley Interscience, New York, 1966)
10. D.R. Poirier, G.H. Geiger, *Transport Phenomena in Materials Processing* (TMS, Warrendale, 1994)
11. J.W. Goodwin, R.W. Hughes, *Rheology for Chemists, An Introduction* (Royal Society of Chemistry, London, 2000)
12. J.D. Ferry, *Viscoelastic Properties of Polymers* (Wiley, New York, 1970)
13. J. Campbell, *Casting* (Butterworth-Heinemann, Oxford, 1991). Chap. 2
14. Y.D. Kwon, Z.H. Lee, The effect of grain refining and oxide inclusion on the fluidity of Al-4.5Cu-0.6 Mn and A356 alloys. Mater. Sci. Eng. **A360**, 372–376 (2003)
15. A. Assar, N. El-Mahllawy, M.A. Taha, Fluidity of stir-cast Al-10% Cu alloy. Aluminum **57**, 807–810 (1981)
16. M. Suéry, *Mise en Forme des alliages métalliques a l'état semi solide* (Lavoisier, France, 2002)
17. M.C. Flemings, Behavior of metal alloys in the semi-solid state. Metal. Trans. A **22A**, 952–981 (1991)
18. D.H. Kirkwood, Semisolid metal processing. Int. Mater. Rev. **39**, 173–189 (1994)
19. Z. Fan, Semisolid metal processing. Int. Mater. Rev. **47**(2), 49–85 (2002)
20. H.V. Atkinson, Modelling the semisolid processing of metallic alloys. Progr. Mater. Sci. **50**, 341–412 (2005)
21. M.J. Stefan, Versuche Uber Die Scheinbare Adhasion SitzberMth. Naturw. Kl. Bagar Akad. Wiss Munchen **69**, Part 2 (1874)
22. X. Yang, Y. Jing, J. Liu, The rheological behavior for thixocasting of semi-solid aluminum alloy (A356). J. Mater. Proc. Tech. **130–131**, 569–573 (2002)
23. W. Nan, S. Guangji, Y. Hanguo, Rheological study of partially solidified Tin-Lead and Aluminum-Zinc alloys for stir casting. Mater. Trans. JIM **31**(8), 715–722 (1990)
24. R. Ghomashchi, *An Introduction to Engineering Materials* (University of South Australia, Australia, 1999)
25. R. Zehe, First production machine for rheocasting. Light Met. Age **57**(9), 62–66 (1999)
26. A. Beaulieu, L. Azzi, F. Ajersch, S. Turenne, F. Pineau, C.A. Loong, Numerical modeling and experimental analysis of die cast semi-solid A356 alloy, in *Proceeding of M.C. Flemings on Solidification and Materials Processing*, ed. by R. Abbaschian, H. Brody, A. Mortensen (TMS, Warrendale, 2001), 261–265
27. A. Wahlen, Modelling the processing of aluminum alloys in the semi-solid state. Mater. Sci. Forum **396–402**, 185–190 (2002)

28. M. Modigell, J. Koke, Rheological modelling on semi-solid metal alloys and simulation of thixocasting processes. J. Mater. Proc. Tech. **111**, 53–58 (2001)
29. C.G. Kang, H.K. Jung, A study on thixoforming process using the thixotropic behavior of an aluminum alloy with an equiaxed microstructure. J. Mater. Eng. Perform. **9**(5), 530–535 (2000)
30. M.C. Flemings, R.G. Riek, K.P. Young, Rheocasting processes. AFS Int. Cast Metals J. **1**(3), 11–22 (1976)
31. K. Brissing, K. Young, Semi-solid casting machines, heating systems, properties and applications. Die Casting Eng. **44**(6), 34–41 (2000)
32. T.Z. Kattamis, T.J. Piccone, Rheology of semi solid Al-4.5%Cu-1.5%Mg alloy. Mater. Sci. Eng. **A131**, 265–272 (1991)
33. D. Brabazon, D.J. Browne, A.J. Carr, Experimental investigation of the transient and steady state rheological behavior of Al-Si alloys in the mushy state. Mater. Sci. Eng. **A356**, 69–80 (2003)
34. J.Y. Chen, Z. Fan, Modeling of rheological behavior of semisolid metal slurries, part 1—theory. Mater. Sci. Technol. **18**, 237–242 (2002)
35. P.A. Joly, R. Mehrabian, The rheology of partial solid alloys. J. Mater. Sci. **11**, 1393–1403 (1976)
36. O. Lashkari, R. Ghomashchi, F. Ajersch, Deformation behavior of semi-solid A356 Al-Si alloy at low shear rates: the effect of sample size. Mater. Sci. Eng. **A444**, 198–205 (2007)
37. O. Lashkari, R. Ghomashchi, Deformation behavior of semi-solid A356 Al-Si alloy at low shear rates: effect of morphology. Mater. Sci. Eng. **A454–455**, 30–36 (2007)
38. O. Lashkari, R. Ghomashchi, Deformation behavior of semi-solid A356 Al-Si alloy at low shear rates: effect of fraction solid. Mater. Sci. Eng. **A486**, 333–340 (2008)
39. C.L. Martin, P. Kumar, S. Brown, Constitutive modeling and characterization of the flow behavior of semi-solid metal alloy slurries-II. Structural evolution under shear formation. Acta Metall. Mater. **42**(11), 3603–3614 (1994)
40. O. Lashkari, R. Ghomashchi, The implication of rheology in semi-solid metal processes: an overview. J. Mater. Proc. Tech. **182**, 229–240 (2007)
41. O. Lashkari, R. Ghomashchi, Evolution of primary α-Al particles during isothermal transformation of rheocast semi solid metal billets of A356 Al-Si alloy. Canad. Metall. Quart. **53**, 47–54 (2014)
42. H.V. Atkinson, D. Liu, Coarsening rate of microstructure in semi-solid aluminium alloys. Trans. Nonferrous Met. Soc. China **20**, 1672–1676 (2010)
43. M. Perez, J.C. Barbe, Z. Neda, Y. Brechet, L. Salvo, M. Suery, Investigation of the microstructure and the rheology of semi-solid alloys by computer simulation. J. Phys. IV France **11**, 93–100 (2001)
44. E. Tzimas, A. Zavaliangos, Evaluation of volume fraction of solid in alloys formed by semisolid processing. J. Mater. Sci. **35**, 5319–5329 (2000)
45. Y. Zhu, J. Tang, Y. Xiong, Z. Wu, C. Wang, D. Zeng, The influence of the microstructure morphology of A356 alloy on its rheological behavior in the semisolid state. Sci. Tech. Adv. Mater. **2**, 219–223 (2001)
46. S. Nafisi, D. Emadi, R. Ghomashchi, Semi solid metal processing: the fraction solid dilemma. Mater. Sci. Eng. **A507**(1–2), 87–92 (2009)
47. J.Y. Chen, Z. Fan, Modeling of rheological behavior of semisolid metal slurries, part 3—transient state behavior. Mater. Sci. Technol. **18**, 250–257 (2002)
48. G. Chai, T. Roland, L. Arnberg, L. Backerud, Studies of dendrite coherency in solidifying aluminum alloy melts by rheological measurements. in *Second International Conference, Semi-Solid Processing of Alloys and Composites* (Cambridge, 1992), 193–201
49. O. Lashkari, R. Ghomashchi, The implication of rheological principles for characterization of semi-solid Al-Si cast billets. J. Mater. Sci. **41**, 5958–5965 (2006)
50. Z. Fan, J.Y. Chen, Modelling of rheological behavior of semisolid metal slurries, part 4—effects of particle morphology. Mater. Sci. Technol. **18**, 258–267 (2002)

51. S. Jabrane, B. Clement, S. Ajersch, Evolution of primary particle morphology during rheoprocessing of Al-5.2%Si alloy. in *Second International Conference, Semi-Solid Processing of Alloys and Composites* (Cambridge, 1992), 223–236
52. M. Silva, A. Lemieux, H. Blanchette, X. Chen, The determination of semi-solid processing ability using a novel rheo-characterizer apparatus. In *10th International Conference on Semi-Solid Processing of Alloys and Composites* (Aachen, Germany, 2008) (published in Solid State Phenomena, vol. 141–143, 2008, 343–348)
53. E.J. Zoqui, M. Paes, M.H. Robert, Effect of macrostructure and microstructure on the viscosity of the A356 alloy in the semi-solid state. J. Mater. Proc. Tech. **153–154**, 300–306 (2004)
54. M. Mada, F. Ajersch, Rheological model of semisolid A356-SiC composite alloys. Part I. Dissociation of agglomerate structures during shear. Mater. Sci. Eng. A **212**, 157–170 (1996)
55. M. Mada, F. Ajersch, Rheological model of semisolid A356-SiC composite alloys. Part II. Reconstitution of agglomerate structures at rest. Mater. Sci. Eng. A **212**, 171–177 (1996)
56. T.Y. Liu, P.J. Ward, D.H. Kirkwood, H.V. Atkinson, Rapid compression of aluminium alloys and its relationship to thixoformability. Int. Mater. Rev. **34**, 409–417 (2003)
57. M. Modigell, L. Pape, M. Hufschmidt, Kinematics of structural changes in semisolid alloys by shear and oscillation experiments. in *8th International Conference on Semi-Solid Processing of Alloys and Composites* (Limassol, Cyprus, 2004)
58. S. Sannes, L. Arnberg, M.C. Flemings, Orientational relationships in semi-solid Al-6.5wt% Si, in *Light Metals*, ed. by W. Hale (TMS, Anaheim, 1996), 795–798
59. Z. Fan, J.Y. Chen, Modeling of rheological behavior of semisolid metal slurries, part 2—steady state behavior. Mater. Sci. Technol. **18**, 243–249 (2002)
60. L. Arnberg, A. Bardal, H. Sund, Agglomeration in two semisolid type 6082 aluminum alloys. Mater. Sci. Eng. **A262**, 300–303 (1999)
61. M. Hirai, K. Takebayashi, Y. Yoshikawa, Effect of chemical composition on apparent viscosity of Semi-solid alloys. ISIJ Int. **33**(11), 1182–1189 (1993)
62. M. Hirai, K. Takebayashi, Y. Yoshikawa, R. Yamaguchi, Apparent viscosity of Al-10mass% Cu semi-solid alloys. ISIJ Int. **33**(3), 405–412 (1993)
63. H. Wang, C.J. Davidson, J.A. Taylor, D.H. St John, Semisolid casting of AlSi7Mg0.35 alloy produced by low temperature pouring. Mater. Sci. Forum **396–402**, 143–148 (2002)
64. W. Mao, C. Cui, A. Zhao, J. Yang, X. Zhong, Effect of pouring process on the microstructure of semi solid AlSi7Mg alloy. Mater. Sci. Eng. **17**(6), 615–619 (2001)
65. K. Xia, G. Tausig, Liquidus casting of a wrought aluminum alloy 2618 for thixoforming. Mater. Sci. Eng. **A246**, 1–10 (1998)
66. S. Midson, K. Young, Impact of casting temperature on the quality of components semi-solid metal cast form alloys 319 and 356. in *5th AFS International Conference of Molten Aluminum* (1998), 409–422
67. O. Lashkari, S. Nafisi, R. Ghomashchi, Microstructural characterization of rheo-cast billets prepared by variant pouring temperatures. Mater. Sci. Eng. **A441**, 49–59 (2006)
68. D. Doutre, G. Hay, P. Wales, J.P. Gabathuler, SEED: a new process for semi solid forming. in *Light Metals Conference, COM 2003*, ed. by J. Masounave, G. Dufour (Vancouver, Canada), 293–306
69. D. Bouchard, F. Pineau, D. Doutre, P. Wales, J. Langlais, Heat transfer analysis of swirled enthalpy equilibration device for the production of semi-solid aluminum. in *Light Metals Conference, COM 2003*, ed. by J. Masounave, G. Dufour (Vancouver, Canada), 229–241
70. H.I. Lee, R.D. Doherty, E.A. Feest, J.M. Titchmarsh, Structure and segregation of stir-cast aluminum alloys, The Metals Society. in *Proceeding of International Conference, Solidification Technology in Foundry and Cast House* (UK, 1983), 119–125
71. D. Brabazon, D.J. Browne, A.J. Carr, J.C. Healy, Design, construction, and operation of a combined rheocaster/rheometer. in *Fifth International Conference on Semi-Solid Processing of Alloys and Composites* (Golden, 1998), 21–28

72. O. Lashkari, R. Ghomashchi, F. Ajersch, Rheological study of 356 Al-Si foundry alloy prepared by a new innovative SSM process. in *EPD Congress*, ed. by Mark E. Schlesinger (TMS, Warrendale, 2005), 149–156
73. P.R. Prasad, A. Prasad, C.B. Singh, Calculation of primary particle size in rheocast slurry. J. Mater Sci. Lett. **14**, 861–863 (1995)
74. E.A. Vieira, A.M. Kliauga, M. Ferrante, Microstructural evolution and rheological behavior of aluminum alloys A356, and A356 + 0.5% Sn designed for thixocasting. J. Mater. Proc. Tech. **155–156**, 1623–1628 (2004)
75. V. Laxmanan, M.C. Flemings, Deformation of semi-solid Sn-15 Pct Pb alloy. Metal. Trans. A **11A**, 1927–1937 (1980)
76. L. Azzi, F. Ajersch, Development of aluminum-base alloys for forming in semi solid state. in *TransAl Conference* (Lyon, France, 2002), 23–33
77. H. Mirzadeh, B. Niroumand, Fluidity of Al-Si semisolid slurries during rheocasting by a novel process. J. Mater. Proc. Tech. **209**, 4977–4982 (2009)
78. S. Nafisi, O. Lashkari, R. Ghomashchi, A. Charette, Effect of different fraction solids on the fluidity of rheocast 356 Al-Si alloy. in *Multi Phase Phenomena and CFD Modeling and Solidification in Materials Processes* (TMS, North Carolina, 2004), 119–128
79. Y. Murakami, K. Miwa, M. Kito, T. Honda, N. Kanetake, S. Tada, Effects of injection conditions in the semi-solid injection process on the fluidity of JIS AC4CH aluminum alloy. Mater. Trans. **54**(9), 1788–1794 (2013)
80. J. Wannasin, D. Schwam, J.A. Yurko, C. Rohloff, G. Woycik, Hot tearing susceptibility and fluidity of semi solid gravity cast Al Cu alloy. in *Ninth International Conference on Semi-Solid Processing of Alloys and Composites* (Busan, Korea, 2006) (published in Solid State Phenomena, vol. 116–117, 2006, 76–79)
81. M. Forte, D. Bouchard, A. Charette, Fluid flow investigation of die cast tensile test bars. in *Ninth International Conference on Semi-Solid Processing of Alloys and Composites* (Busan, Korea, 2006) (published in Solid State Phenomena, vol. 116–117, 2006, 457–463)
82. T. Haga, H. Fuse, Semi-solid die casting of Al-25%Si. in *13th International Conference on Semi-Solid Processing of Alloys and Composites* (Muscat, Oman, 2014) (published in Solid State Phenomena, vol. 217–218, 2015, 436–441)
83. F. Kolenda, P. Retana, G. Racineux, A. Poitou, Identification of rheological parameters by the squeezing test. Powder Tech. **130**, 56–62 (2003)
84. J.A. Yurko, M.C. Flemings, Rheology and microstructure of semi-solid aluminum alloys compressed in the drop forge viscometer. Metal. Trans. A **33A**, 2737–2746 (2002)
85. M. Suéry, M.C. Flemings, Effect of strain rate on deformation behavior of semi-solid dendritic alloys. Metal. Trans. A **13A**, 1809–1819 (1982)
86. Y. Fukui, D. Nara, N. Kumazawa, Evaluation of the deformation behavior of a semi-solid hypereutectic Al-Si alloy compressed in a drop-forge viscometer. Metal. Trans. A **46A**, 1908–1916 (2015)
87. J.H. Han, D. Feng, C.C. Feng, C.D. Han, Effect of sample preparation and flow geometry on the rheological behavior and morphology of micro phase-separated block copolymers: comparison of cone-and-plate and capillary data. Polymer **36**, 155–167 (1995)
88. W.M. Gearhart, W.D. Kennedy, Cellulose acetate butyrate plastics. Indus. Eng. Chem. **41**(4), 695–701 (1949)
89. G.H. Dienes, H.F. Klemm, Theory and application of the parallel plate plastometer. J. Appl. Phys. **17**, 458–464 (1946)
90. O. Draper, S. Blackburn, G. Dolman, K. Smalley, A. Griffiths, A comparison of paste rheology and extrudate strength with respect to binder formulation and forming technique. J. Mater. Proc. Tech. **92–93**, 141–146 (1999)
91. S. Turenne, N. Legros, S. Laplante, F. Ajersch, Mechanical behavior of Aluminum matrix composite during extrusion in the semisolid state. Metal. Trans. A **30A**, 1137–1146 (1999)
92. M. Ferrante, E. de Freitas, Rheology and microstructural development of Al-4wt%Cu alloy in the semi solid state. Mater. Sci. Eng. **A271**, 172–180 (1999)

93. M. Ferrante, E. de Freitas, M. Bonilha, V. Sinka, Rheological properties and microstructural evolution of semi-solid aluminum alloys inoculated with Mischmetal and with Titanium. in *Fifth International Conference on Semisolid Processing of Alloys and Composites* (1998), 35–42
94. H. Blanchette, Development a quality control method for Aluminum semi solid billet obtained from the SEED process. M. Eng Thesis, University of Quebec, Canada, 2006
95. O. Lashkari, R. Ghomashchi, A new machine to characterize microstructural evolution of semi solid metal Billets through viscometry. Mater. Design **28**(4), 1321–1325 (2007)
96. J.D. Sherwood, Squeeze flow of a Herschel-Bulkley fluid. J. Newtonian Fluid Mech. **77**, 115–121 (1998)
97. J.D. Sherwood, D. Durban, Squeeze flow of a power-law viscoplastic solid. J. Non-newtonian Fluid Mech. **62**, 35–54 (1996)
98. J.P. Gabathuler, Evaluation of various processes for the production of billets with thixotropic properties. in *Proceeding of 2nd Conference of Semi solid Materials* (1992), 33–46
99. D.J. Lahaie, M. Bouchard, Physical modeling of the deformation mechanisms of semisolid bodies and a mechanical criterion for hot tearing. Metal. Trans. B **32B**, 697–705 (2001)
100. S. Nafisi, R. Ghomashchi, The microstructural characterization of semi-solid slurries. JOM **58**(6), 24–30 (2006)
101. Oxford instruments www.ebsd.com
102. OIM Analysis Tutorials, Ametek, Inc.
103. B. Niroumand, K. Xia, 3D study of the structure of primary crystals in a rheocast Al-Cu alloy. Mater. Sci. Eng. A **283**, 70–75 (2000)
104. J. Alkemper, P.W. Voorhees, Three-dimensional characterization of dendritic microstructures. Acta. Mater. **49**, 897–902 (2001)
105. T.L. Wolfsdorf, W.H. Bender, P.W. Voorhees, The morphology of high volume fraction solid-liquid mixtures: an application of microstructural tomography. Acta. Mater. **45**(6), 2279–2295 (1997)
106. Y. Ito, M.C. Flemings, J.A. Cornie, Rheological behavior and microstructure of Al-6.5wt%Si alloy. in *Nature and Properties of Semi-Solid Materials*, ed. by J.A. Sekhar, J. Dantzig (TMS, Warrendale, 1991), 3–17
107. Private communication with Prof. Xiangjie Yang, School of Mechanical Engineering, Nanchang University, China, 2016
108. L. Salvo, M. Suery, A. Marmottant, N. Limodin, D. Bernard, 3D imaging in material science: application of X-ray tomography. Comptes Rendus Phys. **11**, 641–649 (2010)
109. K.M. Kareh, In situ synchrotron tomography of granular deformation in semi solid Al Cu alloys. PhD Thesis, Imperial College London, Sep 2013
110. G.C. Gu, R. Pesci, L. Langlois, E. Becker, R. Bigot, M.X. Guo, Microstructure observation and quantification of the liquid fraction of M2 steel grade in the semi solid state, combining confocal laser scanning microscopy and X-ray microtomography. Acta Mater. **66**, 118–131 (2014)
111. W. Xu, M. Ferry, N. Mateescu, J.M. Cairney, F.J. Humphreys, Techniques for generating 3-D EBSD microstructures by FIB tomography. Mater. Character. **58**, 961–967 (2007)
112. Z. Zaafarani, D. Raabe, R.N. Singh, F. Roters, S. Zaefferer, Three dimensional investigation of the texture and microstructure below a nanoindent in a Cu single crystal using 3D EBSD and crystal plasticity finite element simulations. Acta Mater. **54**, 1863–1876 (2006)
113. G. Vander Voort, *Metallography, Principles and Practice* (ASM International, Materials Park, 1999)
114. A. Apaydin, K.V. Prabhakar, R.D. Doherty, Special grain boundaries in Rheocast Al-Mg. Mater. Sci. Eng. **46**, 145–150 (1980)
115. R.D. Doherty, H.I. Lee, E.A. Feest, Microstructure of stir-cast metals. Mater. Sci. Eng. **A65**, 181–189 (1984)

116. N. Limodin, L. Salvo, M. Suery, M. DiMichiel, In situ investigation by X-ray Tomography of the overall and local microstrucutral changes occurring during partial remelting of an Al158 wt Cu alloy. Acta Mater. **57**, 2300–2310 (2009)
117. R.T. DeHoff, F.N. Rhines, *Quantitative Microscopy* (McGraw-Hill, New York, 1968)
118. D. Emadi, L.V. Whiting, S. Nafisi, R. Ghomashchi, Applications of thermal analysis in quality control of solidification processes. J. Thermal Anal. Calorimetry **81**, 235–242 (2005)
119. N. Saunders, A.P. Miodownik, *CALPHAD (Calculation of Phase Diagram): A Comprehensive Guide* (Elsevier, Oxford, 1998)
120. B. Sundman, *ThermoCalc Version L User's Guide, Division of Computational Thermodynamic* (Department of Material Science and Engineering, Royal Institute of Technology, Stockholm, 1996)
121. S. Nafisi, R. Ghomashchi, A. Charette, Effects of grain refining on morphological evolution of Al-7%Si in the swirled enthalpy equilibration device (SEED). in *66th World Foundry Congress* (Turkey, Sep 2004), 1253–1263
122. R.A. Martinez, M.C. Flemings, Evolution of particle morphology in semisolid processing. Metal. Trans. A **36A**, 2205–2210 (2005)
123. R.I. MacKay, M.B. Djurdjevic, J.H. Sokolowski, Effect of cooling rate on fraction solid of metallurgical reactions in 319 alloy. AFS Trans. **108**, 521–530 (2000)
124. D. Apelian, Semi-solid processing routes and microstructure evolution. in *7th International Conference on Semi-Solid Processing of Alloys and Composites* (Tsukuba, Japan, 2002), 25–30
125. S. Nafisi, R. Ghomashchi, Grain refining of conventional and semi-solid A356 Al-Si alloy. J. Mater. Proc. Tech. **174**, 371–383 (2006)

Chapter 5
Rheocasting: Low Pouring, SEED, and EMS Techniques

Abstract This chapter concentrates on understanding the mechanisms of nucleation and growth in three different rheocasting processes of low pouring, SEED, and electromagnetic stirring processes. The application of thermal analysis, parallel plate viscometry, and quantitative metallography verified the mechanisms responsible for microstructural evolution during the mentioned SSM processes.

5.1 Introduction

As discussed in Chap. 2, there is a range of SSM processes to produce billets having globular or near globular primary particles. If it is required to classify the entire SSM processes, it may be logical to use criteria such as temperature and fluid flow as the main parameters to induce globularity in resulted solidification structure through thermal or mechanical or thermomechanical disintegration of dendritic structure during solidification. In this regard three well-researched processes, which may be classified as mainly thermal (Low Pouring), and a combination of both mechanical and thermal (thermomechanical) of Electromagnetic Stirring, EMS and Swirled Enthalpy Equilibration Device, SEED have been selected to discuss in this chapter.

5.2 Low Pouring Temperature Technique

Pouring temperature directly affects the nucleation and growth of primary crystals such as α-Al in hypo-eutectic Al–Si alloys, where nucleation controls the quantity and size of the grains formed and growth determines the grain morphology and distribution of alloying elements within the matrix. The driving force for nucleation is the degree of undercooling created during solidification and growth is controlled by the temperature gradient and solute concentration in the liquid. However, both processes are affected by the rate of heat extraction. The final structure is therefore dependent on the nucleation density, growth morphology, fluid flow, and diffusion and transport of solutes. Therefore, close control of casting conditions, such as pouring temperature, cooling rate, nucleation sites, and temperature gradient, results in formation of the desired as-cast structure.

S. Nafisi, R. Ghomashchi, *Semi-Solid Processing of Aluminum Alloys*,
DOI 10.1007/978-3-319-40335-9_5

From semi-solid standpoint, microstructural variation resulted from pouring temperatures is turned into various viscosities during the application of the external force to the slurries. This is because the flow characteristics, as specified by the viscosity, are dependent on the metallurgical parameters including the fraction solid and its morphology (e.g., dendritic, rosette, or globular), solid particle size and distribution, chemical composition of the alloy, and pouring temperature [1–3]. The effect of solid phase characteristics on the viscosity and rheological behavior of SSM billets is also discussed in details in Chap. 4 of this book.

In the current section, effect of pouring temperature is specifically reported and also a correlation between superheat, morphology of primary α-Al particles, and subsequent rheological behavior of the slurries is established.

5.2.1 *Experimental Setup*

Commercial 356 alloy with the chemical composition given in Table 5.1 is used for the entire experiments. Melt preparation and experimental setup were the same as described in Chap. 4 (a coated cylindrical steel mold of 75 mm diameter and 250 mm long, superheat range of 0–80 °C). In all cases, two K type thermocouples were installed at the mold center and near the wall with their tips at 80 mm from the bottom of the mold. After pouring, solidification continued up to the point where the melt temperature at the center of the billet reached 593 ± 2 °C. It is expected to have a solid fraction of 0.3–0.35 at this temperature according to equilibrium lever rule and Scheil's equation. The melt temperature at the vicinity of mold wall was registered at about 591 ± 1 °C. The billets were then taken out of the mold, still in the mushy zone, and quenched quickly in cold water. For the rheological tests, the temperature of the furnace which was installed on the parallel plate compression test machine was kept at 594 ± 1 °C and the applied dead weight varied between 2.1 and 5.1 kg (see Sect. 4.2).

5.2.2 *Thermal Analysis*

The cooling curves of Fig. 5.1 represent temperature variations of the liquid near the mold wall and the center accordingly. As indicated, there are two different cooling regimes for the wall and central areas at the beginning of each trial, but an almost uniform cooling rate is reached later on for the entire test series. The melt

Table 5.1 Chemical composition of the melts (wt. %)

Si	Mg	Fe	Mn	Cu	Ti	B	Sr	Al
6.9–7.1	0.3–0.31	Max 0.09	Max 0.001	Max 0.001	0.12–0.13	Nil	Nil	Bal.

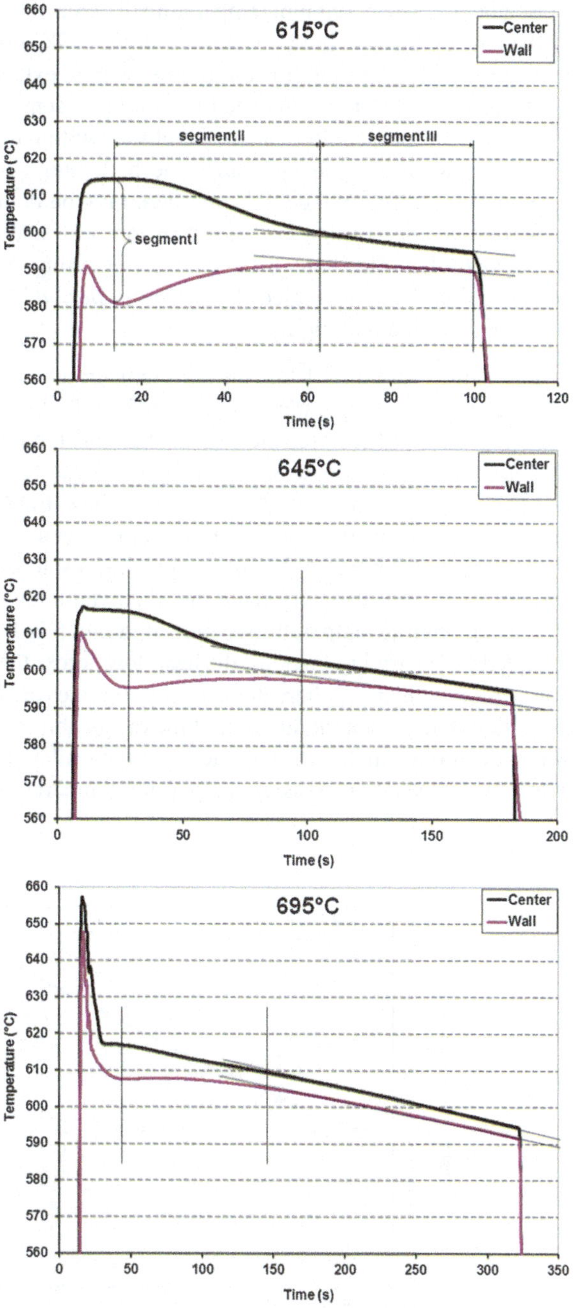

Fig. 5.1 Typical wall and center cooling curves for 615, 645, and 695 °C pouring temperatures [4]

temperature is always higher at the center of the mold immediately after pouring, which is expected due to the chilling effect of the mold wall. There are three distinct regions identified on the cooling curves of Fig. 5.1 as follows:

- Segment I: The maximum temperature difference between the wall and center of the cast billet, seen at the beginning of the bulk solidification.
- Segment II: The solidification time which extends between the maximum temperature difference between the wall and center of the billet up to the point where a uniform cooling rate is established within the bulk liquid. This is labeled the "*metastability period,*" since the bulk liquid has not yet reached thermal equilibrium. Thermal equilibrium here may be defined as uniform cooling rate across the entire billet, i.e., near the mold wall and mold center.
- Segment III: The time between the beginning of uniform cooling rate and quenching. This is labeled the "*stability period,*" due to uniform cooling rate throughout the bulk liquid.

It is evident in Fig. 5.1 that the melt temperature near the mold wall is different for three pouring temperatures of 615, 645, and 695 °C with 615 °C being heavily undercooled. Furthermore, the rise in the wall temperature following pouring is greatest, ~10 °C, for the lowest pouring temperature of 615 °C. Such temperature rise can only be justified by either a rapid phase transformation and releasing the latent heat of fusion, or rapid heat flow from the central regions toward the wall due to the large temperature difference, segment I, established at the beginning of bulk solidification. It is also noticeable that the magnitude of segment-I decreases with increasing pouring temperature as plotted in Fig. 5.2. It is believed that segment-I is regarded as the driving force for the heat flow across the bulk liquid, so its reduction must prolong the time where uniform cooling rates are established within the mold. This should be evident in duration of the metastability period (segment II), the time where the molten metal cooling rates near the wall and center of the mold approach a uniform value. By increasing the pouring temperature the extent of metastability

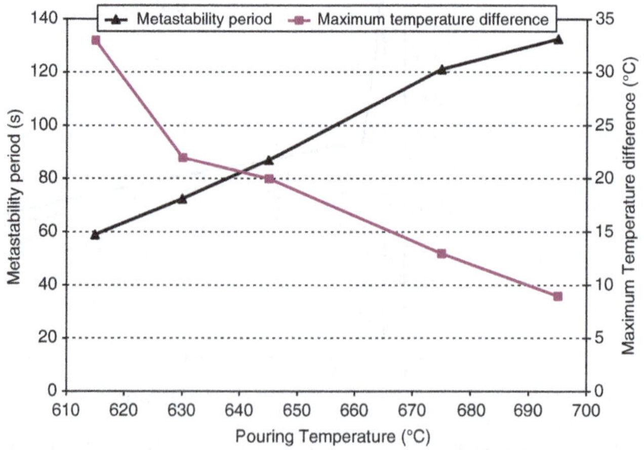

Fig. 5.2 Effect of pouring temperature on the maximum temperature difference (segment I) and metastability period (segment II) [4]

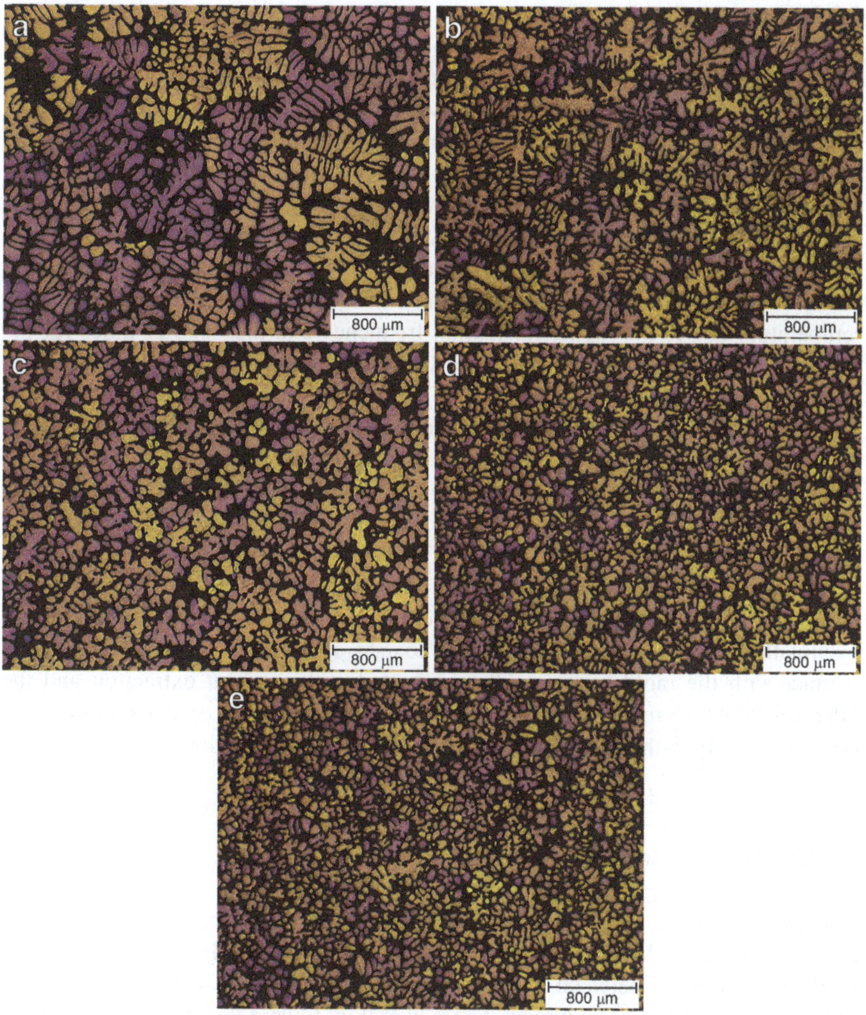

Fig. 5.3 Polarized light micrographs showing the microstructural evolution associated to superheat variation; (**a**) 695, (**b**) 675, (**c**) 645, (**d**) 625, and (**e**) 615 °C

period increases, Fig. 5.2. The optical micrographs shown in Fig. 5.3 clearly illustrate the effect of pouring temperature on the resulting microstructure.

Based on the aforementioned observations, two hypotheses put forward to explain the fine-grained globular structure seen in Fig. 5.3. These are based on either nucleation or growth processes as the dominant factors on the formation of the primary α-Al particles.

5.2.2.1 Nucleation-Based Hypothesis

At low superheat, the liquid is rapidly cooled by contacting the mold wall to a temperature at which nucleation occurs with a large number of nuclei, big bang, or copious nucleation. This results in shifting up the melt temperature adjacent to the mold wall. The volume of liquid that is cooled to this extent increases with decreasing the pouring temperature. However, the temperature rise near the mold wall is not sufficient enough to pose any threat to solidification by remelting the nuclei even for the lowest pouring temperature, nor the total heat content of the melt (Fig. 5.1). The rate of temperature rise near the wall is greater for lower pouring temperature due to the higher nucleation frequency and thus shorter times for the metastability period. The central parts however solidify in normal manner, but assisted by the floating nuclei, due to fluid flow during pouring and natural convection from the wall region to eventually establish a uniform cooling rate. The probability of such nuclei to survive is much greater at lower pouring temperature and the uniform and multi directional cooling, associated to the cylindrical shape of the mold, promotes the formation of fine-grained equiaxed as-cast structure as clearly seen in Fig. 5.3 for 615 °C pouring temperature.

By increasing the superheat, the liquid volume which is cooled to the nucleation temperature is reduced and copious nucleation mechanism is diminishedwhich eventually leads to the formation of a coarser structure. The nuclei formed in contact with the mold grow rapidly associated with the heat extraction and the other nuclei which move with the liquid into the central parts could not survive and dissolve rapidly within the hotter liquid within the central region.

5.2.2.2 Growth-Based Hypothesis

The formation of fine-grained equiaxed structure at lower pouring temperatures is also dependent on the growth characteristics of the nuclei. The large temperature gradient established between the wall and center during the early stages of solidification (e.g., 615 °C) encourages faster heat flow toward the mold wall. In other words, the molten alloy near the mold wall acts as a heat sink. Furthermore, the greater thermal and fluid convection promote grain multiplication and structural evolution. However, the mold coating coupled with the resulting air gap and assisted by the mold thin wall reduces the rate of heat dissipation through the mold wall into the surrounding to increase the melt temperature near the wall. The outcome is a uniform temperature distribution to establish within the bulk liquid which occurs at much shorter time at the lower pouring temperature of 615 °C. This is supported with its shortest "*metastability period*," segment II. As a result, the heat flow toward the mold wall diminishes and the growth rate is reduced much sooner for the 615 °C pouring temperature. This is coupled with multidirectional heat flow due to the mold geometry and promotes the formation of equiaxed and globular structure, Fig. 5.3, i.e., 615 °C.

As pouring temperature increases, the mechanism of copious nucleation is less pronounced and there are fewer nuclei within the bulk liquid which eventually lead to the formation of coarser structure. The amount of temporary rise in temperature or recalescence also decreases with increasing pouring temperature which supports the hypothesis of lower nucleation frequency (it has to be noted that the temperature rise observed at different pouring temperatures in Fig. 5.1 may be explained with respect to the liquid convection due to the large temperature variation at the vicinity of the wall and center of the billet). Such temperature variation is largest at the lowest pouring temperature. It is important to reemphasize the fact that there is longer metastability period at higher temperature, e.g., 695 °C, which encourages better development of primary α-Al particles and the formation of more classical dendritic morphology as illustrated in Fig. 5.3.

5.2.3 Structural Analysis

Effect of superheat on the microstructural evolution of quenched billets is shown in Fig. 5.3. With increasing the superheat, the morphology of primary α-Al particles becomes more dendritic. As the superheat reduced, the morphology changes to rosette, equiaxed, and eventually with ripening process and remelting of thinner arms and growth of thicker arms, the structure becomes globular. Polarized light microscopy facilitates the characterization of morphological variation due to the superheat. Samples with higher pouring temperatures highlight dendritic growth while by decreasing the superheat there is a clear transition from dendritic to globular form (Fig. 5.3). As the pouring temperature approaches 615 °C, the melt is no longer superheated. The mold itself (due to its thin wall) reduces the rate of heat extraction from the mold wall and thus establishes a shallow temperature gradient within the molten alloy. Such low temperature gradient (G) encourages the widespread nucleation of primary α-Al particles within the melt and consequently the formation of equiaxed and eventually globular structure as shown in Fig. 5.3 (625 and 615 °C). It should be stressed that by superheat reduction not only the growth morphology changes but also the size of grains and globules reduces.

For better characterization and further to highlight the correlation between microstructure with pouring temperature, quantitative metallography was carried out on polished specimens using automatic image analysis system (Fig. 5.4). The image processing results may be summarized as follows:

- Area/perimeter ratio and equivalent circular area diameter decreased with lowering the pouring temperature, while number density increased (Fig. 5.4a–c). The findings are in line with optical micrographs presented in Fig. 5.3, where as pouring temperature decreased, the morphology of primary α-Al phase becomes finer and less dendritic.

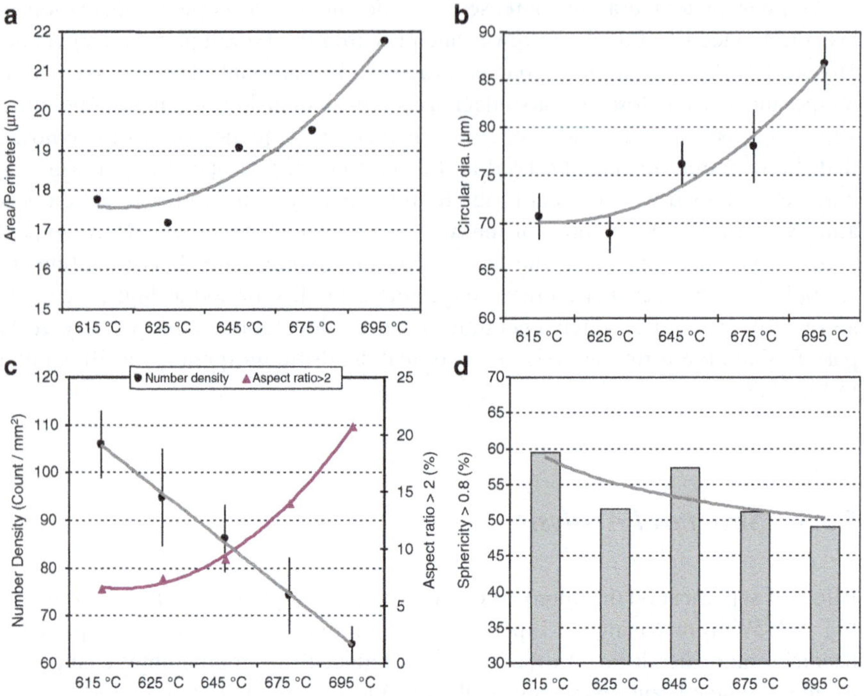

Fig. 5.4 Image analysis data obtained from the conventional cast slugs [4]

- With decreasing the superheat, equiaxed solidification appears to be dominant and growth is in all directions, so globularity and percentage of particles having sphericity >0.8 increased (Fig. 5.4d). This is in line with the percentage of particles with aspect ratio greater than 2 which decreases with lowering the pouring temperature. For better clarification, the distribution of particles with different sphericity values is shown in percentage scale given in Fig. 5.5.

Image analysis technique is based on the digital reproduction of the microstructure which may not be exactly the same as that being studied (discussed in Sect. 4.3). Therefore, it should always be kept in mind that what image analysis cannot do, for example, in the quenched conventional billets; it does not differentiate between the branches of dendrites intersected by the plane of polished and globules, Fig. 5.6. Image analysis only detects the differences between individual particles and ignoring the origin of isolated particles. It means that the way dendrites intersect the polish surface may generate individual isolated particles. Consequently, image analysis will result in a lower value for circular diameter or rise in the sphericity values which does not display the actual dendritic structure. However, even with considering this deceptive point, the data shows fairly good correlation with the variation of pouring temperature.

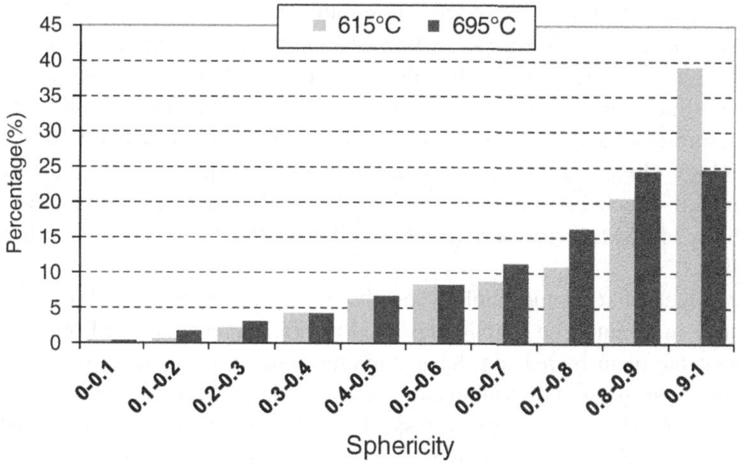

Fig. 5.5 Histogram distribution showing the variation of sphericity percentage for 615 °C and 695 °C [4]

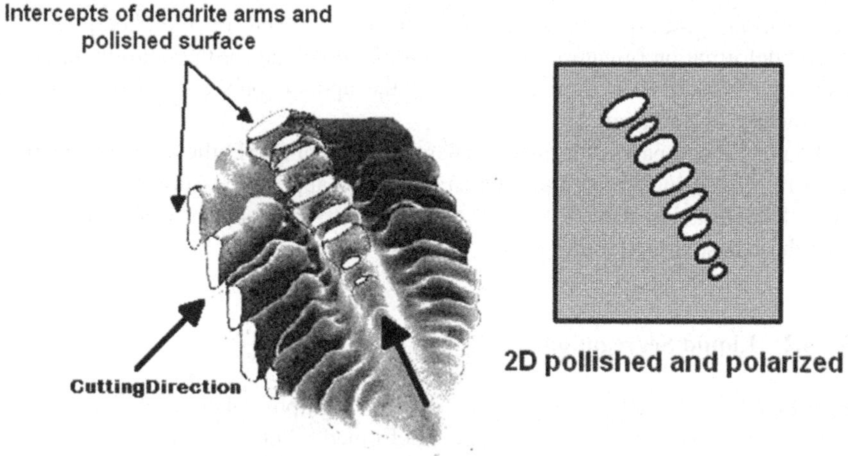

Fig. 5.6 3D view of a dendrite-polished surface intersection showing the dendrite branches look like rounded individual particles

5.2.4 *Rheological Study*

5.2.4.1 Strain–Time Relationship

Formation of the globular microstructure, as shown in the case of 615 °C—Fig. 5.3, has always been the main goal of SSM studies as it has the lowest viscosity number among different morphologies (readers referred to Sect. 4.2). Increasing the temperature gradient within the melt encourages directional growth of the primary α-Al phase. A combination of equiaxed and columnar (directional) growth within the pouring temperature of 630–645 °C brings about the formation of rosette morphology of higher viscosity values than those of globules but lower than the dendritic structure. Such morphology transforms to fully dendritic feature within the melt temperature of 675–695 °C. Formation of the early skeleton structure in dendritic morphology along with its higher dendrite coherency point, DCP, induces some resistance to mechanical deformation for the mush. Such resistance governs the flowability of SSM slurries and makes the filling of the die harder during shaping of the billets.

The effect of α-Al morphology on the ability of billets to deform is clearly detectable in strain–time graphs presented in Fig. 5.7. The applied pressure is an important factor in increasing the maximum values for engineering strain. However, the maximum and minimum engineering strains are always obtained at 615 °C and 695 °C pouring temperatures, respectively, regardless of the applied pressure. In other words, the structure at 615 °C shows less resistance to plastic deformation and if examined in conjunction with optical micrographs in Fig. 5.3, the obvious conclusion is that, "the more globular and finer microstructure yields more engineering strain and better flow." The pouring temperatures of 630 and 645 °C as the representative of rosette microstructure have shown a moderate resistance to deformation somewhere between the globular and dendritic morphologies. The lowest deformation (greatest resistance to flow) obtained for the pouring temperatures of 675 and 695 °C, irrespective of the applied pressure, where dendritic morphology is dominant.

Figure 5.8 compares deformation characteristics of SSM billets cast at 695, 630, and 615 °C pouring temperatures under a constant applied pressure of 4.8 kPa. The flow characteristics of the billet cast at 615 and 630 °C appear to be much better than those cast at higher temperatures.

5.2.4.2 Liquid Segregation

Figures 5.9 and 5.10 present the optical micrographs taken from longitudinal sections from central to wall parts of the deformed billets. In this study, the billets showed almost no liquid segregation for low pouring temperature and some minor segregation at the higher superheats. The absence of liquid segregation for low pouring temperature billets may be attributed to the globular structure of the billets, which allows easy gliding of the primary particles over one another almost at the same speed as the liquid flows through the interparticle channels. This is further assisted by the fine particle size and the low values of shear rate (less than 0.01 s^{-1}). Low shear rates allow solid particles to move easily without excessive collision to bring about homogenous distribution of α-Al phase. The micrographs presented in Fig. 5.9 confirm such hypothesis and show almost no segregation of liquid at different applied pressures. Nonetheless minor liquid segregation is detectable near the wall, particularly in the dendritic structure, Fig. 5.10, which is due to the dendrites interlocking and thus inability to move freely, in contrast to the globules.

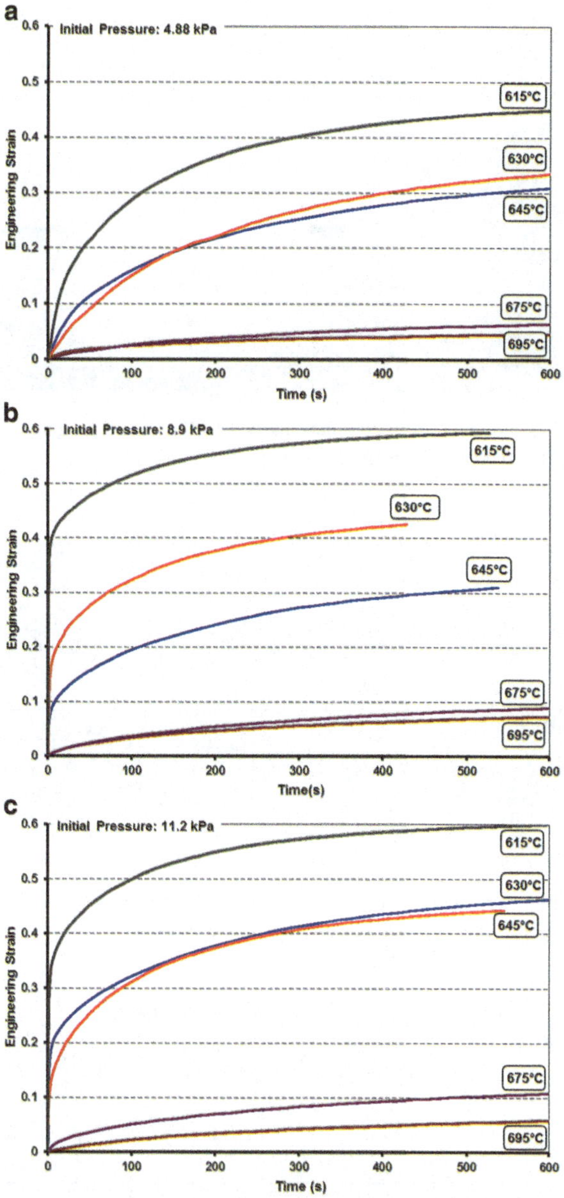

Fig. 5.7 Strain-time graphs obtained at different pouring temperatures and under different applied pressures: (**a**) 4.8 kPa, (**b**) 8.9 kPa, (**c**) 11.2 kPa (reproduced from [4])

The reader should be reminded that such segregation within dendritic morphology has taken place just after 0.1 engineering strain, Fig. 5.10, against 0.6 strain in the case of globular microstructure, Fig. 5.9. This further supports the suitability of globular structure by having lower susceptibility to liquid segregation and better flowability during die casting of SSM billets. Therefore, it is expected the die-cast parts made from billets with globular morphology have more uniform structure and thus properties.

Fig. 5.8 As-deformed billets under 4.8 kPa applied pressure, cast at 695 °C, 630 °C, and 615 °C, from *left* to *right*, respectively [4]

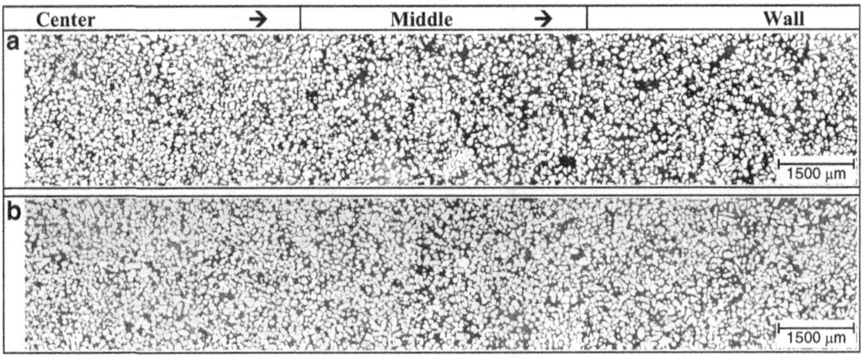

Fig. 5.9 Microstructure of as-deformed SSM billets, maximum strain = 0.6, cast at 615 °C; (**a**) 8.9 kPa, (**b**) 11.2 kPa applied pressures [4]

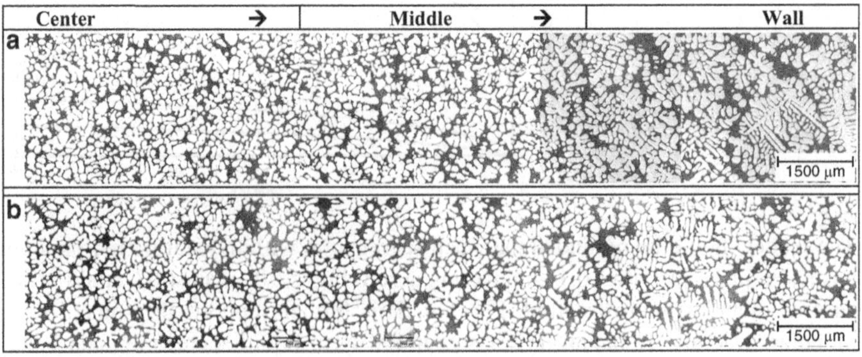

Fig. 5.10 Microstructure of as-deformed SSM billets, maximum strain = 0.1, cast at 695 °C; (**a**) 8.9 kPa, (**b**) 11.2 kPa applied pressures [4]

The main shortcoming of semi-solid materials, the segregation of residual liquid during deformation, which is due to the separation of liquid from the solid phase under pressure is therefore mitigated or even overcome by the refinement and globularization of the solid particles. In such case, application of lower shear rate (lower force) is not only helpful in segregation matter but also it reduces the cost of the HPDC machine.

5.2.4.3 Viscosity

Figure 5.11 shows the graphs where the left-hand side of (4.30) ($\frac{3Vh_o}{8\pi P_o}\left(\frac{1}{h^4}-\frac{1}{h_o^4}\right)=\frac{t}{\eta}$) is plotted against time and only the results of the quasi steady state segments of

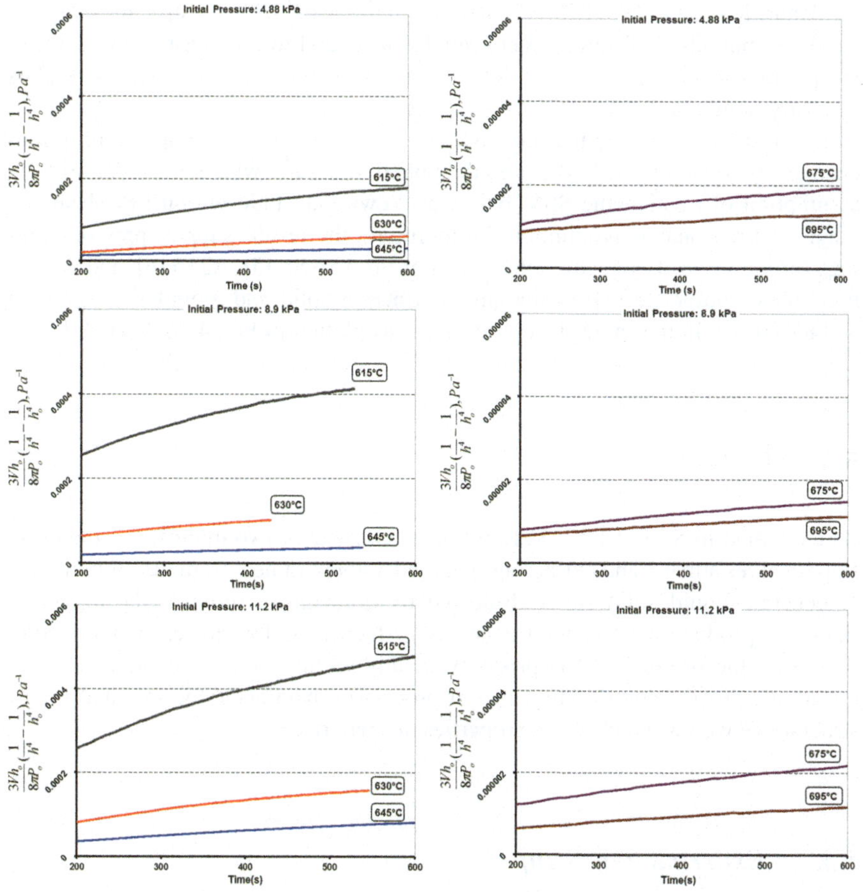

Fig. 5.11 Quasi-Steady state part of the graphs, where $\frac{3Vh_o}{8\pi P_o}\left(\frac{1}{h^4}-\frac{1}{h_o^4}\right)$, Pa^{-1} is plotted against time within 200–600 s after the beginning of each test for viscosity calculation (reproduced from [4])

Table 5.2 Viscosity numbers at different pressures and pouring temperatures, (Log η), (Pa-s) and log shear rate (log $\dot{\gamma}$), (s^{-1}) [4]

Initial applied pressure, P_o	Log η, [log $\dot{\gamma}$] (695 °C)	Log η, [log $\dot{\gamma}$] (675 °C)	Log η, [log $\dot{\gamma}$] (645 °C)	Log η, [log $\dot{\gamma}$] (630 °C)	Log η, [log $\dot{\gamma}$] (615 °C)
4.8 kPa	9, [−3.9]	8.6, [−3.95]	7.3, [−3.54]	7, [−3.33]	6.6, [−3.09]
8.9 kPa	9, [−3.82]	8.6, [−3.83]	7.3, [−3.18]	6.6, [−2.87]	6.3, [−2.74]
11.2 kPa	9, [−3.84]	8.6, [−3.82]	7, [−3.12]	6.6, [−2.86]	6.3, [−2.68]

the graphs (~200 s after the beginning of each compression test) are presented for viscosity calculation. Viscosity is calculated from inverse slope of such graphs. The calculated values of viscosity and shear rates are listed in Table 5.2 for all trials. There is almost three orders of magnitude difference between the viscosity numbers for dendritic (cast at 695–675 °C), and globular (cast at 615 °C) structures and two orders of magnitude difference between dendritic and rosette (cast at 630–645 °C) morphologies. In other words, billets having dendritic structure have the highest viscosity numbers.

It is further noticeable that the viscosity numbers for each temperature did not change significantly with the initial applied pressure which may confirm the assumption of treating the SSM billets as Newtonian fluid within low shear rate range as a reasonable assumption. Furthermore, the results support previous findings for the viscosity of semi-solid alloys of Pb–15%Sn [5], A356 [6], and Al–SiC particulate composite [7] having similar fraction solid and globular morphology and tested within the same shear rate range, as plotted in Fig. 4.16, Sect. 4.2.

5.3 SEED

As explained in Sect. 2.1.12, SEED process consists of two main steps: the initial step involves stirring and extracting specified amount of heat from the molten metal to generate liquid–solid slurry, followed by drainage of some of the remaining liquid to produce a compact free-standing feedstock for rheocasting operation (based on the original SEED process). In this section, microstructural evolution during the process and the influence of processing parameters on the final microstructure as well as mechanical properties are reported.

5.3.1 Experimental Setup

The SEED process is explained in Chap. 2. Table 5.3 shows the chemical composition of commercial 356 alloy that was used for this study [8, 9].

Table 5.3 Chemical composition of 356 foundry alloy (wt. %)

	Si	Mg	Mn	Ti	Fe	Cu	Zn	Al
A356	6.5–7.5	0.2–0.45	0.1	0.2	0.2	0.2	0.1	Bal.

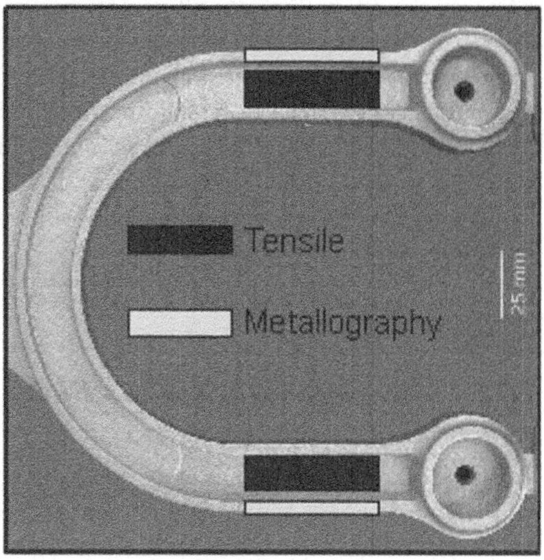

Fig. 5.12 Cast part and sampling location (metallographic analysis was carried out on the inner surface of the white section) [8]

The prepared melt was poured in the mold at various temperatures between 625 and 690 °C. The mold was initially swirled for a certain time "x" and then left to rest for a specific time "y." After that, the bottom enclosure of the mold was opened and the remaining liquid drained for specific duration of time "z" (x–y–z are swirling, resting, and drainage periods, respectively).

The total processing time was increased by changing the swirling (x) and draining duration (z) only and having the pausing time constant (y) as 10 s. These conditions are as follows:

- t_1: 45-10-45 (total time 100 s)
- t_2: 60-10-60 (total time 130 s)
- t_3: 75-10-75 (total time 160 s)

The semi-solid billets (slugs) were cast in a 600-t Buhler machine equipped with a U-shaped mold. The cast parts were quenched in water upon removal (ejection) from die cavity. During the study, the final temperature of the semisolid billet was measured for each condition to ensure repeatability and to minimize the error. Finally, the cast pieces were X-rayed to confirm part quality. The castings were then heat treated to obtain a T5 temper condition (6 h at 170 °C). Samples were taken for microstructural analysis and tensile testing. Figure 5.12 shows the cast piece and sampling locations.

5.3.2 Cooling Curves

Figure 5.13 demonstrates the cooling curves at the center of the mold during the SEED and conventional (no stirring, no drainage) routes to show the thermal history in both processes (pouring temperature 645 °C). As shown, the temperature increases rapidly with pouring the molten alloy in the mold, but the final temperature is slightly lower than the initial pouring temperature. The solidification of the alloy begins after a short thermal equilibration as seen in Fig. 5.13.

As time elapses, the swirling causes the homogenous temperature distribution in the sample. During the drainage step, the cooling rate appears to have increased. The higher cooling rate during drainage could be attributed to the fluid flow; the cooler liquid at the top of the billet is flowing downward while the warmer liquid at the bottom is draining out [10].

5.3.3 Microstructure, As-Cast Billets

Swirling is expected to break down the dendrites, remove or reduce the slope of temperature gradient across the melt and die wall and thus brings about finer and more equiaxed α-Al particles. Such effect is clearly seen in the optical micrographs

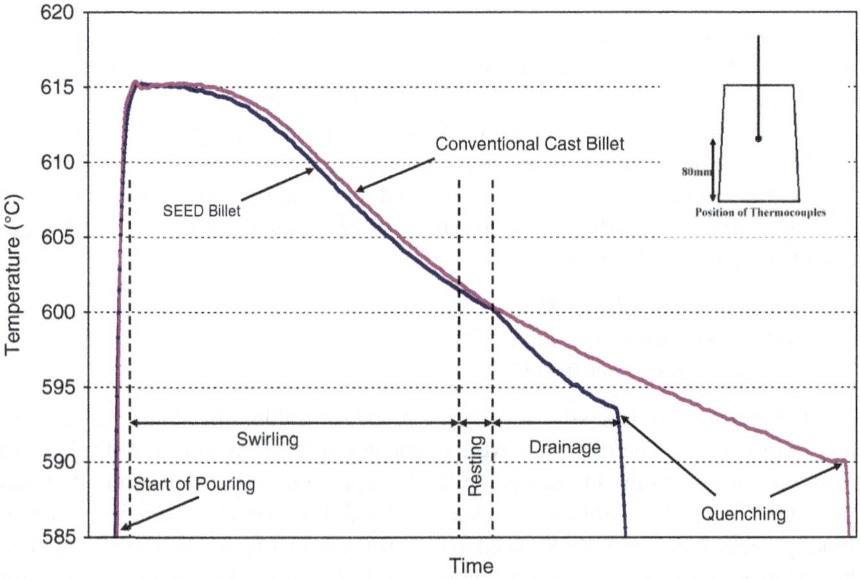

Fig. 5.13 Typical cooling curves for SEED (*blue curve*) and Conventional (*pink curve*) castings (cast temperature 645 °C) [10] (reproduced with permission from the Canadian Institute of Mining, Metallurgy and Petroleum)

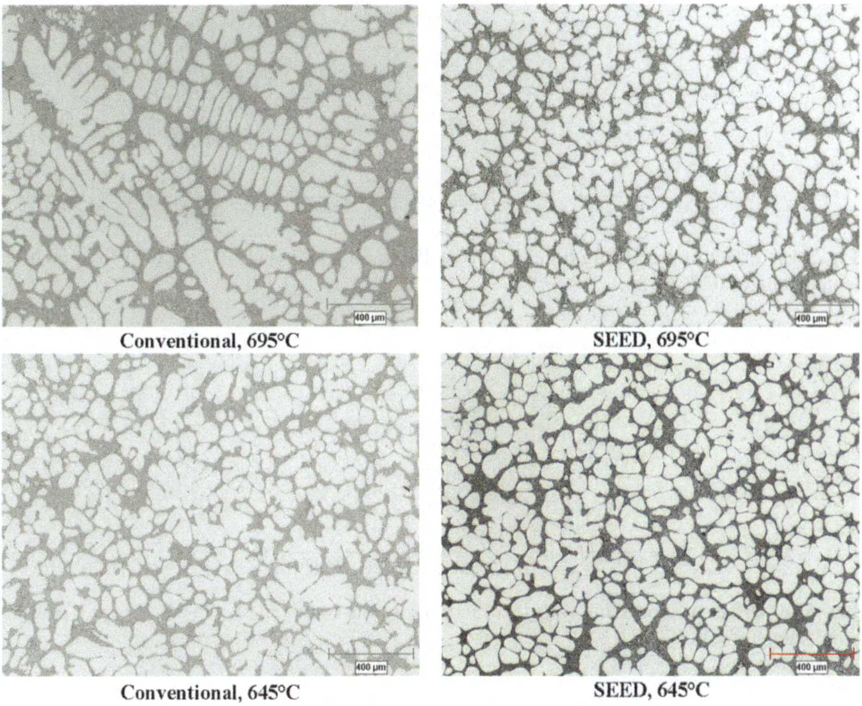

Fig. 5.14 Typical SEED and Conventional cast 356 alloy, quenched billets microstructure cast at 695 and 645 °C

of slugs in Fig. 5.14. It should be emphasized that for conventional microstructures, the isolated α-Al particles are the branches of dendrites intersected by the polished surface not the individual globules (see Fig. 5.6). The SEED process, however, appears to be quite capable in transforming the primary α-Al dendrites to rosette and globular morphologies and its effectiveness is particularly distinct for higher pouring temperatures. The swirling function not only brings about temperature homogenization but also causes fragmentation of dendrites to form α-Al globules (the resting period stabilizes the thermal gradient in the system and allows the remaining liquid to reach a more stable condition. During the draining stage, the fraction solid in the slug increases and a free-standing slug is resulted) [10].

5.3.4 Optimization of Processing Parameters

Figure 5.15 shows the effect of processing time on the microstructure using the initial pouring temperature of 625 °C. As emphasized earlier in this chapter, low pouring temperature significantly modifies the morphology of SSM slurry. Therefore, the low pouring overshadows the processing time variation (it seems that the

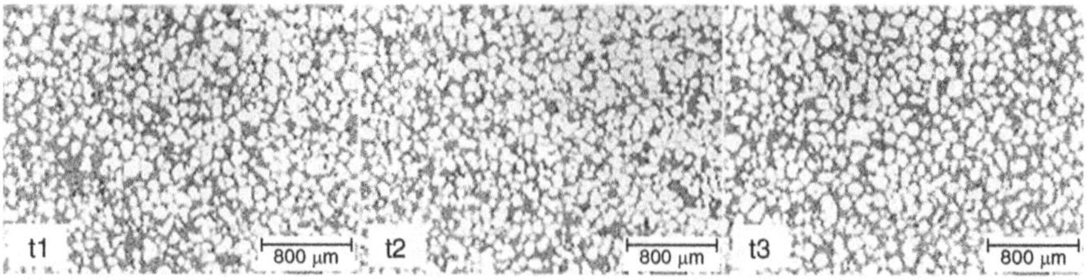

Fig. 5.15 Effect of processing time (t_1: 45-10-45, t_2: 60-10-60, and t_3: 75-10-75) for a pouring temperature of 625 °C [8]

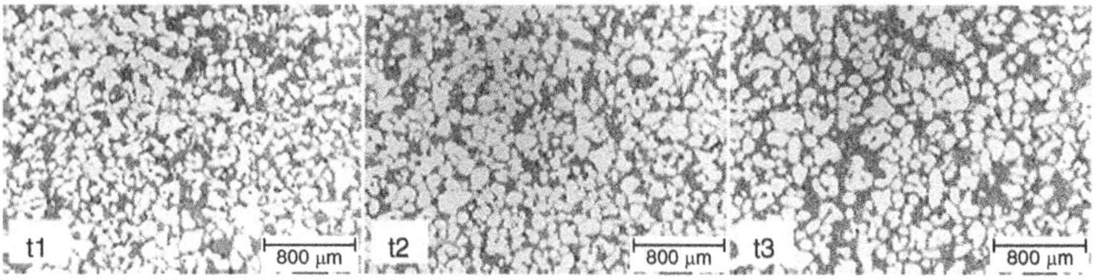

Fig. 5.16 Effect of processing time (t_1: 45-10-45, t_2: 60-10-60, and t_3: 75-10-75) for a pouring temperature of 635 °C [8]

processing time variations have no significant effect on the morphology of primary α-Al particles). As a side issue, it appears that extended swirling / overall time results in more spherical particle.

With increasing the pouring temperature, 635 °C, the morphology of the primary α-Al particles deviated slightly from the globular shape (Fig. 5.16). This is attributed to both the nucleation kinetics and growth directionality of the primary α-Al particles that took place within the melt having a steeper temperature gradient and slightly directional cooling. This caused the primary α-Al particles to depart from the typical globular shape before processing of the semi-solid slug. In line with the findings of 625 °C, globules in t_3 samples seem to be more spherical which is mainly due to the longer swirling/overall time. The solid fraction is slightly reduced with increasing pouring temperature since the overall temperature within the mushy zone at the end of swirling/resting/drainage cycles is expected to be higher compared to 625 °C, i.e., more liquid fraction (see Fig. 5.25). This is more pronounced in Fig. 5.17 with a pouring temperature of 660 °C. This becomes more complicated by considering the drainage of solid/liquid phases.

Figure 5.17 shows the effect of total processing time on the microstructure at a pouring temperature of 660 °C. The increase in processing time at this temperature changed the morphology and the amount of primary α-Al dendrites. A higher pouring temperature greatly impacted the nucleation kinetics and growth behavior. The highest pouring temperature (in contrast to Figs. 5.15 and 5.16) reduced the nucleation rate and promoted steeper temperature gradient within the melt; thus changing the primary α-Al phase morphology from rosette-like to dendritic. For all

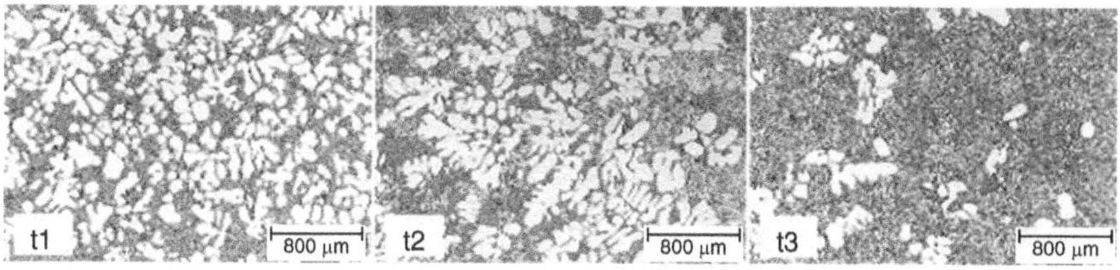

Fig. 5.17 Effect of processing time (t_1: 45-10-45, t_2: 60-10-60, and t_3: 75-10-75) for a pouring temperature of 660 °C [8]

Table 5.4 Typical mechanical properties versus specific conditions (T5 temper condition 6 h at 170 °C) [8]

Condition		UTS (MPa)	YS (MPa)	Elongation (%)
625 °C, t_2	max	281	206	7.9
	min	252	174	2.7
	ave	265	187	5.1
635 °C, t_3	max	276	203	11.5
	min	249	173	9
	ave	263	184	10
660 °C, t_3	max	282	221	2
	min	230	205	1
	ave	256	214	1.5

three pouring temperatures, the fraction of primary α-Al particles appears to have reduced with increasing processing time, t_3. This is partially related to the longer swirling time which promotes dendrites remelting but encourages better distribution of primary phase within liquid. Also it is noteworthy that liquid segregation phenomenon which is active during high pressure die casting is more pronounced at higher pouring temperature. That is why the liquid fraction at 660 °C appears greater than the others.

In general, image analysis confirmed that [8]

- Increase in the superheat leads to larger primary α-Al particles
- Primary α-Al particles are more spherical at lower pouring temperature
- Total processing time has some influence on the shape and percentage of the primary α-Al particles

The mechanical properties at different processing conditions were demonstrated in Table 5.4 for the T5 temper condition (6 h at 170 °C). The mechanical properties and particularly the ductility were significantly affected by the change in the SEED process parameters. The main cause for the changes in the elongation was mostly associated with the microstructure. In addition, it is worth mentioning that the transitions from dendritic to globule structure reduce the defects such as oxide, cold shut, and shrinkage porosity which are obviously detrimental to the mechanical properties.

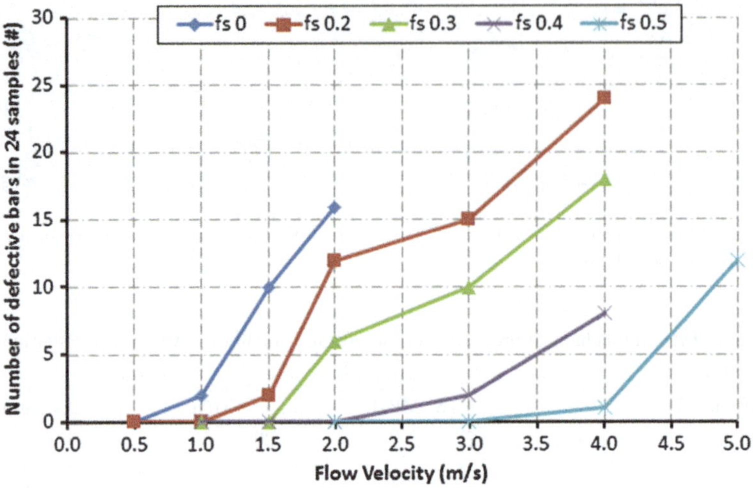

Fig. 5.18 Flow/turbulence-related defects versus fraction solid and flow velocity, SLC process modified and grain-refined A356 alloy, followed by heat treatment to T6 temper condition (solution heat treated at 540 °C for 4 h, quenched in hot water and precipitation aged at 170 °C for 4 h) (adapted from [11])

Furthermore, it may be concluded that the changes in morphology as well as solid fraction could affect the rheological behavior of the slurry and, therefore, its overall filling and feeding capability. Jorstad et al. [11] showed that molten metal in squeeze casting process could flow at 0.5 m/s without turbulence-related defects but 1 m/s was too fast and encountered porosity and blisters. As the fraction of solid was gradually increased, the allowable flow velocity is increased as well. In another way, larger solid fraction results in flow stability and therefore higher flow velocities could be reached without any X-ray visible defects and/or blisters following heat treatment. It was shown that at 50 % solid (f_s=0.5), flow could exceed 3 or 4 m/s without creating defects (Fig. 5.18).

Table 5.4 shows the main results for mechanical properties obtained for all three processing times. The best strength to elongation relationship was gained for the specimens with 635 °C pouring temperature. This could be attributed to the formation of more rosette morphology at 635 °C.

5.3.5 Chemical Uniformity

Solidification of alloy systems is completely different from that of pure metals. In fact, the transformation in a pure metallic system is a heat flow phenomenon while in alloy systems, it is a combination of heat flow and solutes redistribution. Alloy systems solidify over a temperature range with a number of changes taking place simultaneously within the alloy system such as crystallization, solid movement, solute redistribution, constituents' segregation, and coarsening (ripening). In the case of solute redistribution, there are three classical approaches proposed to predict alloy solidification [12]:

- Equilibrium solidification or complete mixing in solid and liquid
- No diffusion in the solid with complete mixing in liquid, Scheil's solidification
- No diffusion in solid and limited liquid diffusion

In real situations, solidification is never slow enough to approach the first condition and therefore the most likely practical cases involve the others. In the case of alloy systems having partition coefficient less than 1 ($k < 1$), the solute atoms are rejected out of the solidifying liquid into the solid–liquid interface to induce solute buildup in the liquid adjacent to the solid. This leads to microsegregation during solidification. On further solidification, and particularly during its final stages, a highly solute enriched liquid region develops which may cause the formation of wider solute-rich area known as "macrosegregation." It also contributes to the formation of different phases or compounds such as intermetallic phases.

The segregation pattern is a key factor in near net shape processes. In SSM processing, the uniformity of alloy distribution within the primary billet is important since the billet is the feedstock for subsequent forming process. By minimizing the chemical variation within the billet, the properties' uniformity of complicated shaped pieces would be assured.

Variation of chemical composition within the microstructure could be studied theoretically, using equilibrium phase diagrams and lever rule or by commercial software packages, and experimentally in macro/micro scales by various methods such as optical emission spectroscopy (OES), scanning electron microscopy (SEM), and electron probe microanalysis (EPMA) through point and line scans.

The fraction solid of the primary phase at a specified location within the liquid–solid zone is given by a mass balance, which in the simplest case, neglects ripening and diffusion in the solid. For this case, the interrelationship between weight fraction of the liquid in the mushy zone, f_l and liquid composition, C_l, assuming complete diffusion within liquid, is given by the Scheil's equation, for constant partition ratio (k) (reformulated from [12]):

$$f_l = \left(\frac{C_l}{C_0}\right)^{\frac{-1}{1-k}} \tag{5.1}$$

where C_0 is the alloy composition. Since the temperature T in the mushy zone and C_l are interdependent through the equilibrium liquidus line, (5.1) may be rearranged as:

$$f_l = \phi^{\frac{-1}{1-k}} \tag{5.2}$$

where ϕ is dimensionless parameter: $\frac{T_m - T}{T_m - T_l}$ and T_m and T_l are the melting point of the pure solvent and the liquidus temperature of the alloy with composition of C_0 respectively.

Equation (5.2) illustrates the correlation of local weight fraction solid to the temperature in the liquid–solid zone; it is often employed in experimental studies to calculate directly the local fraction solid from temperature measurements.

These equations are also employed to calculate the instant chemical composition of a given alloy frozen at a predetermined fraction solid. The chemical analysis of solidified alloy at a distinct fraction solid and temperature renders the concentrations of liquidus and solidus in the binary phase diagram either for equilibrium or nonequilibrium solidification. Equation (5.3) shows a simple relationship between the chemical composition and fraction of phases at a given temperature:

$$\%B = f_s \times C^{T_0}_{\text{@solidus}} + f_l \times C^{T_0}_{\text{@liquidus}} \tag{5.3}$$

where %B is the concentration of the alloying element at temperature T_0, and f_s,f_l, $C^{T_0}_{\text{@solidus}}$, $C^{T_0}_{\text{@liquidus}}$ are the fractions solid and liquid, and the % of element "B" at solidus and liquidus temperatures, respectively.

In this section, a novel experimental procedure is introduced to investigate the constituent concentration of the solidifying alloy. It will also be shown that the pouring temperature plays a key role on the billet chemical uniformity.

5.3.5.1 Methodology

Binary Al7Si and commercial 356 ingots with the chemical compositions given in Table 5.5 were used. Melting procedure was the same as the one mentioned in Sect. 4.1. The prepared melt was poured into the cylindrical steel mold with a range of superheats from ~15 to 75 °C. For the entire tests, at a specific time after pouring (45–50 s), the bottom plug is opened and a portion of liquid drained. The billet was then quenched in cold water after 1 min following drainage. The application of such testing procedure rendered a range of drained molten metal weight and quenching temperatures.

The quenched slugs then were sectioned from top, middle, and bottom as shown in Fig. 5.19. Since the drained droplets of liquid have different compositions and are not homogenous, the extracted material was remelted in an argon atmosphere, cast in standard scissor mold, and prepared for chemical analysis.

Table 5.5 Chemical composition of tested alloys (wt. %)

	Si	Mg	Fe	Cu	Ni	Ti	Al
Al7.54Si	7.54	Nil	0.07	0.002	0.004	Nil	Bal.
356	6.9–7.1	0.32–0.35	Max 0.1	Max 0.002	Max 0.001	0.12–0.13	Bal.

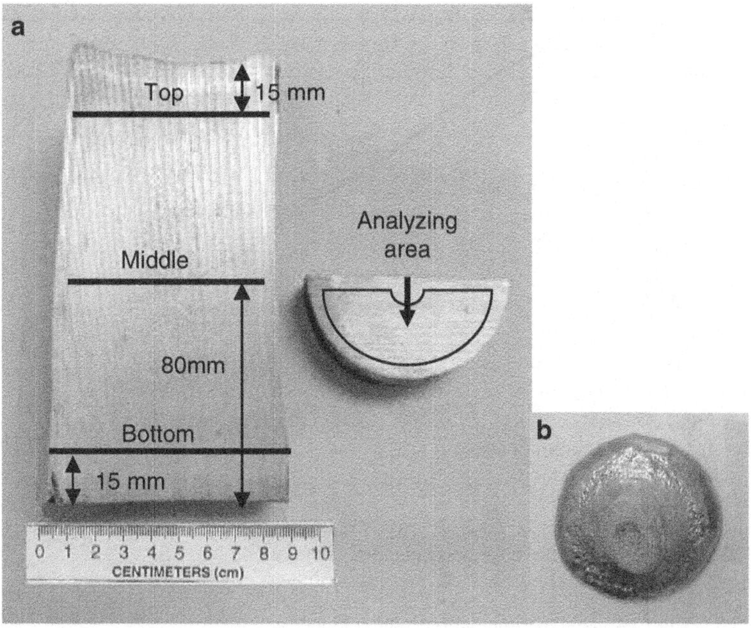

Fig. 5.19 (**a**) Longitudinal section of the billet showing the location of the analyzed areas in the transverse section, (**b**) drained liquid, top view [13]

5.3.5.2 Impact of Partial Draining on Chemical Uniformity

Figure 5.20 shows a typical time–temperature distribution registered during casting of billets poured at 645 °C. During pouring, the temperature at the center of the mold increases rapidly and the peak temperature is lower than the pouring temperature which is associated with the major heat extraction at the initial stages of solidification (see Sect. 5.2.2). By extracting a portion of the liquid, the cooling rate increased slightly which is attributed to the fluid flow and liquid decanting from the bottom.

Binary Al7Si Alloy

Due to the complexity of commercial alloys, the first series of trials were carried out by melting a binary Al–Si alloy, Al–7.54%Si (Table 5.6).

After drainage, the Si concentration both in the billet and drainage is different from the melt which is related to the segregation phenomenon. According to the Al–Si binary phase diagram, the remaining liquid at any stage during solidification becomes more and more enriched with the solute elements. By decanting and extraction of the saturated liquid at the vicinity of solidified particles, the remaining liquid has a different composition. In other words, the remaining liquid solidifies on a different solidification path which is schematically shown in Fig. 5.21.

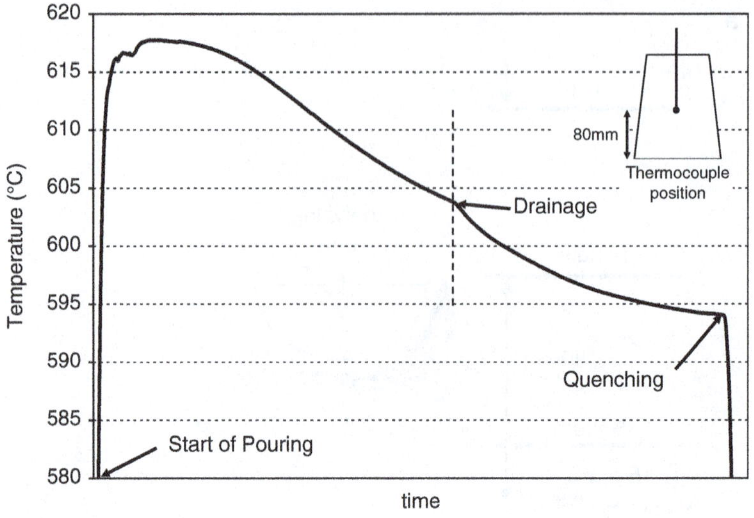

Fig. 5.20 Typical cooling curve for pouring temperature of 645 °C [14]

Table 5.6 Data from the experimental procedure, binary Al–Si alloy

Pouring temp (°C)	T drainage (°C)	T quench (°C)	% Silicon		
			melt	billet	drainage
645	602.1	592.5	7.54	7.25	9.68

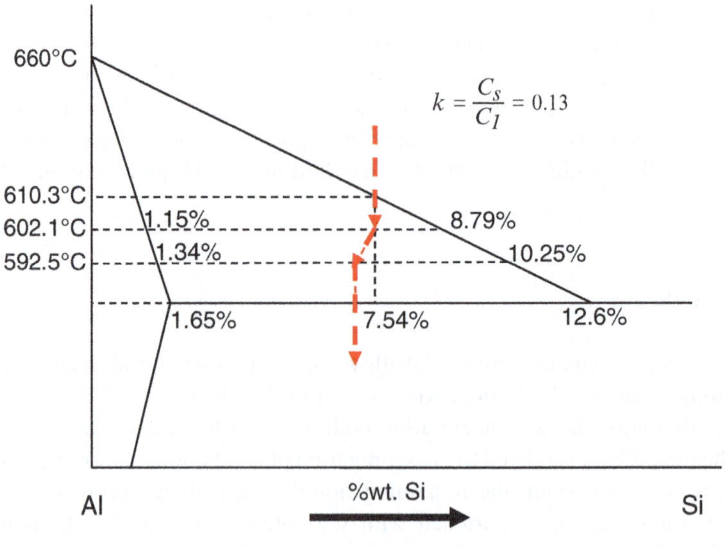

Fig. 5.21 Schematic representation of chemical composition variation due to decanting of Al7.54Si alloy @ 602.1 °C and 592.5 °C for condition provided by pouring at 645 °C [13]

The proposed procedure to verify the liquid enrichment is to drain the liquid from mushy zone at a given temperature range and analyze the decanted liquid and the billet. The broken arrows in Fig. 5.21 approximate the changes in the alloy composition by partially draining the liquid within the temperature range of 602.1–592.5 °C.

According to the equilibrium binary phase diagram, at 602.1 and 592.5 °C, the lever rule ($f_l = \frac{C_0 - C_s}{C_l - C_s}$) yields the liquid fractions of 0.838 and 0.697, respectively, while the Scheil's equation, (5.2), gives liquid fractions of 0.839 and 0.703, respectively. Therefore, it could be concluded that at these temperatures, both Scheil's equation and lever rule calculations are approximately the same. Therefore, by averaging both sets of data, it could be concluded that f_s and f_l are 0.16 and 0.84 for 602.1 °C and 0.3 and 0.7 for 592.5 °C, respectively.

Based on the decanting time of 45–50 s, pouring temperature of 645 °C, and liquid extracting from 602.1 °C to 592.5 °C, about 15 % of the liquid is decanted. The reduction of the liquid volume results in a new equation for solid fraction and increases its portion within the billet. Therefore, the final solid fraction within the billet could be calculated as:

$$f_s^{\text{after-drainage}} = \frac{f_s}{f_s + (f_l - 0.15)} \quad \& \quad f_l = 1 - f_s \tag{5.4}$$

Then at 592.5 °C:

$$f_s = \frac{0.30}{0.30 + 0.55} = 0.353 \quad \& \quad f_l = 0.647$$

With such assumption, the silicon concentration within the samples and drained liquids at equilibrium condition could be as follows:

$$\%\text{Si}_{\text{billet}} = \left(\%f_s^{\text{after-drainage}} \times C^{TO}_{\text{@Solidus}} + \%f_l^{\text{after-drainage}} \times C^{TO}_{\text{@Liquidus}}\right) \tag{5.5}$$

At 592.5 °C:

$$\%\text{Si}_{\text{billet}} = 0.353 \times 1.34 + 0.647 \times 10.25 = 7.1\%\text{Si} \;\; (@\,592.5°\text{C})$$

$$\%\text{Si}_{\text{drainage}}(\text{from binary diagram, Fig. 5.21}) = 10.25\%\ \text{Si}$$

Since the drainage started at 602.1 °C and finished at 592.5 °C, the measured values for Si concentration should be treated as the average composition within such temperature range. In order to be in line with the experimental results, it is reasonable to calculate the average Si content between the two temperatures of 602.1 and 592.5 °C. Therefore, the average contents for silicon calculated from the binary equilibrium phase diagram are 9.52 % Si for the enriched decanted liquid and 7.32 % Si for the billet. As shown in Table 5.6, chemical analysis of the cast piece and decanted liquid of actual Al–7.54%Si binary alloy after 15 % draining

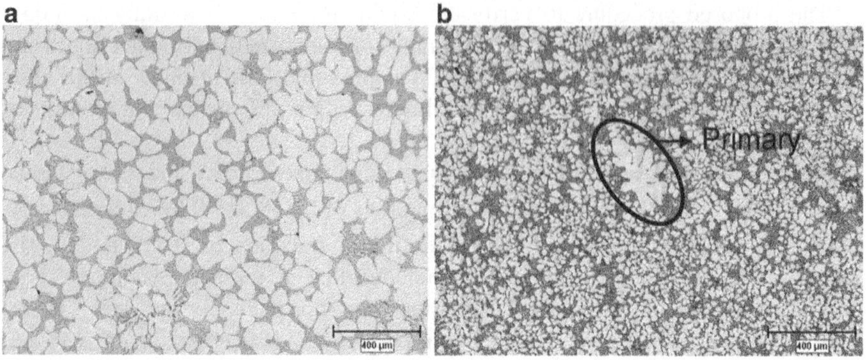

Fig. 5.22 Optical micrographs of middle section, cast piece (**a**) and drained material (**b**), poured at 645 °C

yields 7.25%Si for the cast piece and 9.68%Si for the residual liquid. The results are in good agreement with the calculated values. The small difference between the calculated values and measured data could be attributed to a range of parameters such as the assumption of linearity for the liquidus and solidus lines in the binary phase diagram, trace elements, nonequilibrium solidification of the samples, and lever rule and Scheil equations assumptions. The nonequilibrium solidification is particularly important since it may induce further dissolution of alloy elements within the as-cast billets and lowers the concentration of silicon in the drainage. In addition, the experimental scatter in the measurement is another factor to be considered carefully.

Figure 5.22 shows typical optical micrographs of the middle section of the billet and that of the drained liquid (decanted liquid was quenched in a cold water during the drainage step). For the cast piece (Fig. 5.22a), the growth of the primary α-Al particles has been suppressed by quenching to illustrate the size and morphology of these particles at the respective temperature. The microstructure of the decanted material is quite uniform and finer than the billet structure which is due to rapid quenching of the decanted liquid in water. Infrequently, larger primary particles were seen in the structure of the drained sample. The primary α-Al dendrite observed in Fig. 5.22b has most probably broken away from the solidified network during the drainage stage and carried away with the liquid through the orifice. These atypical features are quite rare and the calculations based on the sole reduction of liquid, not the solid phase, are almost correct and valid.

Commercial 356 Alloy (356)

Similar to the binary Al–Si alloy, commercial 356 was solidified up to about 0.2–0.3 fractions solid and a portion of the remaining liquid was drained. The changes in the concentration of the principal elements were studied by optical

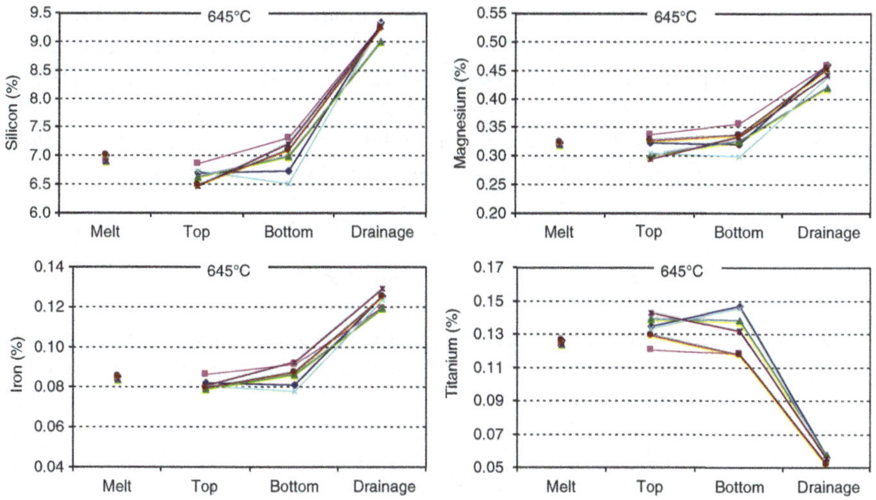

Fig. 5.23 Variation of chemical composition within the melt, *top* and *bottom* of the billets and drainages

emission spectroscopy and the results were compared to those of the theoretical calculations.

Figure 5.23 displays the concentration of Si, Mg, Fe, and Ti within the melt, top, and bottom segments of the billets and the drainage. During the drainage step, the liquid flows downward and since the solubility of alloying elements within the primary α-Al is rather limited, the solute elements are rejected into the remaining liquid. So it is expected to have higher concentration of alloying elements within the partially drained liquid. However, since solidification predominantly starts from the mold top and wall, it is possible to have some enriched liquid entrapped at the top and thus resulting in a slight increase in alloy concentration. At lower pouring temperature such as 645 °C, simultaneous nucleation within the bulk liquid will take place and therefore the possibility of having localized alloy enrichment and segregation is less likely to occur which means more uniform composition.

The results could be summarized as follows:

- There is a slight difference between the composition of the billets at the top and bottom segments for 645 °C pouring temperature.
- Within the drained segment, the concentrations of Si, Fe, and Mg are higher than that of the original alloy composition while for Ti the drained melt is less enriched. This is due to the rejection of solute elements within the liquid at the solidification front during the formation of the primary α-Al particles. For Ti, the trend is reversed which is associated with the peritectic reaction exists in the Al–Ti phase diagram ($k > 1$). As well, Ti initially reacts with molten aluminum to form Al_3Ti particles. These particles act as nucleation sites for primary $\alpha - Al$ due to the peritectic reaction of $L + Al_3Ti \rightarrow (\alpha)Al$. The Ti is expected to be consumed even before the start of decanting process. Therefore, there is not much Ti left in the liquid to flow out of the mold during decanting.

Among several trials, the average Si concentration in the melt was measured 6.96 %, within the billet from top to bottom 6.83 % and for the drainage, 9.29 %. The following paragraphs discuss the theoretical calculations based on Al–Si binary equilibrium phase diagram and (5.4) and (5.5).

Assuming 6.96 % for Si concentration in the Al–Si binary system, the fractions solid at 602.1 °C and 592.5 °C are calculated based on the lever rule:

$$f_s = 0.24 \quad \& \quad f_l = 0.76\ (@602.1\,°\mathrm{C})$$
$$f_s = 0.37 \quad \& \quad f_l = 0.63\ (@592.5\,°\mathrm{C})$$

Fraction solid was also calculated by ThermoCalc software and thermal analysis method which is presented in Fig. 5.24. Comparison between the calculated fraction

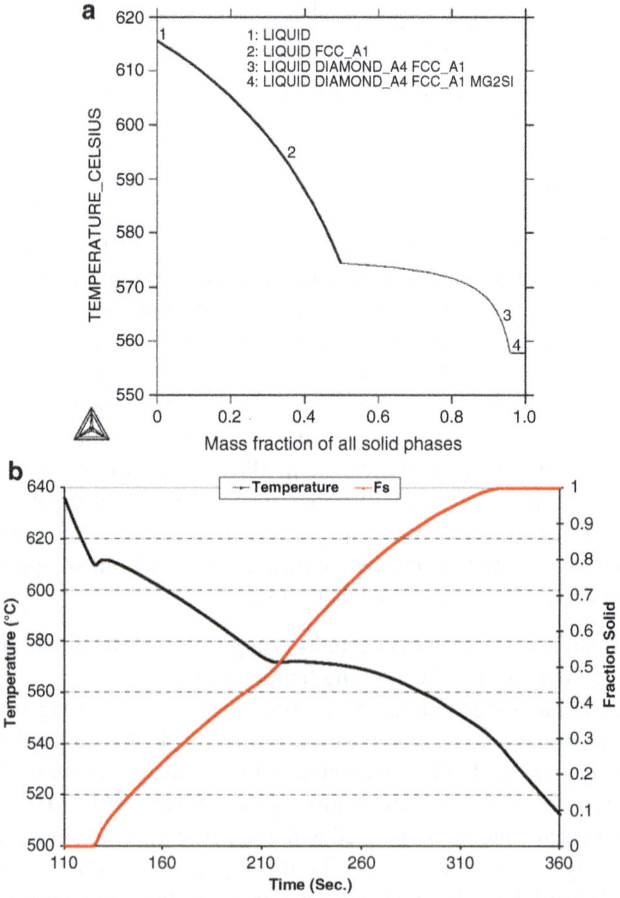

Fig. 5.24 Fraction solid measurements by: (**a**) ThermoCalc software, Al7Si0.35 Mg (0.25 @602.1 °C and 0.35 @592.5 °C) [14], (**b**) thermal analysis, Al7.1Si0.38Mg (0.22 @ 602.1 °C and 0.32 @ 592.5 °C)

solids shows almost the same values. This is an indication that the fraction solid calculated through Scheil's equation, lever rule, or thermal analysis are similar, despite different assumptions considered for these methods.

After drainage of about 15 % fraction volume, the last remaining liquid fraction within the billet at 592.5 °C is calculated at 0.48 (0.63−0.15 = 0.48). Therefore, according to the (5.4), the fraction solid and liquid at the end of drainage are:

$$f_s = 0.43 \quad \text{and} \quad f_l = 0.57 \quad (\text{at } 592.5\,^\circ\text{C})$$

The final Si concentration within the drainage and billet considering (5.5) are as follows:

$$\%\text{Si}_{\text{drainage}} = 10.25 \quad \text{and} \quad \%\text{Si}_{\text{billet}} = 6.41\% \quad (\text{at } 592.5\,^\circ\text{C})$$

The comparison between the experimental result of Si content for the cast piece and drainage (6.96 % and 9.29 %, respectively) and the average calculated Si content above (6.68 % and 9.52 %) is in good agreement. The differences may initiate from the followings:

- Multiple alloying systems are more complex than simple binary alloys. The presence of minor constituent elements within the melt such as Mg, Fe, and Ti can change the equilibrium lines and shift liquidus and solidus up or down. In addition, the minor elements are able to tide up certain number of the main elements such as Si in the form of Mg_2Si and/or formation of iron intermetallics.
- Nonequilibrium solidification during casting process alters the position of solidus, liquidus, and eutectic lines. Besides decanting the hot liquid changes the heat transfer and ultimately the average cooling rate could become greater.
- Scheil's equation which does not consider the back diffusion in the solid.

However despite partial decanting of the remaining liquid, the composition of the final billet is within the A356 alloy range, when compared with the composition of commercial 356 alloy cited by the international standards, Table 5.7.

Investigation on the Chemical Uniformity with Various Pouring Temperatures

Figure 5.25 shows the correlation amongst pouring, drainage, and quenching temperatures with drainage quantity. Higher pouring temperature provides the

Table 5.7 Chemical composition of A356, AA standard (wt. %) [15]

	Si	Mg	Fe	Cu	Ti	Zn	Al
Standard Comp.	6.5–7.5	0.2–0.45	Max 0.2	Max 0.2	Max 0.2	Max 0.1	Bal.
Decanted billet	6.83 ± 0.3	0.32 ± 0.02	0.085 ± 0.005	Max 0.002	0.13 ± 0.01	Max 0.002	Bal.

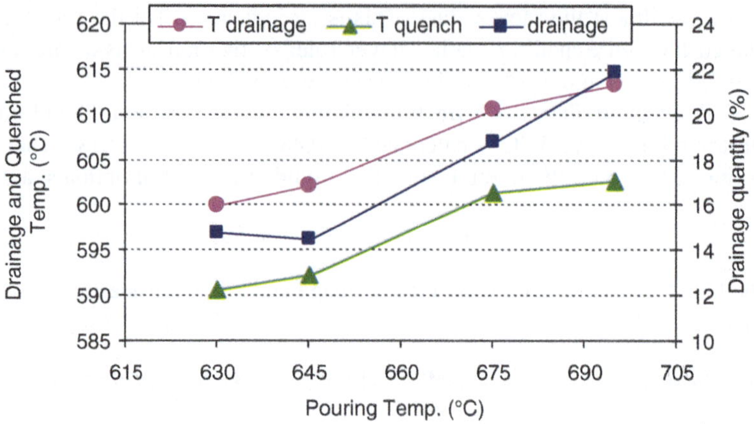

Fig. 5.25 Correlation amongst pouring, drainage, and quenching temperatures, commercial 356 alloy

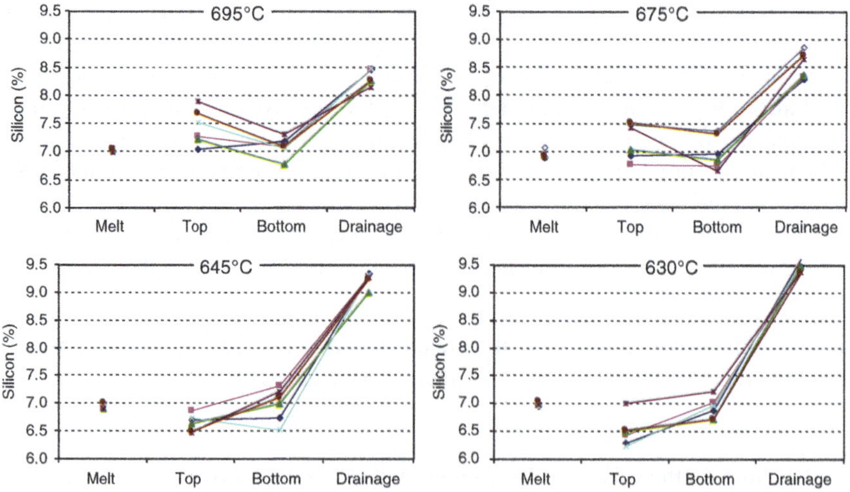

Fig. 5.26 Silicon content in different parts of the as-cast SSM billet with various pouring temperatures (colors showing different casting trials)

mold with higher heat content and increase drainage volume within a constant time frame following pouring (refer to Sect. 5.3.1 for processing time). The higher heat content in a constant processing time promotes less solid fraction. The lower solid fraction is a less effective barrier against liquid flow. The higher heat content also increases liquid fluidity that further simplifies the drainage process.

Figures 5.26, 5.27, and 5.28 show the results of Si, Mg, and Fe analyses for melts, drainages, top and bottom of the billets at various pouring temperatures. Drainage as part of the residual liquid in the mushy zone is enriched in alloying

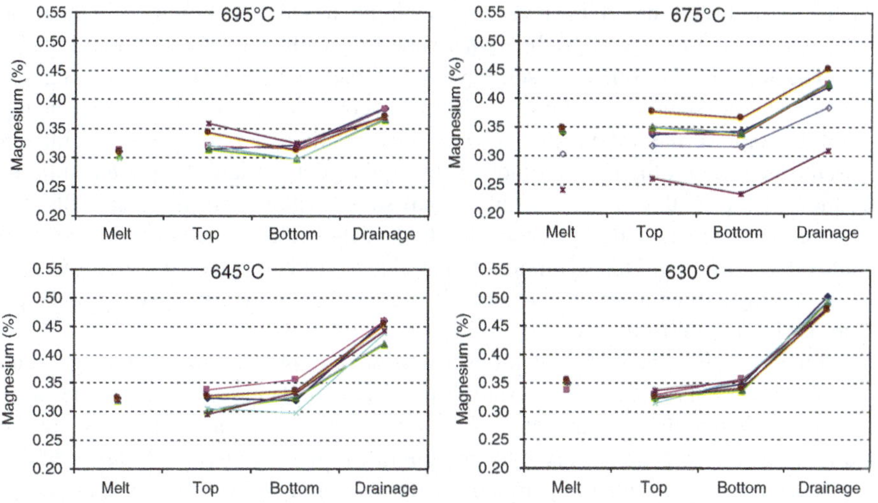

Fig. 5.27 Magnesium content in different parts of the as-cast SSM billet with various pouring temperatures (colors showing different casting trials)

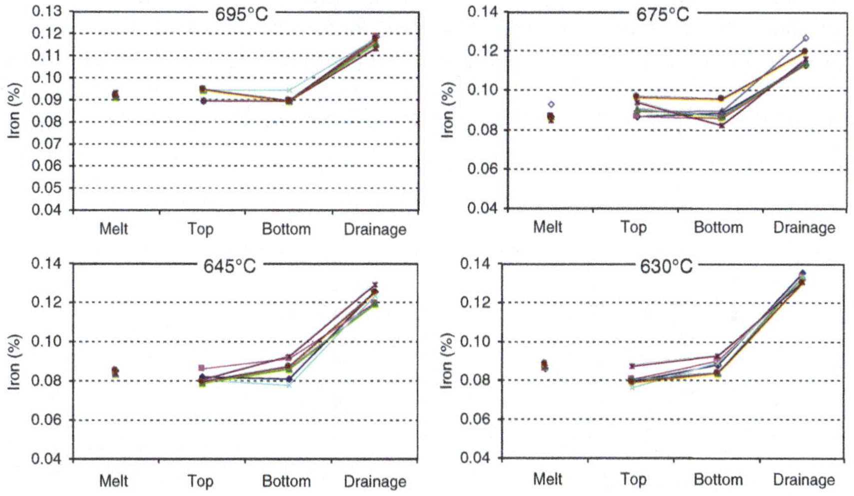

Fig. 5.28 Iron content in different parts of the as-cast SSM billet with various pouring temperatures (colors showing different casting trials)

elements. As solidification proceeds within the mushy zone, according to the equilibrium binary phase diagram, liquid becomes enriched in alloying elements and therefore the analysis shows higher solute contents. Lowering the pouring temperature also increases Si, Mg, and Fe concentrations in the drainage. As mentioned in the experimental setup, the bottom plug is opened at a constant time after pouring and thus lower superheat leads to formation of more α-Al particles and according to the lever rule, the liquid has higher alloying content at the drainage period compared to the higher superheat with less saturated liquid.

The top and bottom analyses have some variations which by lowering the pouring temperature, variations of chemical composition decrease between top and bottom of the billet. This is associated with small temperature gradient established within the melt promoting the equiaxed and multi directional solidification. It means that instead of directional solidification from top and walls, the entire melt starts to solidify almost simultaneously leading to small discrepancy in alloy concentration. This is important in production of cast pieces since by lowering the pouring temperature, the chemical homogeneity increases leading to better properties and performance.

Figure 5.29 shows the titanium variation within the melt, drainage, and top/bottom of the billet. The variation trend is completely opposite to the other elements. While the amounts of Si, Mg, and Fe are at the maximum values in the drained melt, the Ti content is at the minimum. As already discussed, this phenomenon is associated with the peritectic reaction within Al–Ti alloy system and to some extent to the role of Ti as refiner, see Sect. 6.2.1.4. According to the phase diagram, the first solid formed is enriched in Ti and by decreasing the temperature, the Ti level in the remaining liquid gradually decreases. The other reason is Ti capability in generating nucleation sites for the primary α-Al particles due to the formation of Ti intermetallic phases as described elsewhere [16]. A relation also exists between the Ti level in the drainage and pouring temperature. The Ti content in the drainage decreases with lowering the superheat which is an indication of minimizing the possibility of primary α-Al particles extraction during the drainage

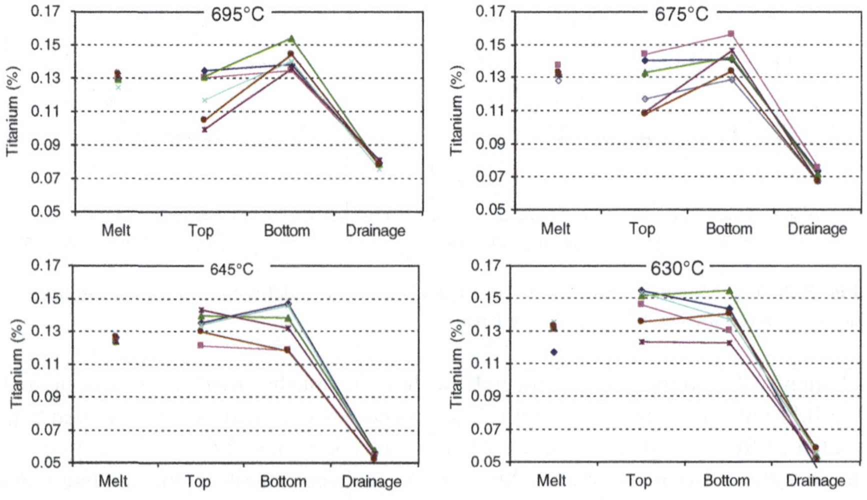

Fig. 5.29 Titanium content in different parts of the as-cast SSM billet with various pouring temperatures (colors showing different casting trials)

period. In another words, the slurry is capable of preserving the primary particles in the billet due to the lower volume of the liquid in the bulk.

At higher superheat, Ti is higher at the bottom comparing to the top portion of the billet. This may be attributed to natural convection in the billet which is greater for high pouring temperature. Higher superheat leads to more liquid extraction from the billet and due to the convection and density variation, the primary particles tend to sediment and concentrate at the bottom of the billet. These primary particles contain Ti and spectroscopy analysis confirms the higher value at the bottom.

5.4 Electromagnetic Stirring, EMS

In the EMS process, as described in Chap. 2, the local shear is generated by electromagnetic field where the solidifying metal acts as the rotor and the stirring action created by liquid movement shears the dendrites formed. Consequently, the processed material could be used as a source of semi-solid material. In this section, the importance of EM stirring, effect of cooling rate, and pouring temperature will be discussed on the grain/globule sizes of α-Al primary particles. The microstructural evolution of silicon and iron intermetallics is already discussed in Sect. 3.3.3.

In order to correlate pouring temperature and stirring effect on the resulted billet structure and quality, two different molds (sand and copper) plus electromagnetic stirring were employed as described in Chap. 4. Figure 5.30 shows the microstructure of the as-cast non-stirred billets from sand and copper molds. Both dendrite cell size and arm spacing (DAS) have a difference due to the various cooling rates of the molds. Higher cooling rate not only reduces the size of primary α-Al grains but also the eutectic constituents become finer. The difference is apparent in the size of the β-iron intermetallic phase (arrowed) and silicon eutectic flakes, where for the copper mold, both becoming thinner and shorter (comparing Fig. 5.30b, d).

5.4.1 Effects of Superheat and Stirring on Grain and Globule Size

5.4.1.1 Sand Mold

The polarized light optical micrographs in Fig. 5.31 present the microstructural evolution due to the variant pouring temperatures combined with stirring application. Primary α-Al particles having dendritic morphology in the unstirred billets appear to have spheroidized by the stirring process. Combining stirring with superheat reduction, the morphology of α-Al phases changes to rosette, equiaxed, and finally globularized with ripening process, remelting of thinner and growth of

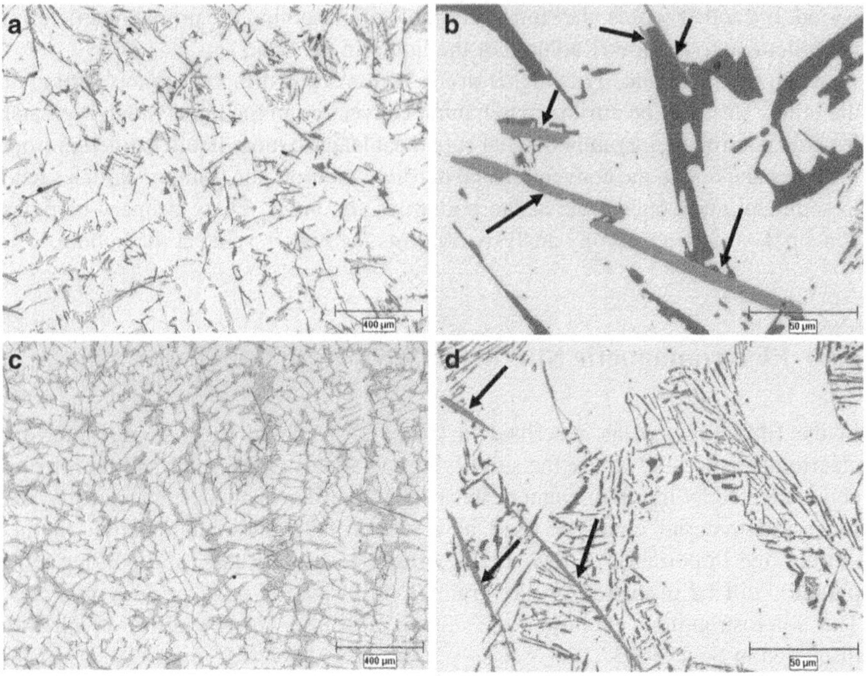

Fig. 5.30 Correlation between microstructure and cooling rate (as-cast, non-stirred): (**a**) and (**b**) sand mold, (**c**) and (**d**) copper mold (pouring temperature 690 °C), iron intermetallics arrowed

the thicker arms. The sand mold itself reduces the heat extraction rate and establishes a shallow temperature gradient across the bulk liquid. Such low temperature gradient promotes nuclei survival with uniform distribution and their eventual freeze off as equiaxed grains.

From SSM processing perspective, microstructural evolution is mainly due to the mechanical and/or thermo-mechanical fragmentation of dendrites and creation of multiple nucleation within the bulk liquid [12]. The application of electromagnetic stirring and the resulting forced convection of the bulk liquid shown to have generated fragmentation of the dendrites either by mechanical disintegration or by dendrite arms root remelting due to the thermal and solutal convections [12, 17, 18]. These fragmented particles are the most favorite nucleation sites for the primary α-Al phase since they are regarded as fresh nucleants having no disregistry with the growing material. This means heterogeneous nucleation is taking place simultaneously throughout the bulk liquid during the course of continuous cooling of the melt under forced convection. The growth mechanism, however, is based on the floating dendrite fragments coarsening with continuously cooled and stirred alloy. With continuous shearing of the melt, dendritic morphology changes to "rosettes" and/or "globules." With further stirring, particle ripening becomes dominant, driven by the reduction of interfacial free energy.

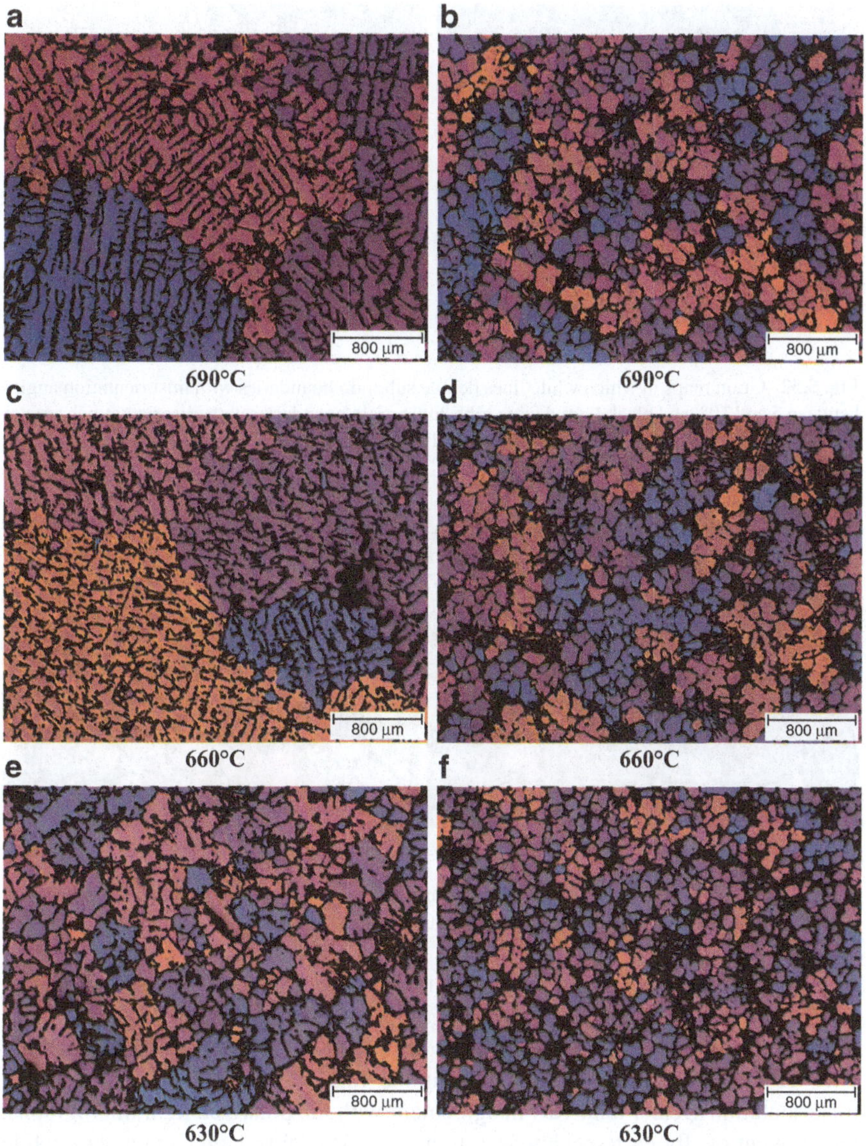

Fig. 5.31 Polarized light micrographs showing the effect of pouring temperature (690, 660, and 630 °C) and stirring on the grain and globule size variations in the sand mold casting, (**a**), (**c**), and (**e**) conventional and (**b**), (**d**), and (**f**) EM-stirred

Results of the EBSD grain mapping of specimens cast at 660 °C pouring temperature are given in Fig. 5.32 [19].[1] Comparable to the polarized optical

---

[1]Specimens were examined using a Hitachi S3000N scanning electron microscope (SEM) operated at 20 kV having an HKL Nordlys EBSD detector attachment and Channel 5 software.

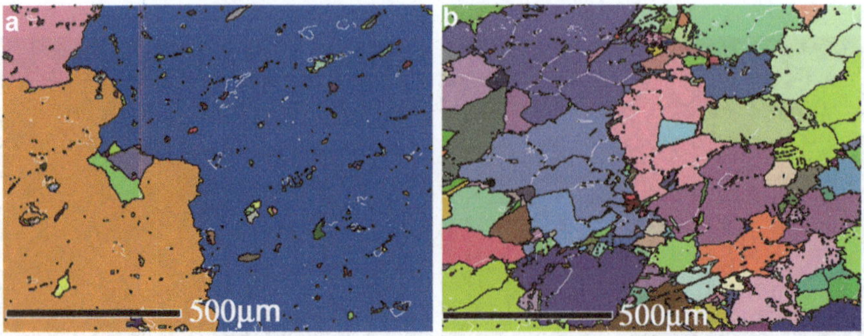

Fig. 5.32 Grain maps in which white lines denote subgrain boundaries with misorientation angles between 5 and 10°, and black lines denote high angle grain boundaries with misorientations greater than 10° (cast at 660 °C pouring temperature) [19]

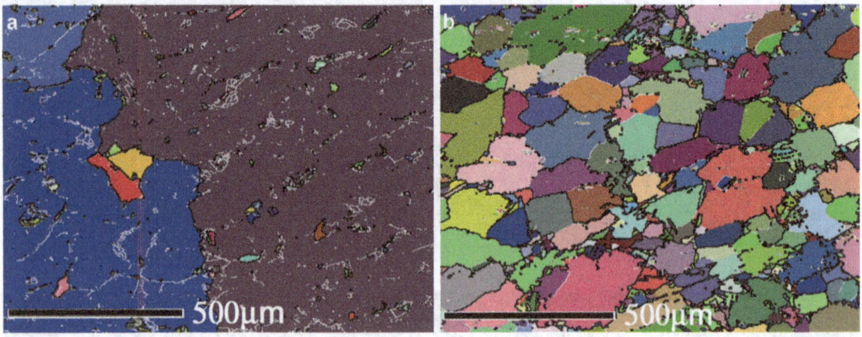

Fig. 5.33 EBSD grain maps, white and black lines indicate grain boundaries with misorientation angles of greater than 1.5 and 10°, respectively (cast at 660 °C pouring temperature) [19]

micrographs, here the grains are shown in unique colors to depict general microstructures, such that no two neighboring grains have the same color. The white lines however denote the subgrain boundaries with misorientation angles ranging from 5 to 10°, and the black lines represent the high angle grain boundaries with misorientation greater than 10°. In agreement with the polarized light micrographs, the conventionally cast specimens exhibit a considerably larger grain size, while stirring has effectively reduced the grain size (the small particles dispersed within the unstirred sample are parts of dendrites/iron intermetallic/eutectic mush cut by the polished surface. This could easily be confirmed with the polarized light micrograph).

Many of the subgrains are not seen in Fig. 5.32 since the maps only reveal subgrains having a misorientation angle greater than 5°. If the minimum misorientation value to define grain boundaries is reduced to 1.5°, the EBSD grain map of the EM-stirred billets show numerous grain or subgrain boundaries as shown in Fig. 5.33 (considering the fact that white and black lines indicate grain

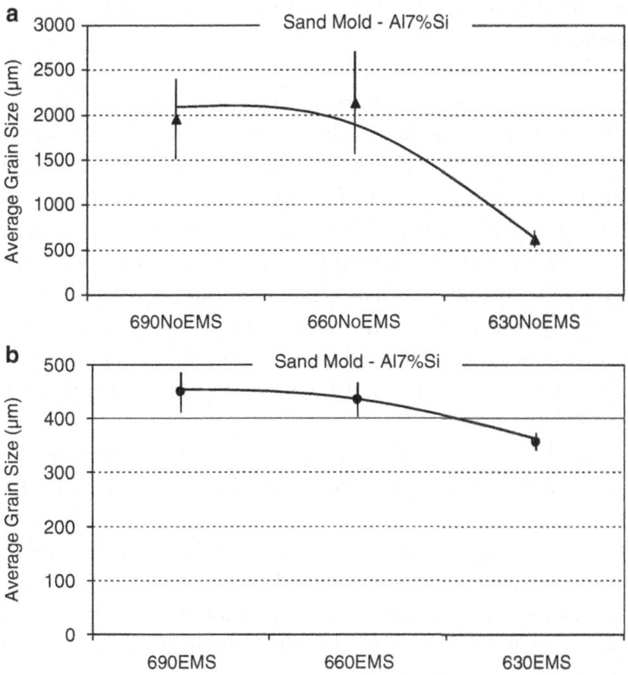

Fig. 5.34 Grain size measurements in: (**a**) conventional, and (**b**) EMS-stirred samples, sand mold casting [19]

boundaries with misorientation angles of greater than 1.5 and 10°, respectively). As with the previous maps, the conventionally cast sample exhibits only a minor change in grain-boundary structure with almost no new subgrains, while the stirred sample possesses a considerably finer grain structure and a larger number of defined subgrain boundaries with the selected lower misorientation limit, 1.5°. Number of subgrain boundaries is a key factor during the reheating process as the reheating time is strongly dependent on the type and quantity of grain boundaries.

Figure 5.34a shows the grain size variation with different superheats in the conventional samples where no stirring was applied. At 690 °C pouring temperature, the structure is fully dendritic and the grain structure is quite coarse with an average size of ~2000 μm.

By reducing the pouring temperature to 660 °C, the grain size was approximately the same and no significant differences detected in the structure. At 630 °C, the columnar dendritic structure changed to equiaxed and the grain size has drastically reduced by almost 70 %. The hefty reduction is due to the overall low temperature of the bulk liquid which is the main drive for establishing a shallow temperature gradient in the mold upon pouring as discussed in details earlier in this chapter. As pointed out, the potential nuclei are therefore able to form almost everywhere within the melt with greater survival rate, due to low temperature of pouring, to promote the formation of equiaxed grains. As per experimental observation, it is believed the starting point of the grain size reduction should lie between 660 and 630 °C temperature range.

Grain size measurements in the stir-cast samples are shown in Fig. 5.34b (note that the grain size was measured using polarized light micrographs, i.e., the measurement was carried out by counting a group of adjacent identical colored α-Al particles as one grain). At first glance, it is clear that the average size is reduced dramatically by stirring. As the superheat decreased, the microstructure becomes more globular and the entire grain size reduction for stirred alloys is ~20 % at the lowest pouring temperature (Fig. 5.34b). This reduction in size is smaller than that of conventional casting and this could be attributed to the effect of vigorous stirring on the particle size which overshadowed the effect of pouring temperature (it is interesting that the standard deviation is tighter for low pouring temperature in both conventional and EM-stirred samples which interprets as the consistency of the process).

In conventional casting, there is always a competition between equiaxed and columnar zones and the outcome depends on the degree and extent of different factors such as temperature gradient, the degree of constitutional undercooling, and the velocity of the columnar front. Hunt [20] analyzed the growth of equiaxed grains ahead of columnar front and defined the columnar to equiaxed transition (CET) parameter and shown that CET is promoted by increasing the alloying content, nucleant particles, and lowering the superheat and critical undercooling during heterogeneous nucleation.

Results of the conventional casting in this work also confirm that by lowering the pouring temperature, a reduction in the columnar growth is observed and consequently equiaxed grains are formed. On the other hand, electromagnetically induced forced flow is capable of refining the grain structure, and promoting a greater proportion of equiaxed zone. The CET in conventional sand casting is occurred at a temperature between 660 °C and 630 °C while by applying the EMS, CET is promoted and raised to the higher pouring temperature, 690 °C. In other words, a comparable microstructure to that of the low pouring temperature conventional casting could be achieved at high pouring temperature with application of vigorous stirring. The added bonus is the formation of more globular particles. This type of the refinement is attributed to the disintegration of dendrites and the subsequent transport and dispersion of the fragments in the bulk liquid.

As explained in Chap. 3, one theory for grain multiplication in the stirred case is bending the dendrite arms plastically under the shear forces created by the stirring action and "dendrite arm fragmentation" was proposed to explain grain multiplication [21, 22]. Plastic deformation is expected to generate dislocations, particularly one sign dislocations to accommodate the curvature with high elastic energy which can be reduced by dislocation migration and rearrangement into lower energy configurations; forming subgrains and ultimately new grain boundaries. These boundaries have a critical role in the degeneration of primary dendrites. The density of dislocations depends on the stirring condition and thermal characteristics of the semi-solid slurry. Due to the thermal condition of the SSM slurry, the dislocations are expected to rearrange themselves into walls (dislocation climb)

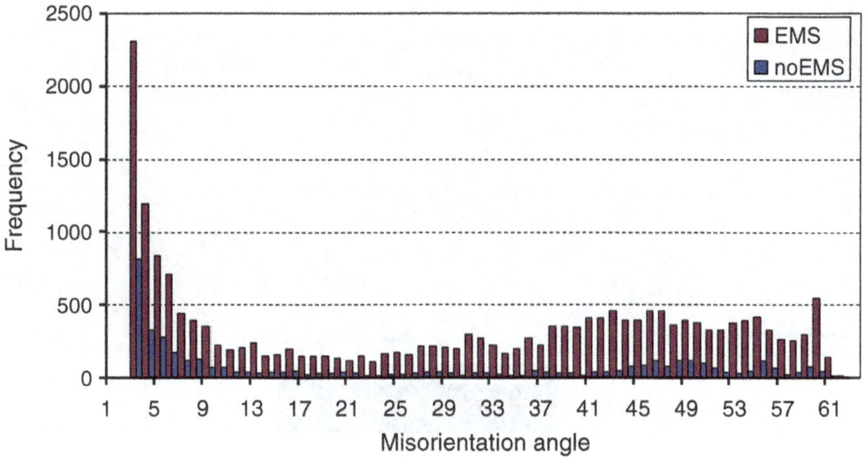

Fig. 5.35 Comparison of frequency of grain boundaries versus degree of misorientation, with/without stirring, 660 °C pouring temperature [19]

with lower total elastic strain energy than the more random dislocation arrangement, i.e., the formation of subgrain boundaries and the accompanying intragranular misorientation.

This theory is confirmed by electron backscatter diffraction (EBSD) analysis. For this objective, specimens of two sets of casting, 660 °C, poured in the sand mold with/without stirring were chosen and prepared for EBSD analysis. Figure 5.35 shows the typical grain boundary misorientation angle histograms for the two specimens [19]. The EM-stirred sample contains a greater concentration of grain boundaries across the misorientation range, particularly in the subgrain boundary region, i.e., <10° misorientation (the histogram is a standard misorientation angle distribution histogram, with the "frequency" on the Y-axis as the number of misorientations). In this case, each measured misorientation across the grain boundary counts, meaning that the misorientation between each neighboring pixel pair straddling a grain boundary counts. So, a perfectly straight "grain boundary" that is 10 pixels long will have 10 pairs of neighboring pixels, more if the boundary is not perfectly vertical or horizontal (since the grain boundary will lie on two or more sides of at least some pixels).

The backscattered electron micrographs with the superimposed grain boundary map in Fig. 5.36 confirm the histograms of Fig. 5.35 where both the fraction of low and high angle boundaries increased in the stirred sample (for the histogram, the sample areas analyzed are the same and therefore the higher density of boundaries could be another indication of a greater concentration of dislocations in the EM-stirred sample).

The TEM micrographs shown in Fig. 5.37 are from conventional and stirred samples. In the conventional condition, dislocations are due to the thermal stresses during liquid to solid transformation and distributed randomly (Fig. 5.37a) while with stirring, not only the density of dislocations is increased, but also there are tangled and rearranged dislocations in the form of subgrain boundaries as depicted

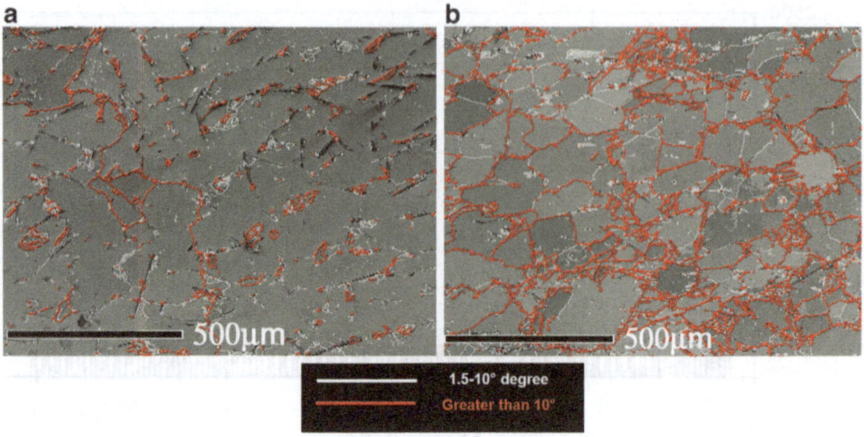

Fig. 5.36 EBSD grain boundary mapping, white lines denote subgrain boundaries with misorientation angles between 1.5 and 10°, and *red lines* denote high angle grain boundaries with misorientations greater than 10° (cast at 660 °C pouring temperature) [19]

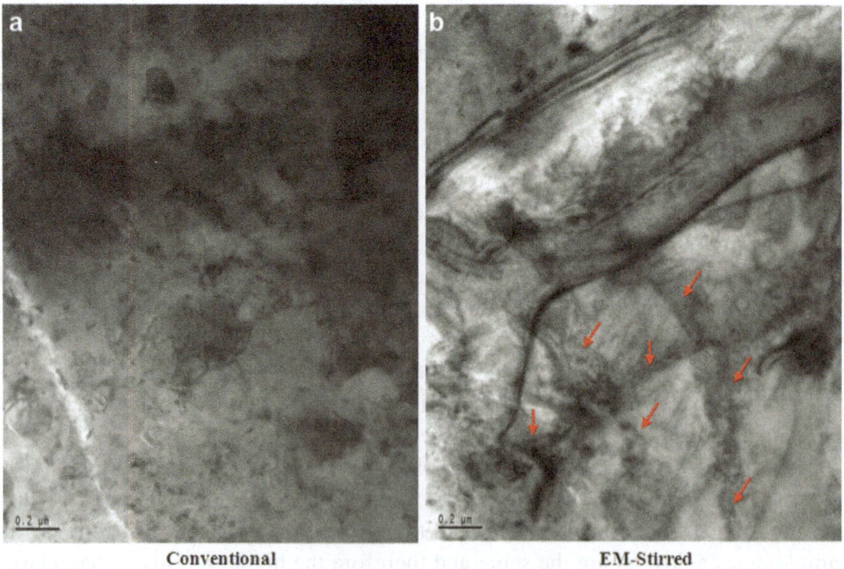

Fig. 5.37 Bright field TEM images showing: (**a**) conventional sample, and (**b**) stirred sample with high density of dislocations and subgrain boundaries, both were poured at 660 °C pouring temperature (*arrows* showing some of the sub grain boundaries) [19]

by the arrows on the micrograph, Fig. 5.37b (the findings emphasize on the ease of operation when EBSD is used in conjunction with bulk specimens in contrast to the complexity of specimen preparation and high degree image interpretation skill required with TEM application).[2]

5.4.1.2 Copper Mold

Figure 5.38 shows the polarized light micrographs of the copper mold cast samples. The comparison between the two sets of micrographs in Figs. 5.31 and 5.38 clearly displays the effect of cooling rate. Thinner, more compacted dendrites, and smaller dendrite arm spacing "DAS" are the benefits of high cooling rate of the copper mold. As for the influence of pouring temperature, similar trends to those of the sand cast are apparent for the copper mold as well, where reduction in pouring temperature encourages the formation of more equiaxed structure, see Figs. 5.38a, c, and e. The key point is to increase the survival rate of nuclei with reducing pouring temperature and changes in the heat flow direction with increasing the number of stable solid particles in the melt which could act as heat sinks. Here the added bonus is the shorter solidification time of copper mold samples which finally leads to smaller dendrite size (and consequently smaller DAS).

By stirring, the morphology of the primary α-Al phase changes to rosette and globular forms. If the micrographs in Fig. 5.38b, d, f are examined closely, the stirred structure is dependent on the pouring temperature, as the pouring temperature controls the starting structure in the first place. Therefore, at higher pouring temperature, stirring has only caused mechanical disintegration of the dendrites and the resulting fragmented segments provide a rosette type structure. Further evolution of rosette structure is hindered by higher cooling rate of copper mold which does not permit complete fragmentation and/or dendrite coarsening to take place. At lower pouring temperature, the morphology of the primary particles is not affected by stirring, since the α-Al particles are already nondendritic. Stirring is more effective at higher superheat as it not only fragments the dendrites, but also removes the concentration gradient and therefore decelerating the growth. This is the main reason why the stirred billets are finer than the unstirred ones.

Another side issue is the sintering/coalescence effect which is apparent in higher magnifications. As shown in Fig. 5.39, particles attached together and form a new one. This phenomenon was elaborated earlier in Sect. 4.3.

The average grain size in unstirred and stirred conditions is presented in Fig. 5.40.

---

[2]In a TEM image, individual dislocations can be seen by the contrast variations due to differences in the intensity of the diffracted beam around the strain fields generated by dislocations. However in the EBSD technique, the local misorientation defines "how dislocations density changes." In TEM, information is collected only from a small area of the sample and may be less representative of the sample as a whole. Moreover, it is also possible to have the nature of the defects altered during sectioning and thin foil specimen preparation for TEM analysis.

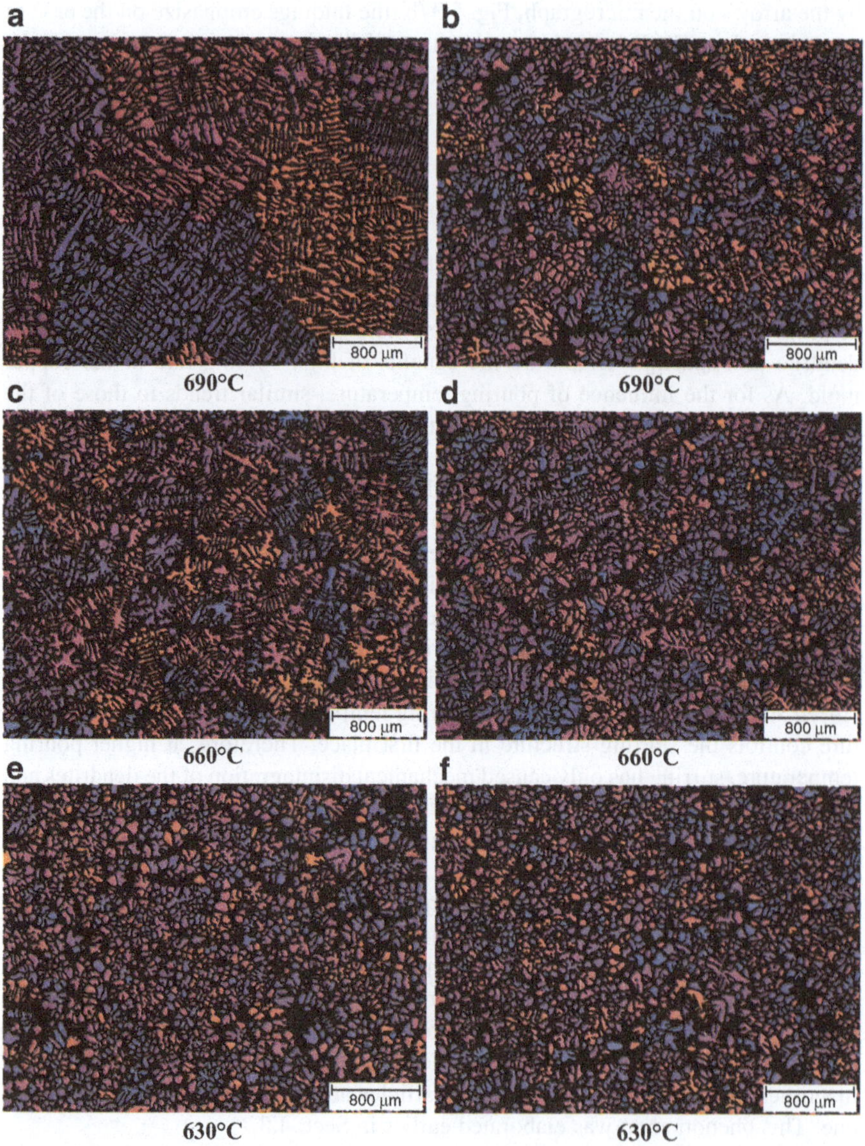

Fig. 5.38 Polarized light micrographs showing the effect of pouring temperature and stirring on the grain and globule size variations in copper mold casting, (**a**), (**c**), and (**e**) conventional and (**b**), (**d**), and (**f**) EMS-stirred

- In conventional castings, there is a sudden reduction in grain size from 690 to 660 °C where reduction in size is about 72 %. By comparing the results with sand mold, Fig. 5.34a, columnar-equiaxed transition, CET, shifted to higher temperatures by increasing the cooling rate. It means that in sand casting, the CET occurs at lower temperatures due to lower cooling rate but with better cooling

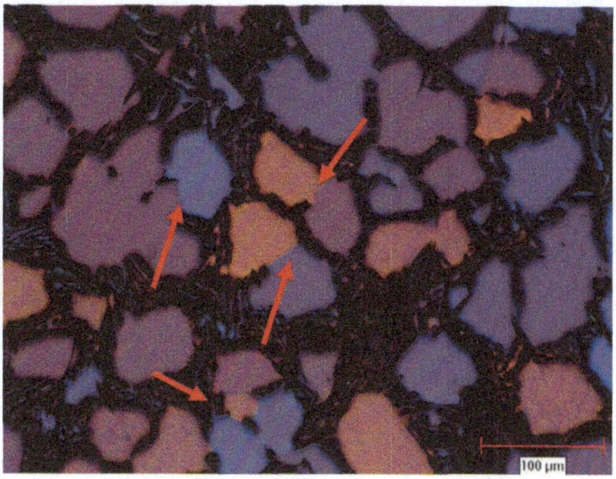

Fig. 5.39 Examples of sintering (*red arrowed*), Copper mold, stirred @630 °C

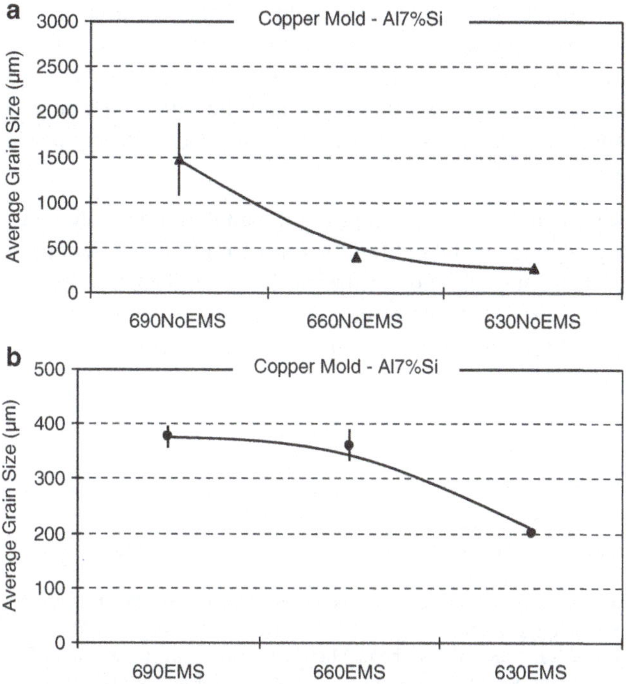

Fig. 5.40 Grain size measurements in: (**a**) conventional and (**b**) EMS-stirred samples, copper mold casting [23]

conditions the CET shifts to higher temperatures, which in this case is about 660 °C. There is also a larger error bar related to the billets cast at 690 °C which indicates a wider range of grain size in the sample while by lowering the temperature, this variation becomes minimum, uniform grain distribution.

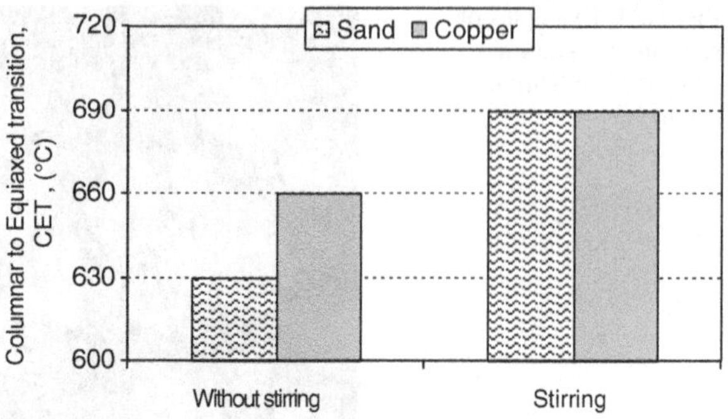

Fig. 5.41 Variation of CET by pouring temperature and application of the stirring

- There is a similar refinement trend in the stirred conditions regardless of the cooling rate, Figs. 5.34b and 5.40b. However, the grain and globule size values shifted to lower ones by increasing the cooling rate.
- Grain size reduction due to the pouring temperature in the EMS treated alloy is not as significant (46 % reduction) as that of the unstirred alloy (81 %). It is believed, stirring and pouring temperature have the same function and that is to increase the number of potential nucleants, former does it mechanically by generating large number of fragments and the latter thermally. Therefore, if low pouring temperature becomes a restricting factor during production, it can be compensated by stirring.

As a general conclusion, by increasing the cooling rate, columnar evolution to equiaxed is accelerated and therefore for similar pouring temperature, the structure is different. By stirring the alloy in the mushy zone, dendritic grains transform to globules and the CET shifts up to higher temperatures, Fig. 5.41 (a similar effect may be proposed for dendrite to globule transition, "DGT").

References

1. M.C. Flemings, Behavior of metal alloys in the semi-solid state. Metal. Trans. A **22A**, 952–981 (1991)
2. D.H. Kirkwood, Semi-solid metal processing. Int. Mater. Rev. **39**(5), 173–189 (1994)
3. Z. Fan, Semisolid metal processing. Int. Mater. Rev. **47**(2), 49–85 (2002)
4. O. Lashkari, S. Nafisi, R. Ghomashchi, Microstructural characterization of Rheo-Cast Billets prepared by variant pouring temperatures. J. Mater. Sci. Eng. A **441**, 49–59 (2006)
5. V. Laxmanan, M.C. Flemings, Deformation of semi-solid Sn-15%Pb alloy. Metal. Trans. A **11A**, 1927–1937 (1980)
6. J.A. Yurko, M.C. Flemings, Rheology and microstructure of semi solid aluminum alloys compressed in drop forge viscometer. Metal. Trans. A **33A**, 2737–2746 (2002)
7. L. Azzi, F. Ajersch, Development of aluminum-base alloys for forming in semi solid state. in *Trans Al Conference* (June 2002, Lyon, France), 23–33

8. J. Langlais, A. Lemieux, B. Kulunk, Impact of the SEED processing parameters on the microstructure and resulting mechanical properties of A356 alloy castings. in *AFS Transactions* (2002), paper 06-125
9. J. Langlais, A. Lemieux, The SEED technology for semi-solid processing of aluminum alloys: a metallurgical and process overview. Solid State Phenomena **116–117**, 472–477 (2006)
10. S. Nafisi, O. Lashkari, R. Ghomashchi, J. Langlais, B. Kulunk, The SEED technology: a new generation in rheocasting. in *CIM-Light Metals Conference* (Calgary, Canada, 2005), 359–371
11. J. Jorstad, Q.Y. Pan, D. Apelian, Effects of key variables during rheocastings—fraction solid and flow velocity. in *111th Metal Casting Congress* (NADCA, 2007)
12. M.C. Flemings, *Solidification Processing* (McGraw-Hill, New York, 1974)
13. O. Lashkari, S. Nafisi, J. Langlais, R. Ghomashchi, The effect of partial decanting on the chemical composition of semi-solid hypo-eutectic Al–Si alloys during solidification. J. Mater. Proc. Tech. **182**, 95–100 (2007)
14. S. Nafisi, O. Lashkari, J. Langlais, R. Ghomashchi, The impact of partial drainage on chemical composition of 356 Al-Si alloy. Mater. Sci. Forum **519–521**, 1765–1770 (2006)
15. ASTM International Standard Worldwide, Volume 02.02, Aluminum & Magnesium Alloys (2004), 78–92
16. D.G. McCartney, Grain refining of aluminum and its alloys using inoculants. Int. Mater. Rev. **34**(5), 247–260 (1989)
17. A. Hellawell, Grain evolution in conventional and Rheo-castings. in *4th International Conference on Semi-Solid Processing of Alloys and Composites* (Sheffield, England, 1996), 60–65
18. R.D. Doherty, H.I. Lee, E.A. Feest, Microstructure of stir-cast metals. Mater. Sci. Eng. **A65**, 181–189 (1984)
19. S. Nafisi, J. Szpunar, H. Vali, R. Ghomashchi, Grain misorientation in Thixo-billets prepared by melt stirring. Mater Charact **60**, 938–945 (2009)
20. J.D. Hunt, Steady state columnar and equiaxed growth of dendrites and eutectic. Mater. Sci. Eng. **A65**, 75–83 (1984)
21. A. Vogel, R.D. Doherty, B. Cantor, Stir-cast microstructure and slow crack growth. in *Proceedings of the Solidification and Casting of Metals* (The Metals Society, London, 1979), 518–525
22. A. Vogel, B. Cantor, Stability of a spherical particle growing from a stirred melt. J. Cryst. Growth **37**, 309–316 (1977)
23. S. Nafisi, D. Emadi, M.T. Shehata, R. Ghomashchi, Effects of electro-magnetic stirring and superheat on the microstructural characteristics of Al-Si-Fe alloy. J. Mater. Sci. Eng. A **432**, 71–83 (2006)

Chapter 6
Rheocasting-Melt Treatment

Abstract The process of melt treatment (grain refining and modification) is a routine practice for conventional Al–Si alloys casting. For SSM processes, it is worthwhile to see the effect of grain refiner and modification on the primary α-Al dendrites and eutectic silicon. This chapter provides a detailed analysis to verify the effect of melt treatment on the microstructure of as-cast SSM billets. The mechanism of grain refining and modification is examined through thermal analysis, rheological studies, and quantitative metallographic characterization for a range of grain refining master alloys and Sr-based modifiers in different rheocasting processes.

6.1 Melt Treatment in Semi-Solid Casting, A Review

The process of melt treatment which covers the practice of grain refining and modification is routinely carried out in conventional Al–Si alloys casting to refine the primary α-Al dendrites and eutectic silicon to strengthen the as-cast structure with improved toughness. In SSM processes, however, the melt treatment is not a routine practice and the information provided here should verify if it needs to be implemented and the resulted improvements warrant melt treatment. In the following sections, both grain refining and modification of Al–Si alloys using Ti, B, and Sr-based master alloys will be discussed.

6.1.1 Effects of Grain Refining on Semi-Solid Structures

Addition of grain refining agents to molten aluminum is now a prerequisite for successful and economic casting. Grain refinement provides an equiaxed grain structure, uniform mechanical properties, and better machinability; it also reduces shrinkage, increases resistance to hot tearing, and improves feeding [1].

In the SSM processing of aluminum alloys, especially Al–Si alloys, the size of the primary α-Al particles plays an important role in the final mechanical properties. Most of the alloying elements reduce the size of α-Al grains but one of the effective methods for reducing the size of grains is the application of grain refiner. It

S. Nafisi, R. Ghomashchi, *Semi-Solid Processing of Aluminum Alloys*,
DOI 10.1007/978-3-319-40335-9_6

was reported that the best flow behavior for semi-solid metal processing is achieved if the raw material structure consists of finer and more rounded globular grains [2].

In the science of semi-solid, there is not much work reported on the direct incorporation of grain refiners during the processing and preparation of slurries and most of the works are related to the post thermal treatment of the cast billets. As a result, the current section is to cover the research works on the two main groups of SSM processes: thixo and rheocasting.

6.1.1.1 Grain Refining During Thixocasting

A typical grain refining curve shown in Fig. 6.1 presents the grain size of primary phase in relation to the grain refiner addition. At lower addition level, the grain size decreases significantly; however, above a certain level of titanium and/or boron addition, C_1, which in this case is around 600 ppm, only slight improvements in grain size are achieved. Due to the technical and economic reasons, the usual addition level is therefore as low as possible; this is in the range of C_1 for cast alloys and even below C_1 for wrought alloys. If Direct Chill (DC) cast bars of this type are reheated for SSM processing, a long reheating time is required to produce globular grains by dendrite root remelting, ripening, and coalescence as explained in details in Chap. 3. However, some preliminary experiments indicated that higher addition levels of grain refining agents may promote the formation of globular grains during reheating, even though it does not significantly change the as-cast structure [3, 4].

Gabathuler et al. [2] by using different refiners, EMS stirring, and ultrasonic treatment, and observing the microstructure as well as the forces needed to extrude SSM AlSi7Mg0.3 billets through a die (ingot with 75 mm diameter), claimed that that Al5Ti1B and Al2.5Ti2.5B master alloys are not efficient enough for the production of thixo-billet. This is because the desired grain size for an adequate

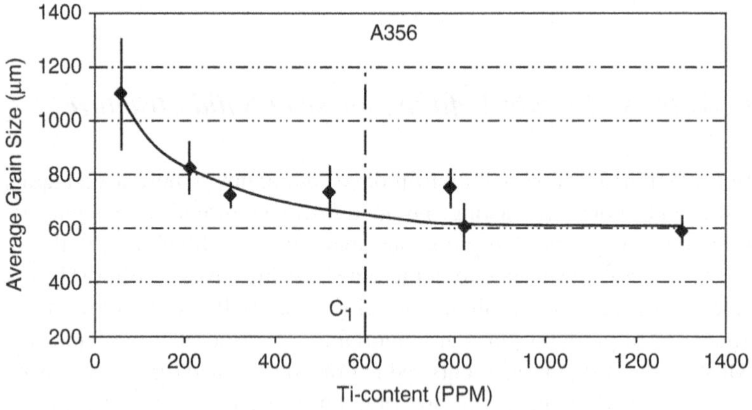

Fig. 6.1 Grain refining of A356 alloy with Al5Ti1B master alloy with different additional levels

Table 6.1 Effect of AlTiB grain refiner on the grain size of DC cast Al7Si0.3Mg [2]

Grain refiner	Concentration [%]	Grain diameter [μm] Near ingot surface	Ingot center	Homogeneity of grain size
Without	–	≃ 5000	≃ 2000	feather crystals
AlTi5B1	0.02	350	500	Good
	0.1	300	500	Good
	0.5	240	420	Very good
AlTi2.5B2.5	0.02	400	450	Good/poor
	0.1	300	350	Good
	0.5	220	300	Very good

thixotropic material is less than 150 μm and the measured grain sizes were never below 200 μm (Table 6.1).

Wan et al. [3, 4] added Al5Ti1B master alloy to the A356 and poured it in a water-cooled conical brass mold (80 mm dia.) and a preheated steel mold (50 mm dia.) to achieve different cooling rates for billet production. The samples were then heated in a medium frequency induction heating furnace to a temperature 5 °C above the eutectic/solidus temperature with varying heating rate and holding period (up to 15 min) and they were quenched after holding. The following results were concluded (Fig. 6.2):

- At a given addition rate, the grain size both in the as-cast and reheated conditions decreases with increasing the cooling rate
- Above approximately 0.025%Ti, the as-cast grain size is almost constant and are equiaxed dendritic
- The minimum reheating time for the formation of globular grains decreases with increasing addition level of refiner
- The grain size remains almost unchanged above the conventional addition percentage but the shape of the primary particles at higher additional levels can be changed from dendritic to globular by reheating within a shorter time
- It was stated that at an addition level of 0.26%Ti, the minimum reheating time is at an acceptable level (5 min) and the globular size is about 120 μm

Mertens et al. [5] examined the aluminum wrought alloy 6082 with different feedstock manufacturing and feeding routes. In the refined as-cast structure, the material already has a globular dendritic matrix. Heating to 635 °C and isothermal holding for 5 min induces a structure suitable for thixoforming with a mean grain diameter of about 90 μm approximately. After holding for 20 min at 640 °C, the α-Al phase is present in an almost spherical form, but the structure is slightly coarser (about 130 μm mean grain diameter) (Fig. 6.3).

Bergsma et al. [6, 7] have observed the combined effects of high solidification rates and grain refining on 76.2 mm diameter billets of A356 and A357. The average grain size before reheating was 120 μm and samples were reheated to

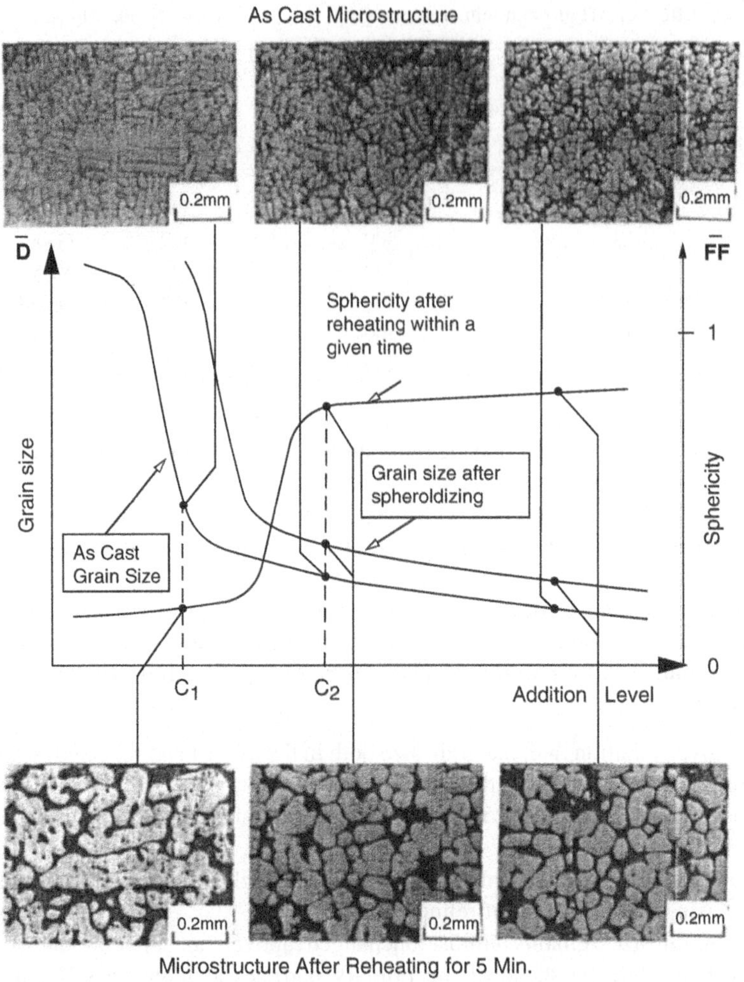

Fig. 6.2 Grain size variation with refiner addition. The microstructures of A356 are shown as example. C_1 = conventional addition level and C_2 = critical addition level where after that, the sphericity remains almost constant [3]

588 °C and then quenched. The results show that combination of refiner addition and thixoforming results in spherical particles formation. Also with increasing the holding time at 588 °C, the particles tend to grow and become more spherical with a shape factor closer to one (Fig. 6.4).

Tahara et al. [8] also reported the effect of adding refiner to Al7Si3Cu-0.1 ~ 1% Fe. In their study, 0.2 %Ti was added to the alloy and poured into a metallic mold with 10 $°Cs^{-1}$ cooling rate. For semi-solid experiments, specimens were reheated at 587 °C for 10 min. The dendritic α-Al particles changed to globules and the grains coarsened slightly to 100 ~ 150 μm from an initial size of about 100 μm.

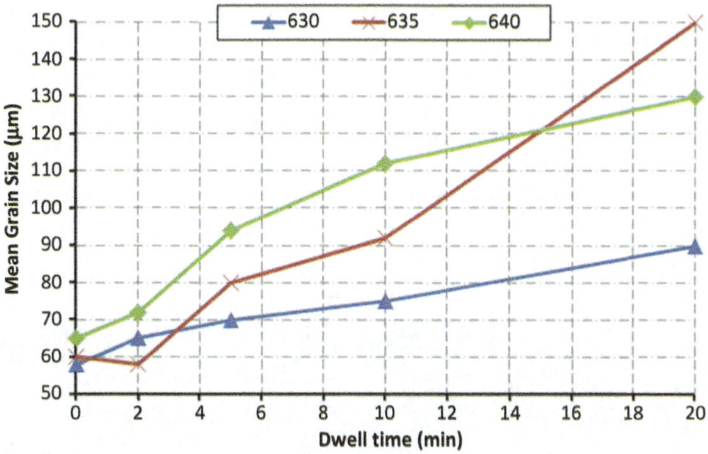

Fig. 6.3 Structural development of the grain refined material as a function of the temperature and holding time (adapted from [5])

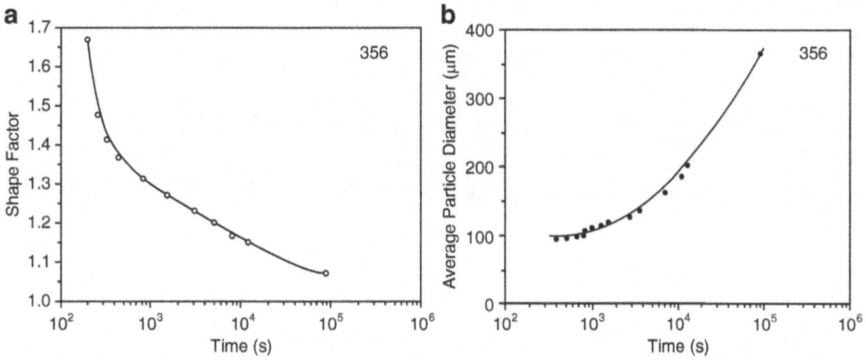

Fig. 6.4 (**a**) Shape factor as a function of holding time, (**b**) average 3D particle diameter versus holding time (shape factor was defined as reverse of sphericity, Sect. 4.3.2.5) [6]

Wang et al. [9–11] with experiments on Al7Si0.35 Mg in 50 mm steel mold at different pouring temperatures have shown that using both low pouring temperature and grain refining could optimize the structure during billet casting, partial remelting, and injecting. They confirmed that the addition of refiner imparts little effect on grain size for the casting condition used in their work but after isothermal holding at 580 °C, it was concluded that the refined samples attained a globular structure faster than the nonrefined alloy as a result of more rosette like structure of the refined sample.

Table 6.2 and Fig. 6.5 show the variation of grain and globule size before and after refiner addition and/or reheating. Grain refiner has an obvious effect on the structure both on the as-cast and reheated structures. The time for globularization

Table 6.2 Selected microstructural parameters measured in the as cast and reheated billets at 580 °C for 15 min [10]

As-cast microstructure			Reheated microstructure	
Casting condition	Structure	Grain size (μm)	Particle size (μm)	Morphology
650 °C	Fine-grained, dendritic	200	102	Spherical globular
650 °C + GR	Fine globular	160	95	Spherical globular

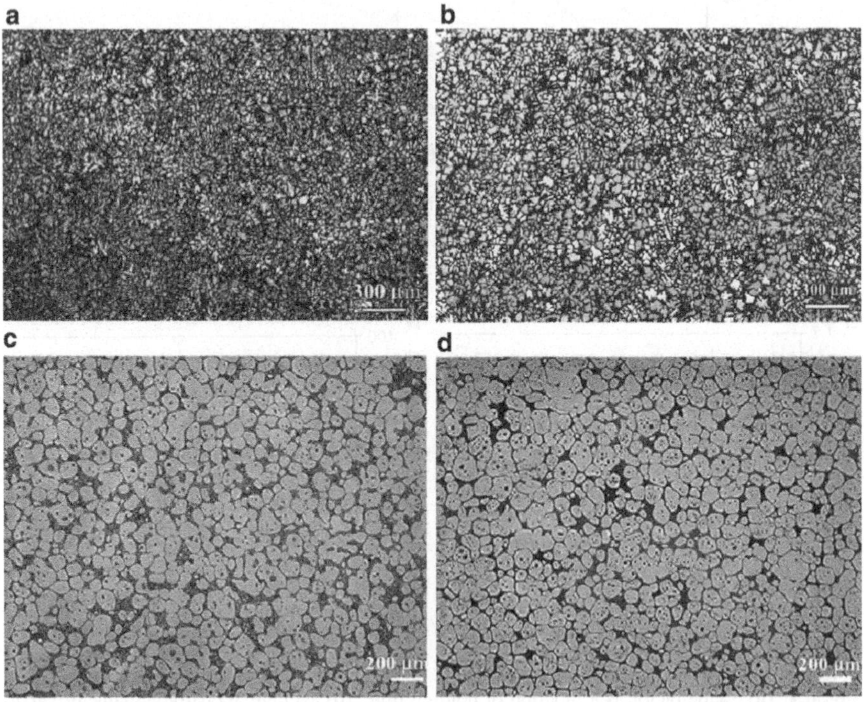

Fig. 6.5 Microstructures of: (**a**) as-cast 650 °C, (**b**) as-cast grain refined 650 °C, (**c**) and (**d**) partial remelted and isothermally held at 580 °C for 15 min from the initial microstructure of (**a**) and (**b**) respectively [10]

depends on the initial grain size and morphology of as-cast structure. As shown in Fig. 6.6, the globularization time decreased from 30 to 10 min when the as-cast grain size changed from 350 to 200 μm. For grain refined structure (shown as "fine globular structure"), the time decreased from 10 to 5 min. Therefore, the smaller grain size billets require lesser reheating time during thixocasting.

Also the shear behavior of the semi-solid material was measured using a direct shear test. Billets were prepared at 650 °C and grain refined with Al5TiB master alloy to reach 0.05%Ti. Samples partially remelted and quickly transferred to the test unit. They were isothermally held at 580 °C for 0.5, 2, 3.5, 5, and 10 min before

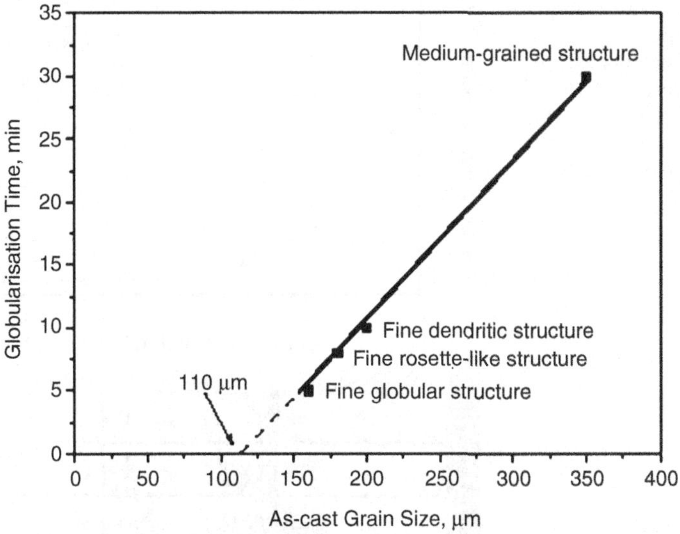

Fig. 6.6 Globularization time for different microstructures isothermally held at 580 °C [10]

shearing. Figure 6.7a shows the shear results for different treatments and holding times. For longer holding times, more globules form which results in lower shear strength. The reduction in shear strength is achieved at longer times for non-grain refined samples; 6 min of reheating to reach to 10 kPa shear strength in contrast to 3 min for the refined specimen to have the same shear strength. The difference in shear strength becomes negligible at longer reheating times [11].

At the next step, grain refined billets were isothermally held at 580 °C for 0.5, 2, and 5 min and then injected into a stepped die as explained in Sect. 2.2.3. All of the castings were completely filled. X-radiographs revealed all castings were free from internal porosity and the quality was reported to be much better than those made from non-grain refined billets (Fig. 6.7b) [11].

Pan et al. [12, 13] used boron refiners in the form of SiBloy trade mark and for comparison, a commercial TiB_2-refined ingot was examined as well. It was found that in non-reheated samples, boron-refined billets have smaller grain size compared to TiB_2 grain-refined sample (Fig. 6.8a, b). Moreover, B-refined billets have more uniform grain size distribution on the billet cross section than commercial Ti-B-refined samples. Figure 6.8 compares typical B-refined and Ti-B-refined billets before and after reheating. One can see the structure of the latter has much more entrapped liquid content, larger α-Al particles, and less spherical particles (comparing Fig. 6.8c, d). Quantitative metallography has shown that the B-refined billets have four times less entrapped liquid. Figure 6.9 shows the difference between Ti-B- and B-refined billets. It was concluded that the optimized boron-refined alloy is an improved candidate billet material for SSM processing.

In another study, EMS billets of 2024 and 7075 wrought aluminum alloys with/without refiner addition (in the form of Al5Ti1B) were thixocast and their

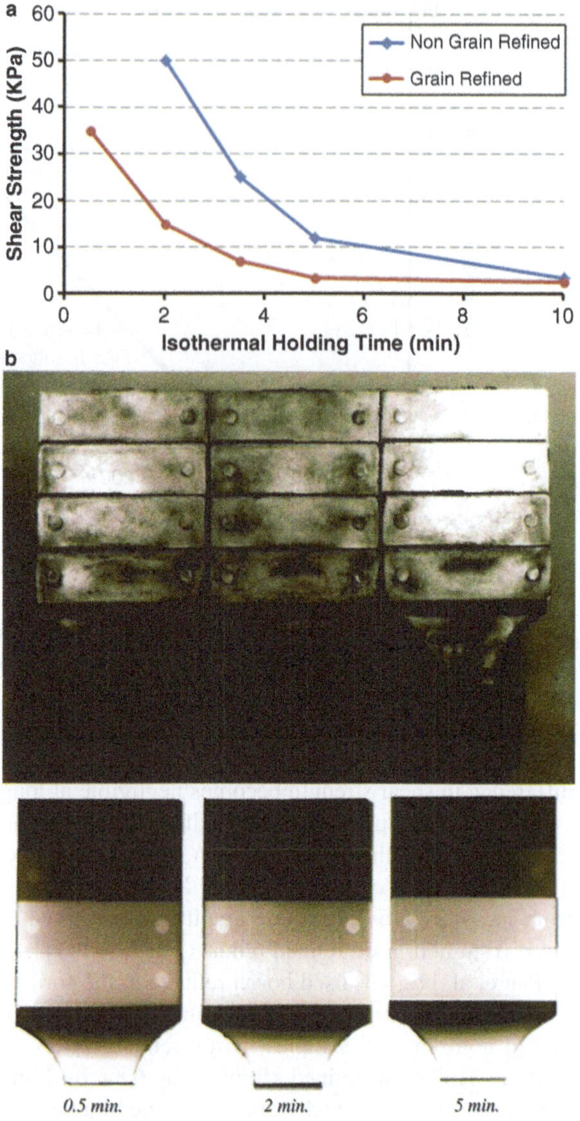

Fig. 6.7 (**a**) Variation of shear strength with isothermal holding time (adapted from [11]), (**b**) semi-solid castings made of grain refined billets and the X-radiographs of the cast pieces [11]

microstructural evolution, mechanical properties, and fluidity were investigated. It was found that by refiner addition, the grain size of as-cast billets and globule size in thixocast samples were decreased. The tensile properties of T6-treated specimens fabricated by EMS were superior to the permanent mold cast ones (for 7075 alloy was claimed even to be better than forged samples). The refined EMS billets had finer globules with uniform liquid distribution leading to better and even flow in the mold [14].

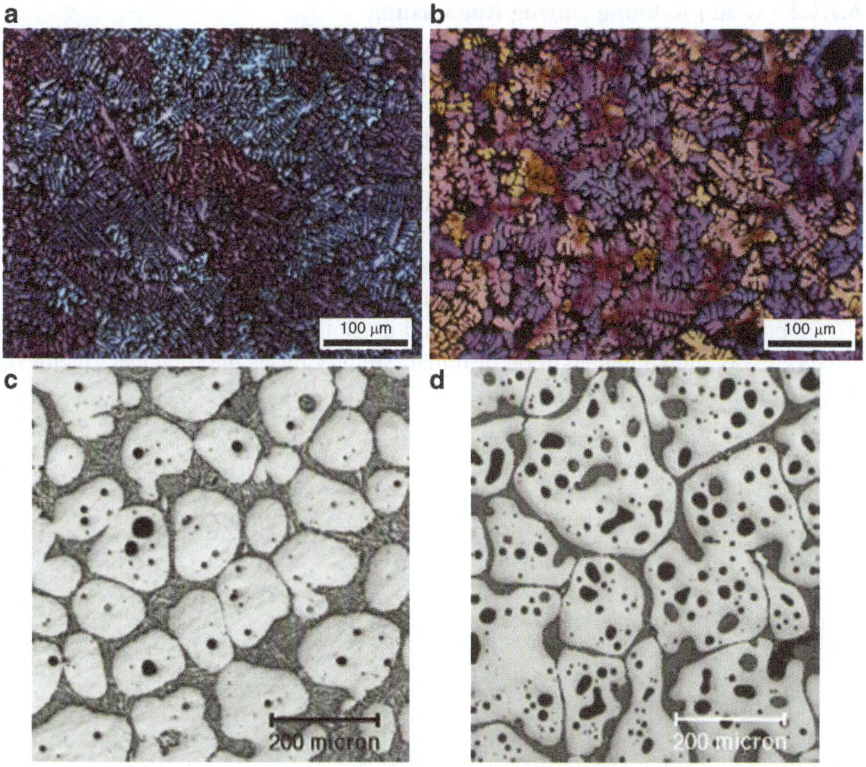

Fig. 6.8 Typical structures of the A356 alloy; (**a**) B-refined through SiBloy, (**b**) as-cast TiB_2 refined, followed by a comparison of semi-solid structures of B-refined billets (**c**) with those of TiB_2 refined (**d**) (reheated temperature: 585 °C) [12]

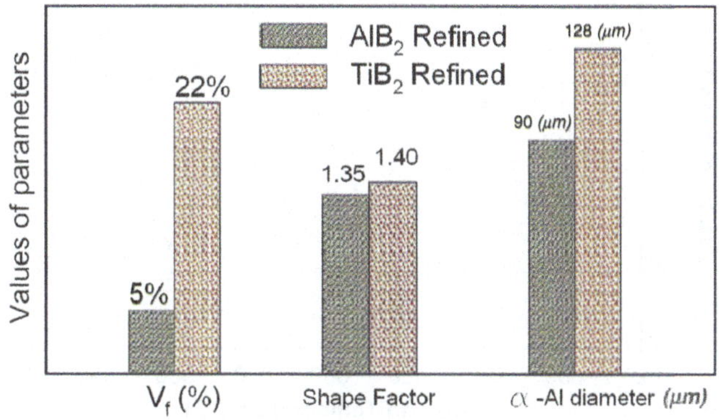

Fig. 6.9 Image analysis results (V_f: entrapped liquid content, shape factor was defined as reverse of sphericity, Sect. 4.3.2.5, and α-Al particle size) [12]

6.1.1.2 Grain Refining During Rheocasting

Grimmig et al. [15] used refiner and modifier to investigate the effect of melt treatment on the final quenched microstructure of A356 SSM material produced via cooling slope process. It was claimed that the addition of refiner did not change the structure and just what supplier added was enough. However, it is clear from the optical micrographs in their paper, Fig. 6.10, that the primary α-Al particles are finer, better distributed, and even α-Al percentage is increased.

In another study, prealloyed ingots of Al7Si0.3 Mg were melted, degassed, and refined by addition of Ti, and Ti-B (all the alloys were modified with Sr within 77–92 ppm range). Melts then were poured in shot sleeves and injected to the die cavity. It was shown that addition of refiner slightly raises the roundness factor of the α-Al particles (Fig. 6.11) [16].

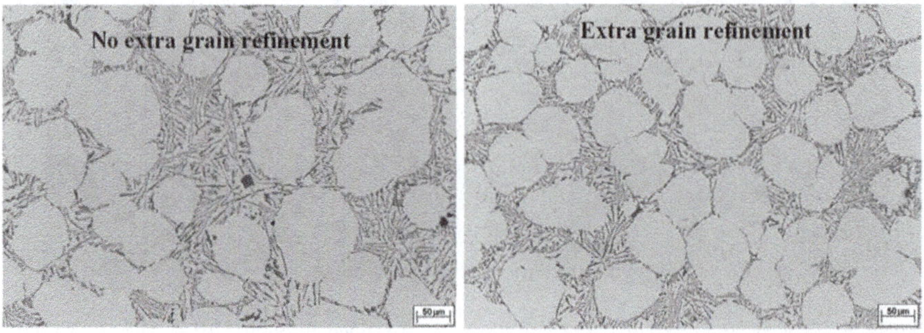

Fig. 6.10 Microstructures with and without inoculation (by addition of Al5Ti1B) in the quenched billets [15]

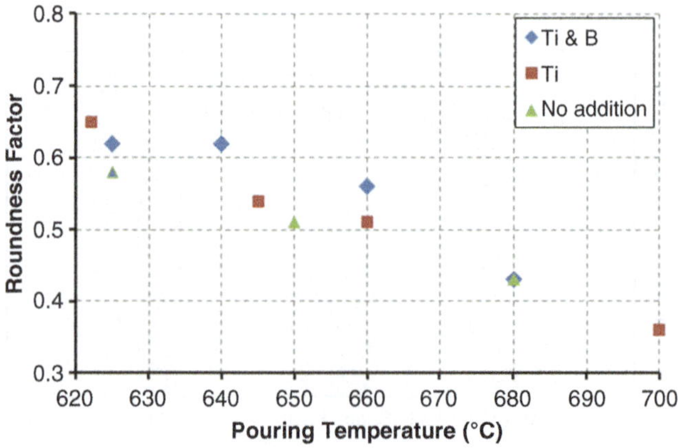

Fig. 6.11 Roundness factor of test specimen (roundness was defined as the ratio of alpha particles over an imaginary circle having the same circumference) (reproduced from [16])

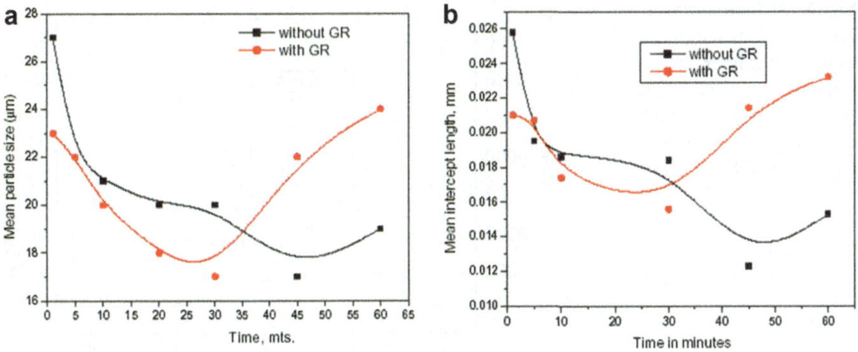

Fig. 6.12 Mean particle size (**a**) and mean intercept length (**b**) versus time for isothermal stirring at 615 °C [17]

In support of the above-mentioned observations, Sukumaran et al. [17] with isothermal rheocasting studies of Al5.2Si alloy have shown that the size of primary α-Al particles obtained with the refiner addition followed by stirring is comparatively smaller than that obtained without refiner. This was attributed to heterogeneous nucleation resulting in the formation of equiaxed grains suppressing the formation of dendritic/columnar structure.

The effects of isothermal stirring at a constant shear rate (210 s^{-1}) on the mean particle size and intercept length are shown in Fig. 6.12. As the isothermal shearing time increases, the particle size decreases initially, reaches a minimum value and increases again. In the case of refined alloy, the time that was required to obtain the minimum particle size is less than that for the nonrefined one. In both figures, the addition of refiner reduces the size of the particles and improves the degree of their roundness.

The relationship between viscosity and refinement efficiency with the addition of various AlTiB master alloys was studied by Yu and Liu [18]. The commercial aluminum alloy melted and treated by AlTiB refiners at 720 °C and the viscosity was measured by using a torsional oscillation viscometer after 30 min holding time (note that the experiments were carried out at liquid state, not within the semi-solid zone). Figure 6.13 shows the results indicating that refiners have a remarkable increase in viscosity of liquid metal compared to unrefined aluminum. The authors claimed that by addition of Al5Ti1B refiner, a Ti transition zone forms at the interface of TiB_2 and results in the increase of the viscosity followed by grain size reduction in solidification [18].

6.1.2 *Effects of Modification on Semi-Solid Structures*

It is well known that the growth mode of the Al-Si eutectic can be modified by the addition of modifiers such as strontium. With the presence of a modifying element, silicon growth mechanism is modified and its morphology changes to fibrous. In

Fig. 6.13 Increase in viscosity from sample 0 (without refiner) by addition of various refiners (second axis in the graph is percentage of viscosity increment) [18]

a

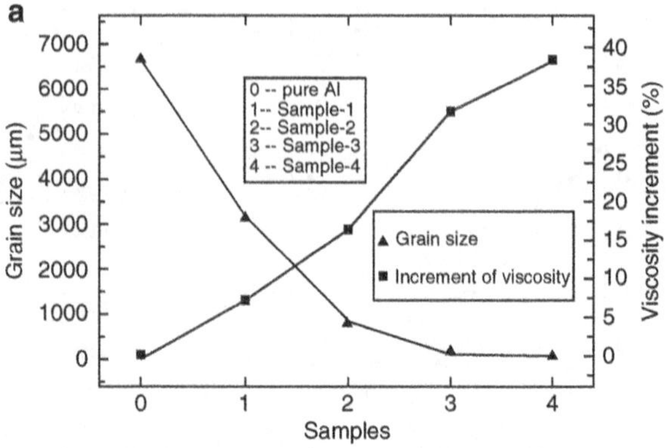

b

Alloys	Viscosity (mPa s)	Increment (%)
Pure Al	0.842277	–
Pure Al+0.3%Al–2.2Ti–IB (Sample-1)	0.901910	7.08
Pure Al+0.3%Al–2.2Ti–IB +0.084%Al–10Ti (Sample-2)	0.981337	16.51
Pure Al+0.3%Al–5Ti–IB (Sample-3)	1.110352	31.83
Pure Al+0.3%Al–5Ti–IB after strong distortion (Sample-4)	1.165869	38.42

general, the structural modification is concluded to be the result of a change in the growth mechanism from layer to continuous growth [19]. The morphological change of the Si phase results in improving mechanical properties, including higher strength and better ductility of the alloy.

One of the main aims of SSM processing, with specific emphasis on Al alloys is to obtain globular structure of the primary phase to improve mechanical properties and thixotropic behavior of the alloy. As a result, there is not much work reported on the other micro-constituents that usually form during solidification of Al alloys such as silicon. Fat-Halla [20] have studied the effect of modification on the stirring of commercial alloys A-S7G03 (7.1%Si, 0.3%Mg) and A-S4G (4.2%Si, 0.1%Mg). As expected, the eutectic silicon in the conventionally cast sample has a flake morphology with relatively large size. With stirring, not only the α-Al particles transformed to rosette and globules but also the eutectic silicon is broken into much smaller sizes. With Sr addition, the eutectic silicon becomes very fine and the optimum modification of stir-cast alloys was achieved at around 0.02%Sr. Higher amount of Sr addition results in overmodification which could have detrimental effects on the mechanical properties [20].

Figure 6.14 shows stress–strain curves for the unmodified and modified stir-cast 356 alloy. The yield strength was reported almost the same value of 110 MPa for both cases. The ultimate tensile strength (UTS) has significantly increased from 160 to 210 MPa with modification. Moreover, the percentage elongation was increased remarkably from 2.75 % for the unmodified alloy to 15 % for the modified

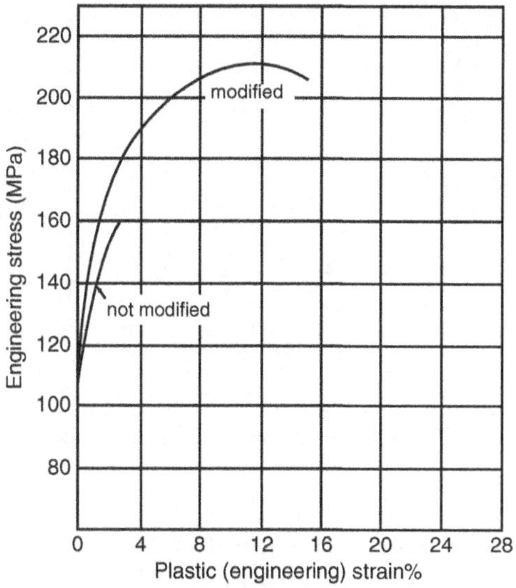

Fig. 6.14 Stress–strain diagram for stir-cast, modified, and unmodified A-S7G03 alloy [20]

one. Furthermore, the fracture surfaces, as examined by SEM, showed dimple and smooth ripple patterns, for the modified stir-cast alloys, reflecting the high ductility [20].

Loue and Suéry [21] performed some experiments with direct chill and permanent mold casting using EMS, and addition of Sr to Al7Si (with 0.3 and 0.6 %Mg) alloy. They have investigated the effect of isothermal holding at 580 °C on the resulting microstructure and shown that with addition of Sr, the morphology of eutectic silicon has changed and the grain size has decreased slightly. It was also found that the kinetics of remelting of the eutectic phase was not altered by the Sr modification. In addition, the presence of Sr in the liquid phase did not change the coarsening characteristics of the solid phase.

Jung et al. [22] investigated the morphological changes of primary Al phase in A356 alloy by using EMS and Sr modification. It was found that the optimum current for globalization is 15A. As well, the optimum addition of Sr was reported to be 50 ppm. Above this amount, the eutectic silicon was well modified but the average size of primary α-Al phase increased. It was suggested that the increase in the α-Al particle size is due to the extended mushy zone resulting from Sr addition. The average size of the eutectic Si for the stirred condition is slightly smaller than that of the non-stirred alloy. This was assumed not directly related to electromagnetic stirring and is probably because of uniform distribution of the modifier due to the stirring (Fig. 6.15) [22].

Grimmig et al. [15] with cooling slope process and addition of 175 ppm Sr to A356 alloy have shown that Sr addition results on changing the Si morphology from lamellar (flake) to fibrous and this effect fades during holding time.

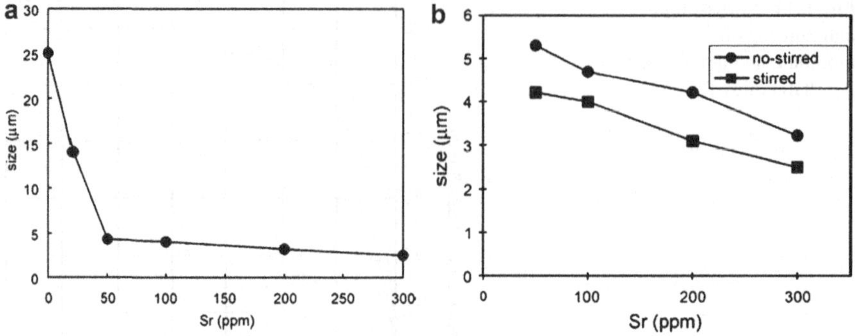

Fig. 6.15 (**a**) Variation of the average size of the eutectic Si stirred at 15A with the amount of Sr, (**b**) Effect of electromagnetic stirring on the average size of the eutectic Si [22]

6.2 Melt Treatment in the Semi-Solid Casting, SEED Technology

One of the most appropriate alloys for SSM processes are aluminum alloys. In this category hypoeutectic Al–Si alloys gained attention due to their superior formability, better mechanical properties, higher strength-to-weight ratio, relative lower melting point, and wider solidification range. Generally, a desired structure in SSM processing is defined as a structure free from dendrites, preferably fully globular (spherical primary particles), and minimum or no entrapped eutectic. Therefore, the formation of globule morphology is of paramount concern while other variables are almost neglected. In the case of hypoeutectic Al–Si alloys, the primary α-Al particle size and eutectic silicon morphology are the main parameters that affect mechanical properties of the selected alloy. Addition of inoculants to the molten aluminum is now a prerequisite for successful and economic casting. Grain refinement in conventional casting provides an equiaxed grain structure, uniform mechanical properties, reduction in solution treatment time, general soundness, and better machinability. It also eliminates shrinkage porosity, increase resistance to hot tearing, and improve feeding. The mechanical properties of Al–Si alloys depend not only on the primary particles size and morphology, but also on the morphology of the eutectic Si particles. Therefore, modification of the Si morphology is the next factor in foundry operations.

In conventional high pressure diecasting, melt treatment (including refining and modification) is rather uncommon. This can be explained by the nature of high cooling rate of the process and consequently finer microstructure. For rheocasting, there are two stages for alloy solidification: one with lower cooling rate during slurry preparation to obtain globular structure and the second step with faster cooling rate, typical of conventional diecasting, during billet injection and shaping (Fig. 6.16). The main step concerning solidification is the slurry preparation (low cooling rate) which involves the nucleation of the primary particles affecting the

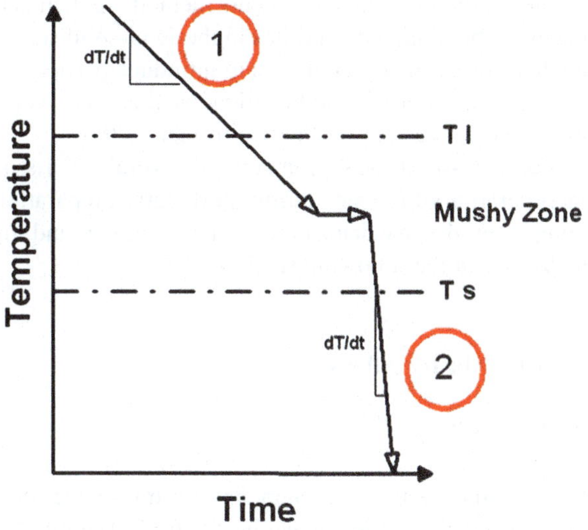

Fig. 6.16 Schematic presentation of various cooling rates in rheocasting; (1) low cooling rate due to the slurry preparation, (2) high cooling rate during diecasting process

entire structure. At this stage, melt treatment (grain refining and modification) is critical as it has an impact on the physical properties of alloy including surface tension and flow characteristics.

6.2.1 Grain Refining

Normally, hypoeutectic Al–Si alloys form coarse columnar, and equiaxed α-Al grains during solidification. The percentage of each zone depends on a multitude of parameters including pouring temperature, cooling rate, temperature gradient in the liquid, and the mold material which also influences thermal gradient established within the molten alloy.

The integrity of hypoeutectic Al–Si cast products is dependent on the fraction, size, and morphology of primary α-Al and this is why in foundry operations a close control over the α-Al formation is of great importance. The quality of the casting is improved by reducing the α-Al grain size and manipulating its morphology, i.e., refinement process. Generally, as mentioned by Kissling and Wallace in 1963, there are three principal methods for achieving grain refinement in Al alloys [23]:

- Rapid cooling (Chilling effect)
- Agitation of the melt and semi-solid metal processing
- Addition of grain refiner to the melt

In the first method, a fine dendritic structure is formed in the casting due to the rapid cooling where large undercoolings may be achieved before solidification initiation. This in turn affects the size of critical nuclei, smaller critical nuclei size, and thus increasing the effective number of nuclei to eventually rendering fine-grained structures. In the second method, refinement is accomplished by some means of breaking up dendrites in the semi-solid state. By far the most convenient method for grain size control is to introduce particles into the melt which nucleate new crystals during solidification as there are certain restrictions due to the mechanics of casting for the two former methods.

For semi-solid casting, in spite of several reports in the literature on the direct incorporation of refiners during SSM slurry preparation, there are still some ambiguities on the mechanism of grain refining and modification which will be addressed in the following section.

6.2.1.1 Master Alloys

Al5%Ti1%B

Optical and electron micrographs of a transverse section of a commercially produced master alloy are shown in Fig. 6.17. The microstructure of Al5Ti1B master alloy consists of Al_3Ti and TiB_2 particles embedded in aluminum matrix. The

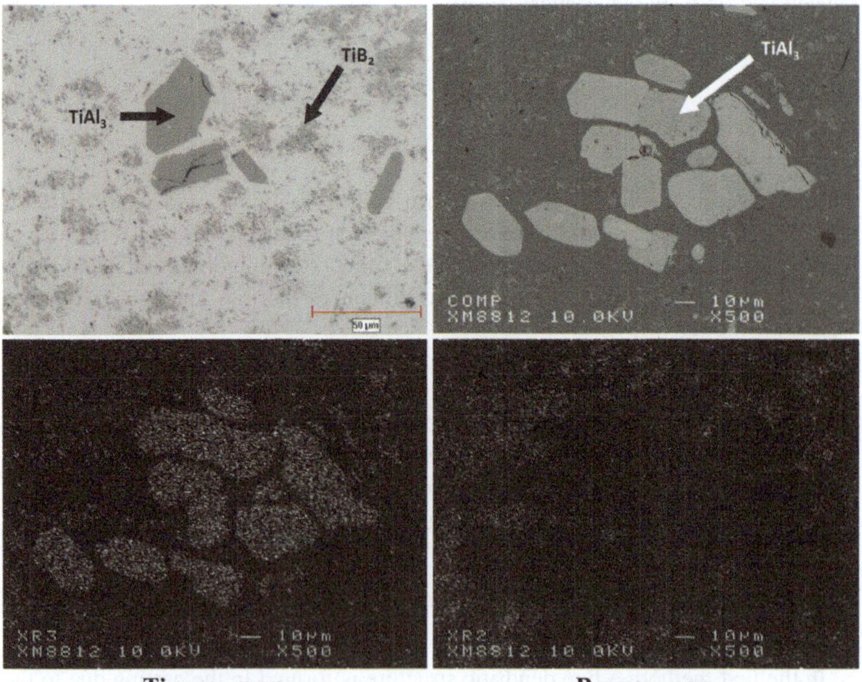

Fig. 6.17 Typical optical and backscattered electron micrographs with X-ray maps of Al5Ti1B master alloy, cross section, rod form [24]

microstructure is almost homogenous with small clusters of TiB_2 particles which are segregated within the aluminum matrix. The average length and width of Al_3Ti particles are 27 ± 15 μm, 11 ± 6 μm, respectively. The presence of the TiB_2 particles is substantial because of the high boron content shown in the boron X-ray map.

Al4%B

Optical and backscattered electron micrographs plus X-ray maps for Al and B are presented in Fig. 6.18. Chemical analysis by microprobe and size measurement confirmed the microstructure of the Al-4%B consists primarily of AlB_{12} particles with the average size of 6 ± 3 μm. The AlB_{12} particles are not distributed well within the master alloy and different sizes of particle agglomeration observed in the structure.

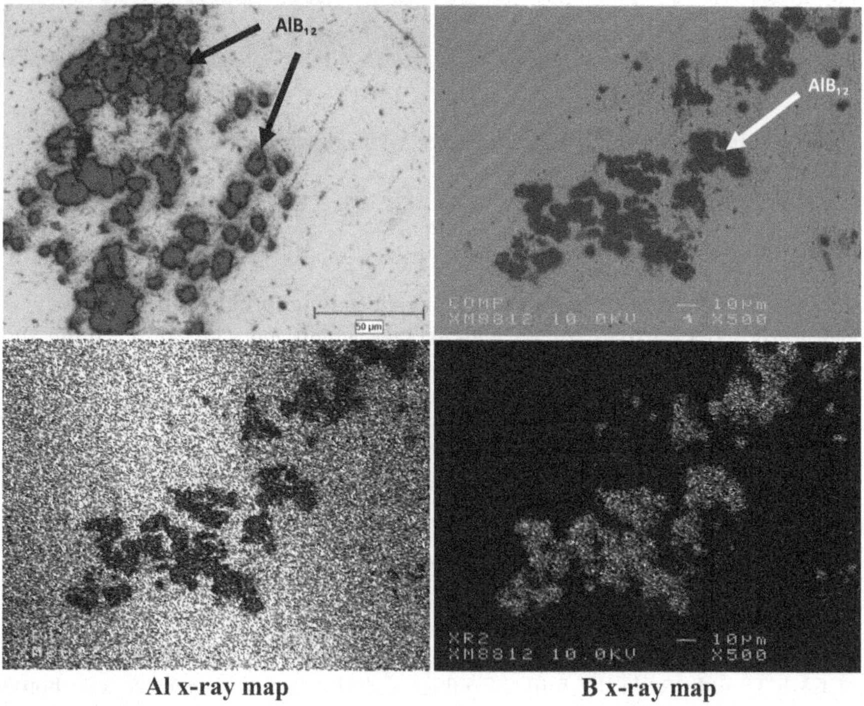

Fig. 6.18 Typical optical and backscattered electron micrographs together with X-ray maps of Al4B master alloy, waffle form [25]

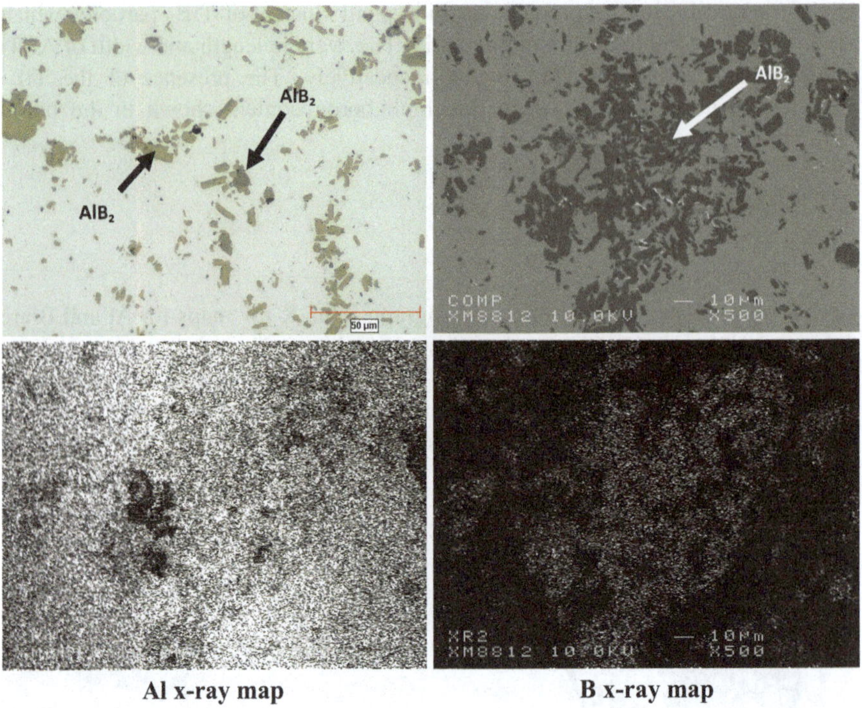

Fig. 6.19 Typical optical and backscattered electron micrographs together with X-ray maps of Al5B master alloy, cross section, rod form

Al5%B

The microstructure of the commercially produced master alloy comprises of fine and nearly blocky AlB_2 particles with average length and width of 9 ± 6 µm and 6 ± 3 µm, respectively. These particles are distributed within the aluminum matrix with some degree of agglomeration as seen in Fig. 6.19.

6.2.1.2 Addition of Ti-B by Al5Ti1B Master Alloy

In different standards such as ASTM [26], commercial 356 alloy contains dissolved titanium in the range of 0.1–0.2 wt%. This Ti has a key role on suppression of the growth of primary α-Al which will be explained later in the form of growth restriction factor (Sect. 6.2.1.4). Therefore, for further clarification on the effect of fresh Ti addition in the form of refiner and also to investigate the sole boron addition, the two Sects. of 6.2.1.2 and 6.2.1.3 report on two different master alloys additions.

The study was performed by addition of Al5Ti1B to the base alloy which was almost free of titanium (maximum 58 ppm Ti in solution), see Table 6.3.

Table 6.3 Chemical analysis of the base alloy (wt. %)

Si	Mg	Fe	Mn	Cu	Ti	B	Sr	Al
6.44–6.53	0.36–0.39	Max 0.07	Max 0.003	Max 0.001	Max 0.0058	Nil	Nil	Bal.

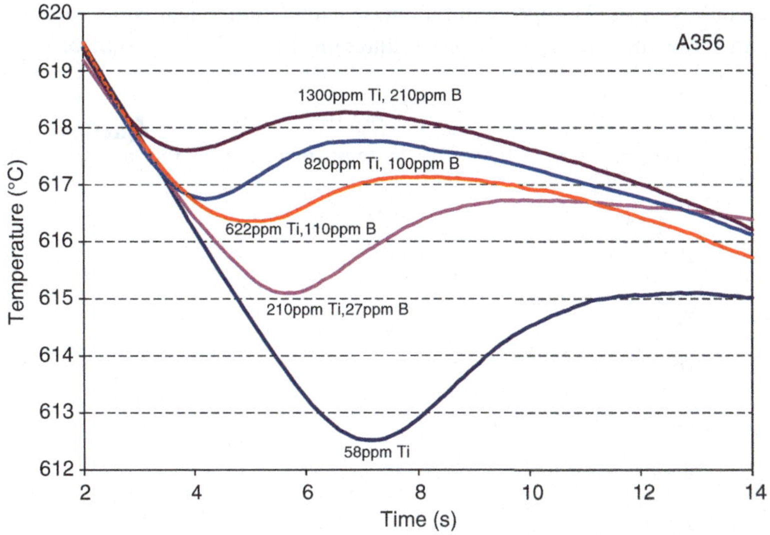

Fig. 6.20 The initial section of cooling curves for untreated and treated A356 alloys (results from central thermocouple, graphite cup) [24]

Conventional Casting

Thermal Analysis

The effect of Al5Ti1B refiner addition to the A356 alloy on the early stages of solidification is illustrated in Fig. 6.20. Small additions of Ti-B shifted the cooling curves up and to the left with the recalescence decreasing with increasing master alloy content. The nucleation event takes place at higher temperatures and at shorter times after pouring. As reported [27–29], an effective refiner does not have any undercooling before the actual growth temperature. In fact, the smaller the recalescence, the more effective is the refiner. In these series of experiments, it is revealed that Al5Ti1B does not remove recalescence completely even at the maximum additional level.

FactSage software[1] was used to predict the influence of Ti on Al7Si1Mg alloy. In this case, an isopleth[2] phase diagram was constructed for Al7Si1Mg alloy

[1]FactSage is a commercial thermodynamic-based software package (http://www.crct.polymtl.ca/factsage)

[2]An isopleth is a type of phase diagram which may come from ternary or higher degree system where the phase relationship is identified for the elements of interest at different temperatures with variation of one element.

with Ti addition, Fig. 6.21. The prediction of forming intermetallic compounds is highly dependent on the selection of the database and the expertise of the user including his/her knowledge of the alloy and possible intermetallic reaction amongst the constituent elements. For example if the intermetallics containing binary Ti–Si or ternary Al–Si–Ti are not selected in the database, then only the intermetallics of Al–Ti appears in the phase calculation which is the case here. In comparison to the binary Al–Ti phase diagram, Fig. 6.21a, the start of peritectic

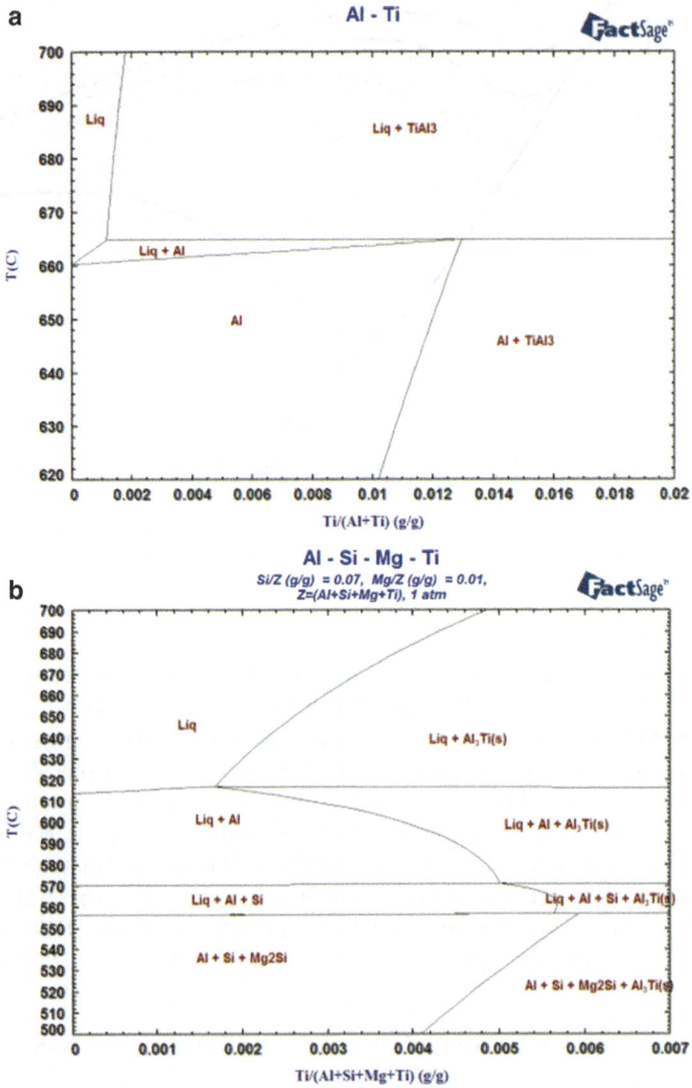

Fig. 6.21 FactSage calculation of: (**a**) Al–Ti binary diagram, (**b**) the isopleth for Al7Si1Mg versus Ti concentration

reaction shifts to the Ti concentration of ~0.17 % Ti for Al7Si1Mg alloy in contrast to ~0.11 % Ti for Al–Ti binary alloys. This has a bearing on the nucleation of α-Al grains due to the formation of Al_3Ti particles as potential nucleant. In actual testing condition, however, other parameters such as alloying and trace elements, cooling rate, and TiB_2 particles (size, morphology, and distribution) in the master alloy may increase solidification temperature range even further and improve refining efficiency. It is also important to indicate that every predictive software calculates the changes based on its logic and database and therefore the result may not be identical.

Figure 6.22 presents selective thermal analysis parameters. By refiner addition, nucleation and growth temperatures of α-Al increased by ~5 and 3 °C, respectively. The increasing nucleation temperature allows new crystals to form ahead of solidification front, rendering an equiaxed fine-grained as-cast structure. Furthermore, the rate (slope of the curve) at which growth temperature increases is less than nucleation temperature. In other words, there are more nuclei with less growth potential and consequently refining mechanism should be active. This is one of the basic principles for a grain refiner to be rated as effective.

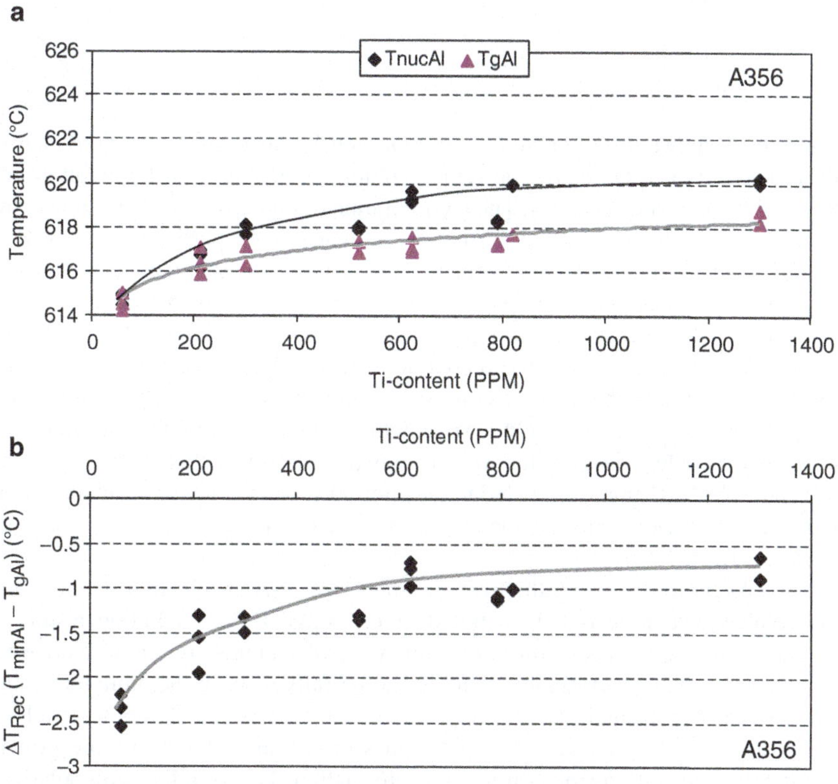

Fig. 6.22 Effect of Ti & B addition on the: (**a**) nucleation and growth temperatures of primary α-Al, (**b**) ΔT_{Rec} [24]

The recalescence becomes approximately constant after ~600 ppm Ti addition and from thermal analysis standpoint; the critical Ti addition for this alloy appears to be about 600 ppm (Fig. 6.22b). Further addition may form Ti-based intermetallics which are detrimental to the mechanical properties of the as-cast products as discussed later. Lowering the recalescence has influence on the growth of primary α-Al particles. The minimum temperature at the beginning of solidification, $T_{min_{Al}}$, shows the temperature where the rate of latent heat liberation is balanced out with the heat extracted from the sample. In untreated alloy, the existence of recalescence means the heat generated with the commencement of solidification could not be transferred out of the mold completely and therefore the heat balance leads to the appearance of recalescence. In refined alloy, nucleation temperature increases and therefore in contrast to the untreated alloy, there are more primaries within the same time interval for the refined alloy. These solid α-Al particles can serve as heat sinks to absorb the heat released from surrounding liquid and therefore leads to lower recalescence compared to the untreated alloy.

There could be a second hypothesis to justify the disappearance of recalescence with inoculation. This could be attributed to the lower growth in the refined alloys which results in lesser heat being evolved to balance out the heat extracted from the solidifying alloy.

Structural Analysis

Figure 6.23 shows the microstructural evolution by addition of Al5Ti1B master alloy. The microstructure without refiner comprises coarse dendritic grains which refine with increasing Ti and B. The incorporation of numerous nucleants due to the refining agent encourages the formation of many α-Al particles with smaller boundary layers due to smaller inter-nuclei spacing. The result is lower growth and multidirectional heat flow to eventually forming equiaxed structure.

Figure 6.24 illustrates the refining effect of the master alloy where with the refiner addition, grain size reduced continuously and the final grain size is around ~590 μm. By comparing average grain size against nucleation and growth temperatures of primary α-Al, one could see that with addition of up to 800 ppm Ti, there is a continuous reduction in grain size but remains almost unchanged in excess of 800 ppm. The lack of grain size reduction after 800 ppm Ti may be attributed to the extra Al_3Ti particles becoming inactive nucleants. This phenomenon was discussed in Sect. 6.1.1.1.

The correlation between the average grain size, α-Al nucleation temperature, and recalescence time is demonstrated in Fig. 6.25. Refiner addition results in decreasing the recalescence time (remember, smaller grains are due to increasing of refiner percentage). Assuming that nucleation fully stops at the minimum value for undercooling, a small value of t_{Rec} signifies that the grains do not have a long growth period. In other words, longer values of t_{Rec} are related to more growth opportunity and consequently larger grains formation. This is in line with polarized

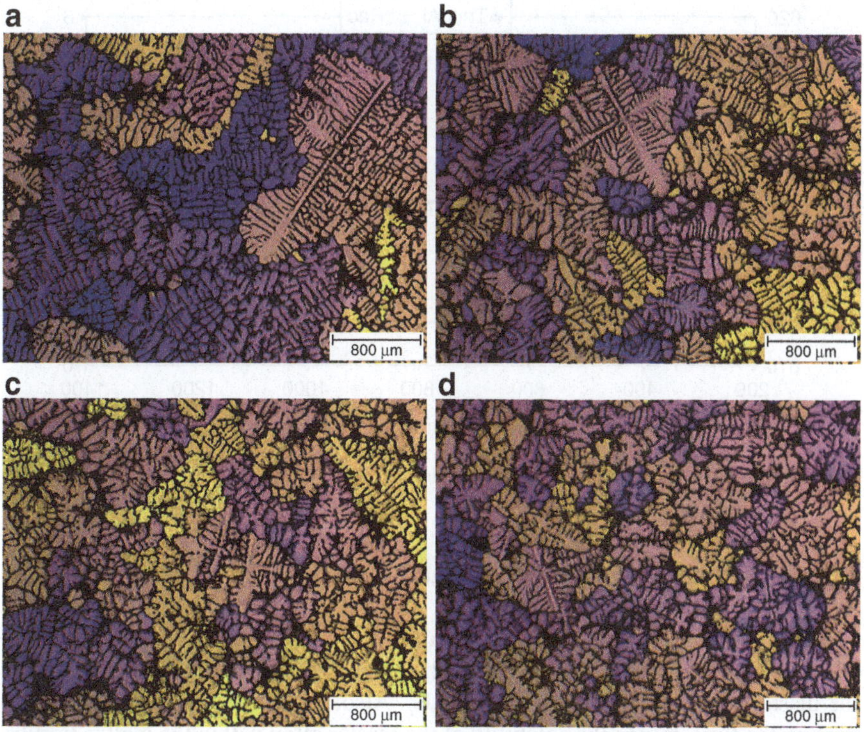

Fig. 6.23 Polarized light micrographs showing the effect of Ti-B addition: (**a**) without grain refiner, (**b**) 210 ppm Ti, 27 ppm B, (**c**) 520 ppm Ti, 73 ppm B, and (**d**) 820 ppm Ti, 100 ppm B

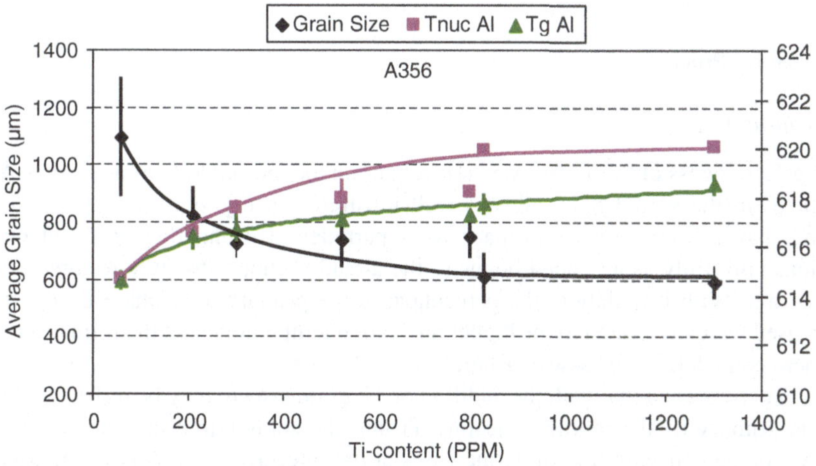

Fig. 6.24 The effect of Al5Ti1B addition on the grain size, $T_{\mathrm{nuc}_{\mathrm{Al}}}$, and $T_{\mathrm{g}_{\mathrm{Al}}}$ [24]

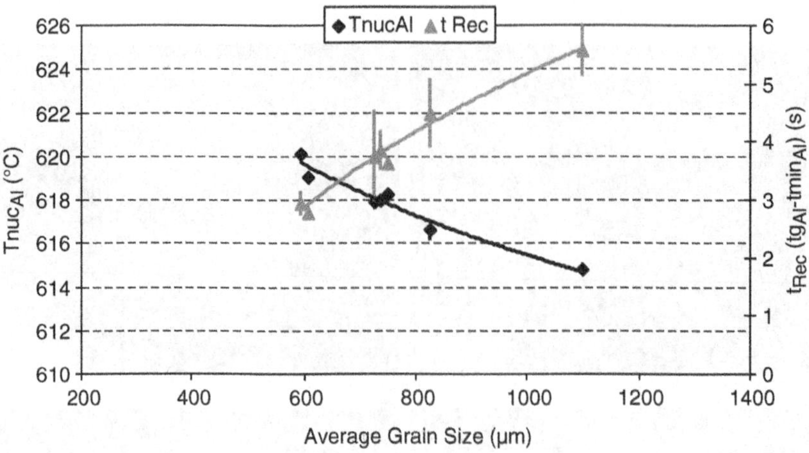

Fig. 6.25 Relation between grain size and recalescence time

light micrographs in Fig. 6.23 and previous findings on the effect of refiner addition to the Al–Si alloys [30, 31].

The ability of the Al5Ti1B master alloy to initiate refining depends on two parameters. First and the obvious one is the potency of the intermetallic particles in the master alloy with an emphasis on TiB_2. Also, after addition of refiner and by passing the time, due to the solubility of the master alloy within the melt, a portion of titanium will dissolve. This portion which comes from the Al_3Ti or the Ti in the α-Al matrix of the master alloy imparts its suppression of the growth to further improve the overall efficiency of the refiner.

Semi-Solid Processing

Structural Analysis

The effectiveness of the added Al5Ti1B refiner on the morphology of primary α-Al particles in the stirred SSM alloy is well illustrated in Fig. 6.26. It appears that refiner has further reduced primary α-Al particles size and dispersed eutectic regions uniformly as an added bonus to the already refined structure resulted due to stirring. With inoculation, the percentage of the primary α-Al phase is slightly increased which is related to the higher nucleation temperature and the formation of numerous nuclei, as discussed earlier.

It may be argued that with the addition of Ti, grain growth may be restricted due to the primary α-Al dendrites rejecting Ti into the solid–liquid interface as they grow. As the concentration of Ti increases at the interface, it may reach the level where new nucleants of Al_3Ti could form within the interface layer. The presence of TiB_2 in the master alloy together with the newly formed Al_3Ti particles may

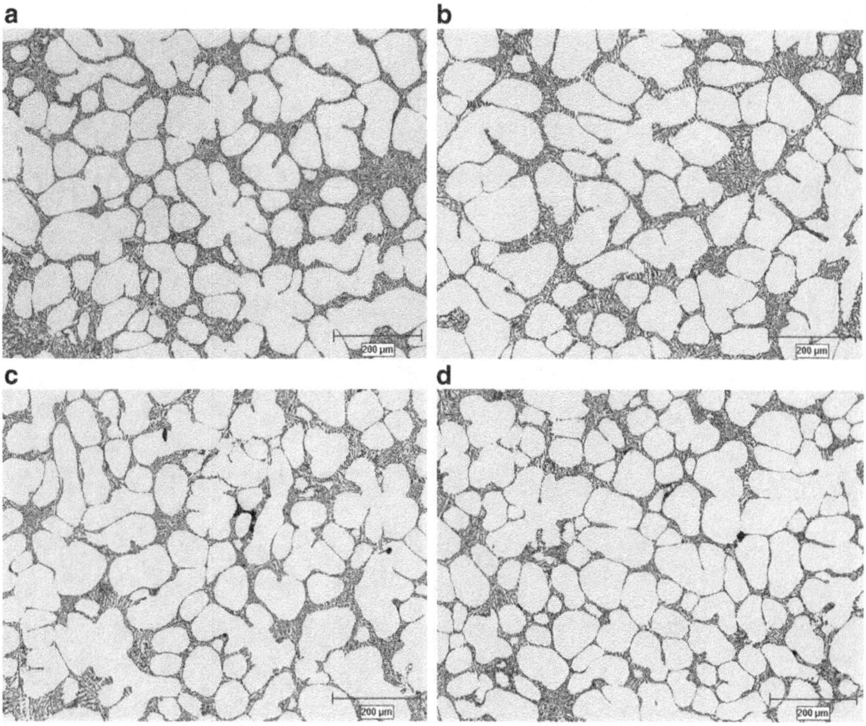

Fig. 6.26 Optical micrographs showing the refiner addition effects in the SEED process: (a) without addition, (b) 210 ppm Ti, 27 ppm B, (c) 520 ppm Ti, 73 ppm B, and (d) 1300 ppm Ti, 210 ppm B

encourage the formation of new α-Al nuclei within the interface. The newly formed α-Al particles are expected to reject Ti into the newly formed interfaces as they grow. The repetition of such mechanism ensures the formation of finer and more equiaxed particles as these globules could not grow much because of the neighboring globules and also higher growth restriction factor due to the titanium addition.

The drawback is the agglomeration of Ti-based compounds with increasing Ti content beyond approximately 1000 ppm which degrades the potency of the nucleants. Basically, the sources for agglomeration are (Fig. 6.27) as follows:

1. The nature of refiner itself where some Ti-based particles may not have the potency to act as nucleation sites nor to go into solution and rejected at the interface to eventually segregate within the eutectic region as the last liquid to solidify.
2. The other source may be attributed to the peritectic reaction. With reference to the Al–Ti binary phase diagram and with computational thermodynamic calculation for approximately the same alloy, Fig. 6.21, the peritectic reaction shifts and this has a contribution on Al_3Ti formation.

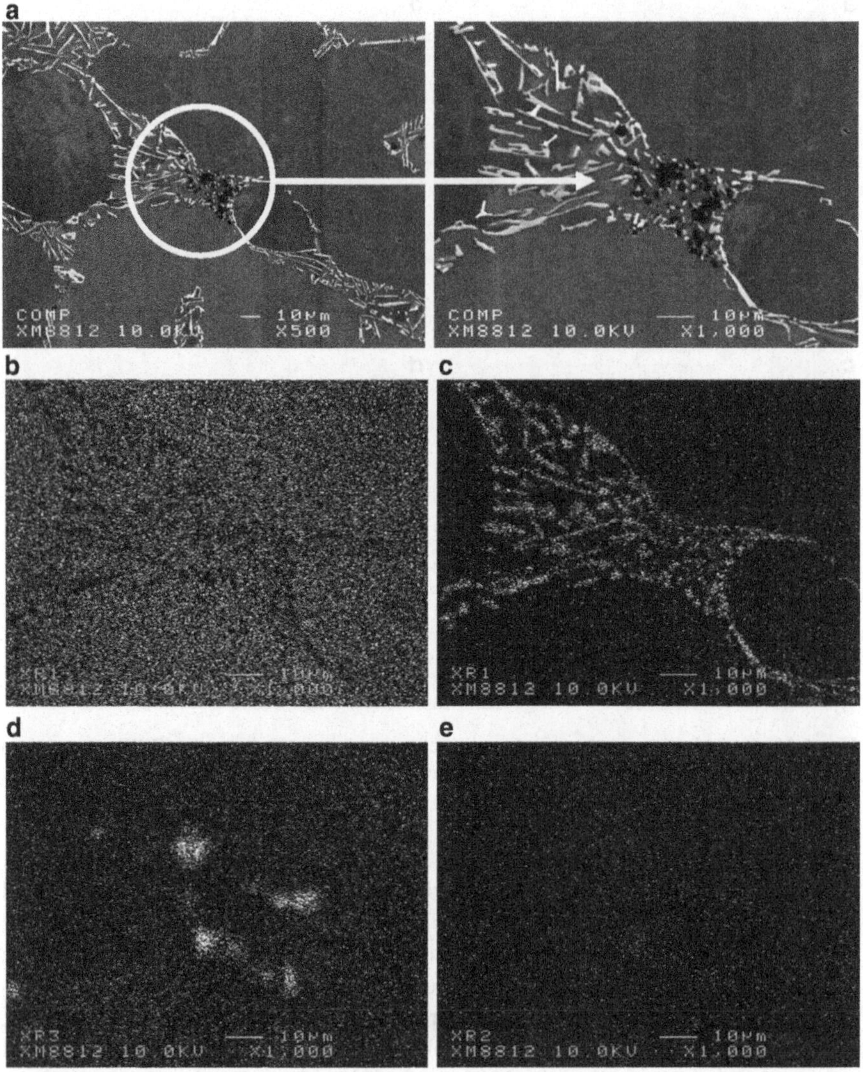

Fig. 6.27 (**a**) Backscattered electron micrographs (different magnifications) and X-ray maps of the intermetallic particles; (**b**) Al, (**c**) Si, (**d**) Ti, and (**e**) B maps (sample with 1300 ppm Ti and 210 ppm B)

Variation of solidification ranges of α-Al particles (ΔT_{α}) and eutectic and post eutectic reactions (ΔT_{eut}) are plotted in Fig. 6.28. By increasing the refiner, α-Al solidification range increases by ~6 °C and addition of refiner has almost no effect on the eutectic solidification range. This issue has a great impact on the rheo-processing as initiation and formation of α-Al globules entirely depends on the SSM processing in this temperature range.

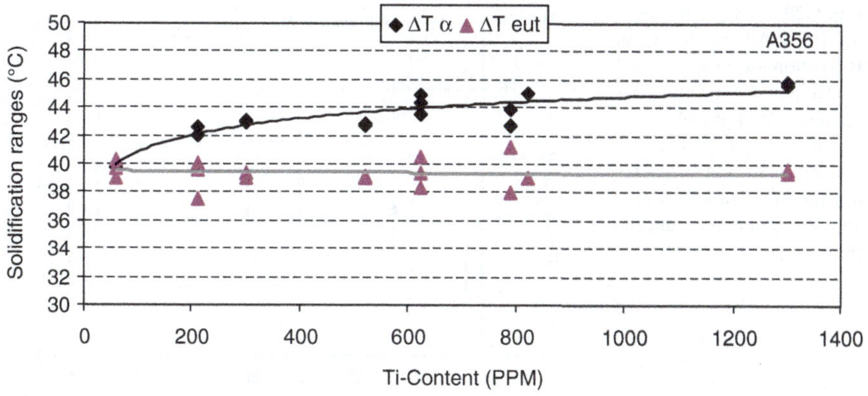

Fig. 6.28 Effect of Ti and B on the solidification range of various phases

Image Analysis

Quantitative results of image processing are shown in Fig. 6.29. With refiner addition, the primary α-Al percentage appears to increase slightly. This is due to the higher number of effective nuclei for the primary α-Al particles and also shifting of liquidus temperature to higher values with eutectic temperature remaining unchanged.

By refiner addition, the equivalent circular diameter decreases. This parameter has a direct relation with number density (the number of primary particles per unit area). More effective nucleation sites result in better distribution of particles in the sample powered by limited growth leading to smaller particle size and greater number of particles. Such increase in number density is in spite of the difficulty in image processing and differentiating between two adjacent α-Al particles especially when they coarsen or there is less eutectic in the structure. Therefore, it is possible to underestimate α-Al number density.

Also as a marginal but critical result, the main criterion of effective SSM processing, fine primary phase particles, preferably less than 100 μm diameter with globular or rosette structure [32], is easily achieved even in untreated alloy.

The percentage of particles having aspect ratio above 2 is a measure of the globularity of the structure. The average trend of area/perimeter and aspect ratio decreased with refiner addition, Fig. 6.29, are indications of having finer and isolated spherical α-Al particles.

The histogram chart in Fig. 6.30a shows the percentage of α-Al particles with sphericity >0.8 rises slightly with increasing refiner addition. Figure 6.30b compares the sphericity values for the alloys with and without refining. The sphericity values closer to 1 are indications of having more spherical particles; these values are greater for the refined alloy.

Image analysis technique is based on the reproduction of the microstructure digitally which may not be exactly the same as the one being studied. Therefore, one should recognize what "image analysis" cannot do. For example, the definition

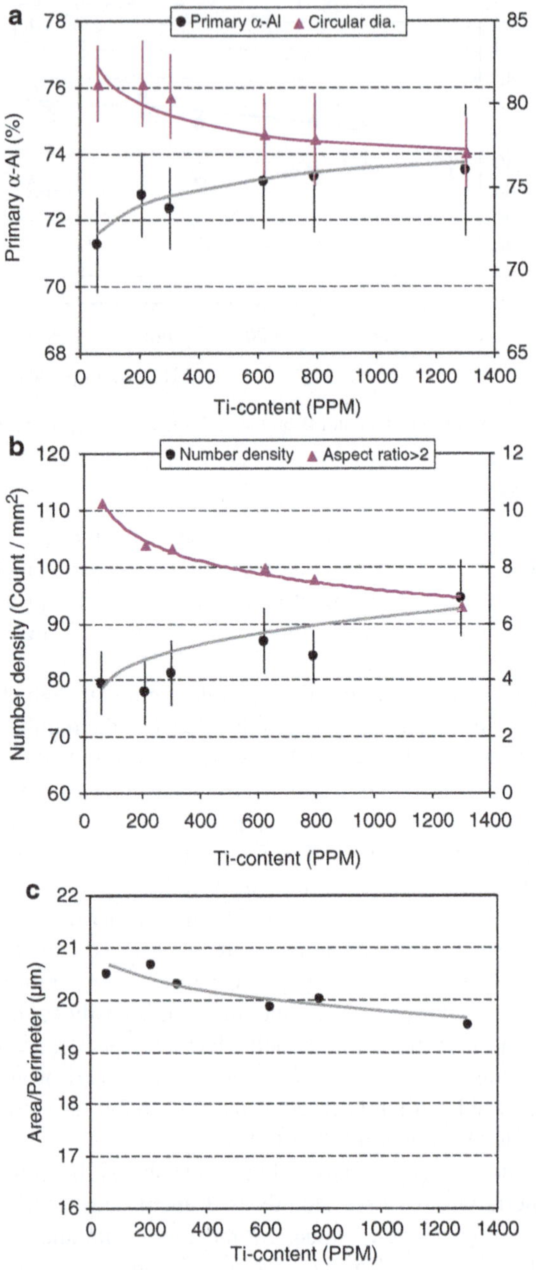

Fig. 6.29 Image analysis results for A356 with Ti and B addition: (**a**) primary α-Al percentage, average circular diameter, (**b**) number density of α-Al particles, percentage of α-Al particles having aspect ratio >2, (**c**) area/perimeter (reproduced from [24])

for particle diameter in the software is based on the equation $(2\sqrt{\text{area}/\pi})$, where area refers to the area of the object being measured. Such measurement makes the assumption that the examined object has a shape close to a circle and therefore the

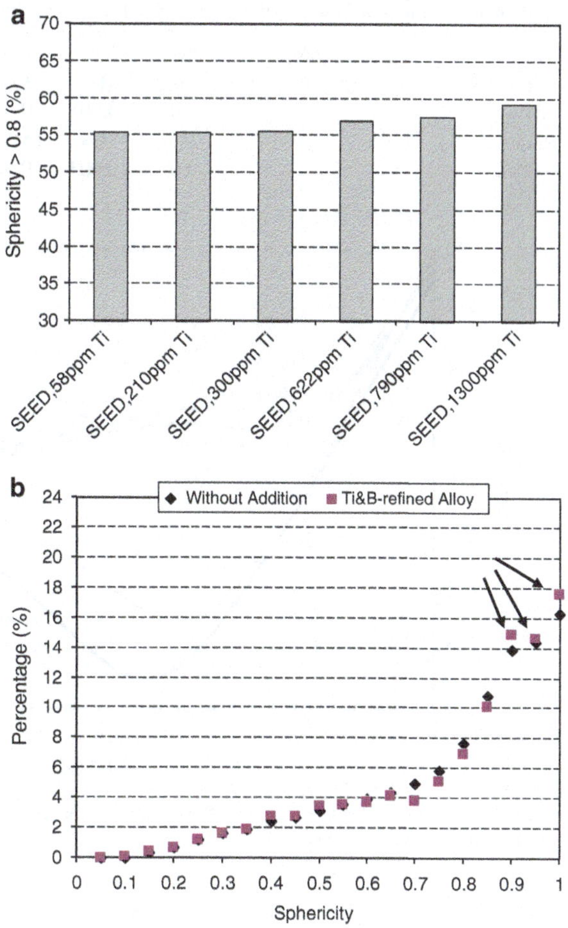

Fig. 6.30 (**a**) The influence of inoculation on the percentages of particles with sphericity >0.8, (**b**) effect of refiner on sphericity values (820 ppm Ti & 100 ppm B) [24]

more the particle is rectangular, the greater is the error in calculation. Considering this point, it is difficult to detect the differences between the average circular diameter of a rosette and globule α-Al particle with the same area (readers are referred to Sect. 4.3).

6.2.1.3 Addition of Boron

Currently, the most widely used industrial refiners for Al–Si alloys are AlTiB master alloys with their well-proven drawbacks such as fading and settling of the so-called potential nucleant particles and lack of grain size uniformity. These factors have negative effects on the as-cast structure and properties of the engineering components.

Table 6.4 Chemical analysis of the grain-refined samples with boron (wt. %) prior to refining

	Si	Mg	Fe	Mn	Cu	Ti	B	Sr	Al
Refined with Al4%B	6.42–6.6	0.36–0.39	Max 0.07	Max 0.003	Max 0.001	Max 0.0058	Nil	Nil	Bal.
Refined with Al5%B	6.5–6.7	0.36–0.4	Max 0.07	Max 0.003	Max 0.001	Max 0.0057	Nil	Nil	Bal.

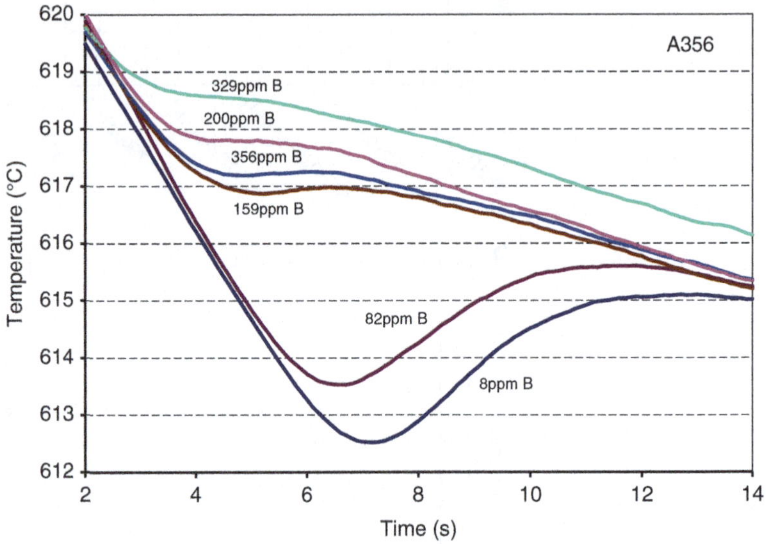

Fig. 6.31 The onset of solidification at the center of cups having different concentrations of boron (Al5B master alloy addition)

In this section, it is shown that boron refiners could act better than Ti-based refiners not only in the conventional casting but also in SSM processing. For this study, two different master alloys were used: Al4%B where the majority of particles are AlB_{12} and Al5B with AlB_2 particles. The base chemical compositions are listed in Table 6.4 and depending on the master alloy additions, the final chemical composition varied. Due to the results similarity, all the reports in this section are related to the Al5B master alloy addition unless otherwise mentioned.

Conventional Casting

Thermal Analysis

Figure 6.31 shows the effect of Al5B refiner addition to the alloy on the early stage of solidification. Small addition of boron has a remarkable effect on the solidification. For instance, the growth temperature is raised about 2 °C by adding only 159 ppm boron and beyond that it remains relatively constant. Furthermore, the

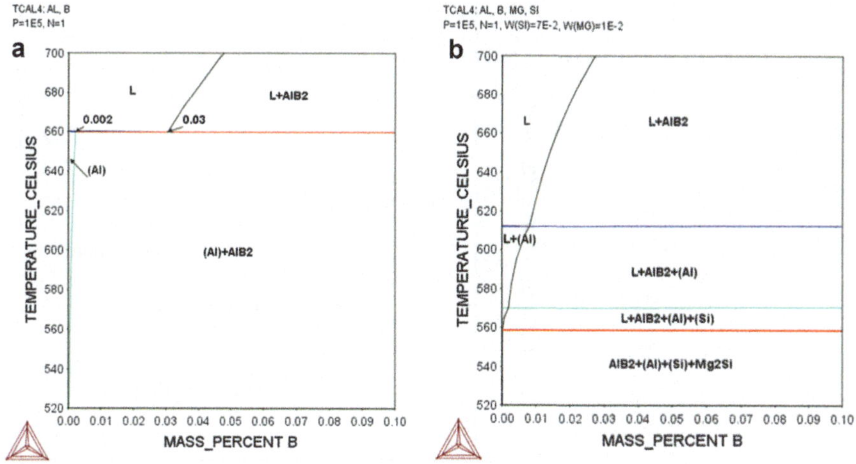

Fig. 6.32 ThemoCalc calculation of Al rich section: (**a**) binary Al-B phase diagram, (**b**) isopleth for Al7Si1Mg versus B

addition of boron eliminates the dip (undercooling) and recalescence resulting in continuous temperature reduction after nucleation. From a thermal analysis point of view, the critical boron content for this alloy seems to be more than 159 ppm but as will be discussed, higher values should be treated cautiously due to the agglomeration and formation of insoluble boride particles.

ThermoCalc software was used as a predictive tool to examine as how the boron addition affects the ternary Al–Si–Mg alloy. For this purpose, aluminum-rich sections of Al-B binary and an isopleth diagram were constructed for the addition of boron to the Al7Si1Mg alloy (Fig. 6.32). In the binary phase diagram, the eutectic reaction leads to the formation of the α-Al plus AlB_2, an active nucleant for the aluminum alloys:

$$Liquid \rightarrow AlB_2 + \alpha\,(Solid)$$

The eutectic reaction takes place at temperature just below the equilibrium freezing temperature of pure aluminum (660 °C) and the resulted AlB_2 particles could act as effective nucleants especially for Al–Si hypoeutectic alloys with melting points well below 660 °C. In Al7Si1Mg, boron addition expanded the mushy zone and brings about the formation of AlB_2 phase. This area is more visible in the isopleth diagram calculated for a constant value of boron, 0.02 %, Fig. 6.33. Furthermore by formation of AlB_2, the liquidus shifts to higher temperature and that is to have earlier nucleation (liquidus is not shown in Fig. 6.33). In actual casting condition, this thermodynamic calculation could be different due to nonequilibrium condition such as higher cooling rate, existence of trace and alloying elements, etc.

With addition of boron up to 329 ppm, the nucleation and growth temperatures increase by ~7 °C and ~3 °C, respectively (Fig. 6.34a). The rising of nucleation

Fig. 6.33 Isopleth of the AlSi1Mg0.02B versus Si

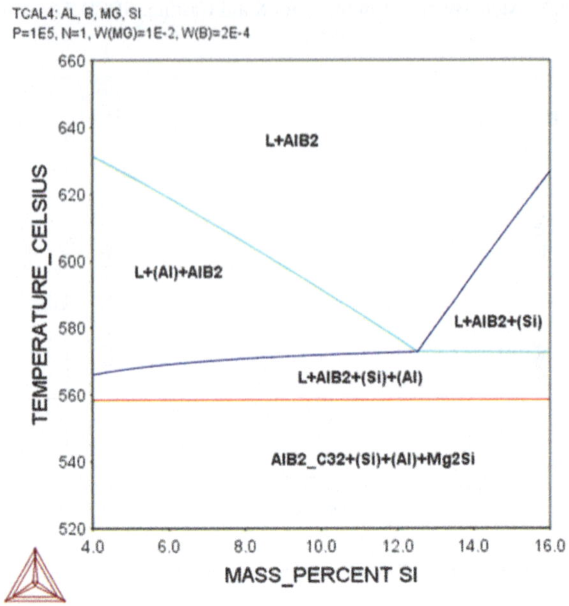

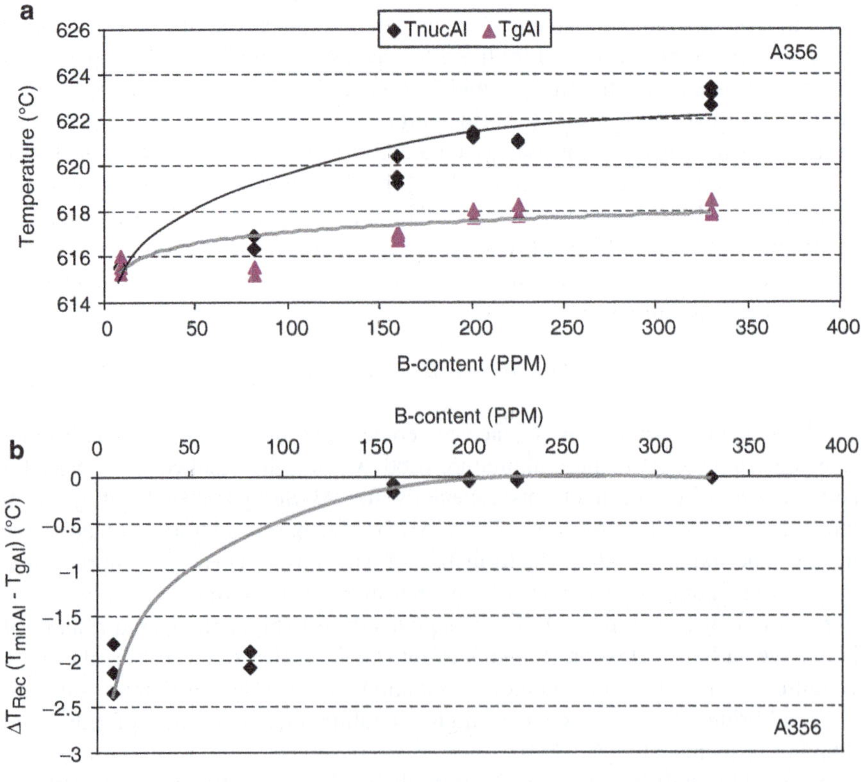

Fig. 6.34 Effect of boron addition on the: (a) nucleation and growth temperatures of primary α-Al, (b) ΔT_{Rec}

temperature allows new crystals, which were not energetically stable before, to form ahead of solidification front, rendering an equiaxed fine grained as-cast structure. The rate (slope of the curve) at which the nucleation temperature ascending is higher than the growth temperature (same reported for Al5Ti1B). This is equivalent to more nuclei generation with less capacity for growth which leads to more equiaxed structure. However, the difference between these two temperatures widens with more boron additions.

For boron content less than 159 ppm, an undercooling of ~2 °C exists, but approaches and remains zero for all other additional levels. Lowering the recalescence has a great influence on the growth of primary α-Al particles. The minimum temperature, $Tmin_{Al}$, shows the temperature where the rate of latent heat liberation is balanced out with the heat extracted from the sample. In untreated alloy, the existence of recalescence means the heat generated with the commencement of solidification could not be transferred out of the mold completely and therefore the heat balance leads to the appearance of recalescence. However, this is not exactly the case for the refined alloy. In the refined alloy, as depicted in Fig. 6.34a, nucleation temperature increases and therefore in contrast to the untreated alloy, there are more primary nuclei within the same time interval for the refined alloy. The same mechanism where the extra particles acting as heat sink as explained in Sect. 6.2.1.2 is applicable here.

With addition of boron up to 329 ppm, the nucleation and growth temperatures increase by ~7 °C and ~3 °C, respectively (Fig. 6.34a). The rising of nucleation temperature allows new crystals, which were not energetically stable before, to form ahead of solidification front, rendering an equiaxed fine-grained as-cast structure. Once more, the rate at which the nucleation temperature ascending is higher than the growth temperature. This is equivalent to more nuclei generation with less capacity for growth which leads to more equiaxed structure. However, the difference between these two temperatures widens with more boron additions.

In commercial applications, Al-B master alloys are used to eliminate trace amounts of transition elements such as chromium, titanium, vanadium, iron, and zirconium from pure aluminum since these elements drastically lower the electrical conductivity. For example, only 0.014 % of vanadium lowers the electrical conductivity by 1%IACS (International Annealed Copper Standard) (e.g., [33]). Boron reacts with these transition elements to form boron compounds which settle at the bottom of the holding furnace. It was shown that AlB_2 particles react slightly faster with transition elements and also form smaller diboride particles than AlB_{12}. The lower boron content per unit volume of the AlB_2 particles in addition to its smaller size contributes to lower settling rate for transition element borides [34].

From thermodynamic standpoint, in reaction competition between different elements, priority depends on the heat of formation values (absolute) and compounds with higher values have greater chance to form. Table 6.5 shows some examples on particle stability. The most important borides for refining purpose are TiB_2/AlB_2 particles. Among different borides, it is shown that TiB_2 has the highest heat of formation value and therefore TiB_2 is the most stable boride. In the case of

Table 6.5 Formation enthalpy of various particles, ΔH_f (kJ/mol) [35]

Compound	ΔH_f (kJ/mol)	
	298 (K)	1000 (K)
AlB_2	−150.996	−165.201
AlB_{12}	−266.102	−288.989
Al_3Ti	−146.44	−181.020
CrB_2	−94.14	−94.11
MgB_2	−92.048	−106.639
MnB_2	−94.14	−99.16
TiB_2	−323.8	−326.647
VB_2	−203.761	−203.823
ZrB_2	−322.586	−325.428

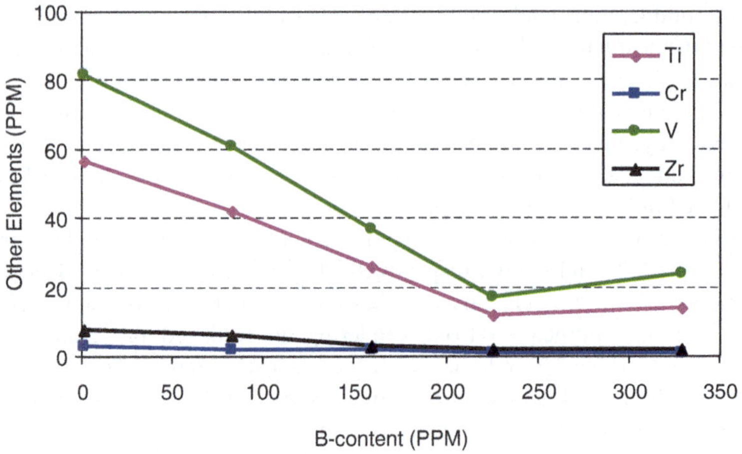

Fig. 6.35 The effect of B addition on the trace amounts of Ti, Cr, V, and Zr

sole boron addition, it should be considered that any residual Ti leads to TiB_2 formation and therefore the quantity of boron to react with aluminum decreased.

According to the above information and in order for Al-B master alloys to be effective, it is critically important to ensure that the molten alloy is free from titanium. This is due to the high affinity of Ti to form TiB_2 and thus diminishing the effectiveness of AlB_2 particles as the preferred nucleation sites for α-Al phase. Figure 6.35 shows that with addition of B, the trace Ti in the alloy reacts with B and consequently reduces the refiner efficacy.

Structural Analysis

Figure 6.36 shows the microstructural evolution due to addition of boron-based refiner. The microstructure in the refined sample is quite different from that of the non-refined with much smaller sizes for the primary α-Al particles, Fig. 6.36c, d.

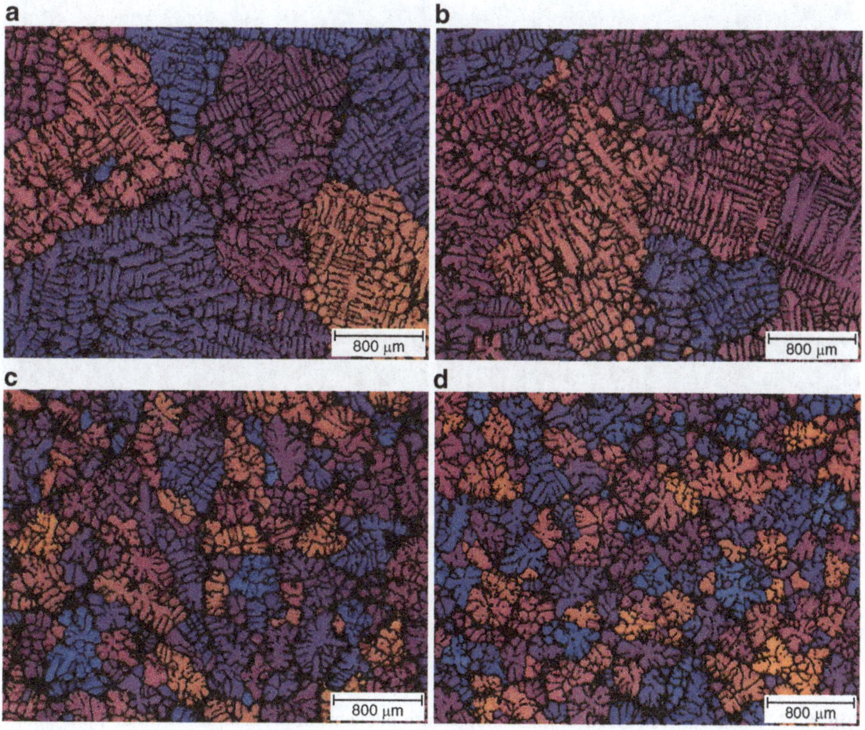

Fig. 6.36 Polarized micrographs showing the effect of B addition: (**a**) without addition, (**b**) 82 ppm, (**c**) 159 ppm, and (**d**) 225 ppm

The B-treated alloy comprises more rosette particles. B refiner encourages the formation of more nuclei in the system, something close to what is termed "copious nucleation" due to a large population of active B-based nucleants introduced into the bulk liquid. Such large number of active nucleants not only restricts the growth of primary α-Al nuclei, but also acts as heat sink to impose a localized multi-directional heat flow within the melt. The outcome would be the formation of small equiaxed primary α-Al particles. Therefore, in the boron-treated alloy, the typical morphology is between rosette and globules.

Identification of nucleants in the as-cast structure is time consuming and tedious which is associated with the detection probability on the polished surface and whether the polished surface intersects the nucleant or not. By boron addition, most of the nucleants in the material expected to be AlB_2 particles. However, the composition of the nucleant depends on the titanium level in the melt. Even with Ti concentration at the ppm levels in the melt, due to the high affinity of B with Ti according to the data presented in Table 6.5, it is expected to have a wide range of nucleant compositions. Figure 6.37 shows a nucleant which has been characterized by backscattered electron micrographs and X-ray maps of the particle. The nucleant is a two-phase compound where a Ti-B particle is encompassed by aluminum boride.

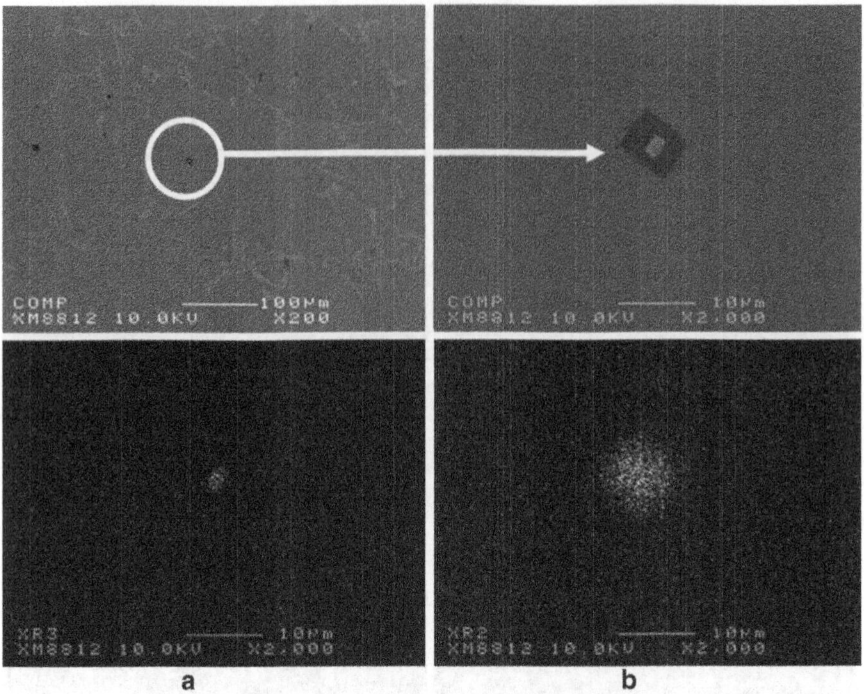

Fig. 6.37 Backscattered electron micrographs of a nucleant in the sample with 720 ppm B, and X-ray maps of: (**a**) Ti, and (**b**) B [25]

Based on such observation and the formation of multiphase nucleants in the system, it is believed that by boron addition, the boron atoms (whether dissolved in the alloy or as AlB_2 or AlB_{12} particles) react with residual titanium atoms in the melt to form stable TiB_2 particles or to partially transform to complex AlTiB-based particles, respectively. The resulted TiB_2 is then expected to act as nucleation site for aluminum diboride. The AlB_2 is then expected to encapsulate TiB_2 particle. For the complex particles, according to Karantzalis et al. [36], the lattice disregistry between Al and TiB_2 or AlB_2 is similar and at 660 °C is 4.53 %. On cooling, α-Al primary particles could easily nucleate on these preexisting particles. In addition, there are still many unreacted AlB_2 particles which would operate as potential nucleants. Figure 6.38 shows the schematic sequence of the aforementioned hypothesis. For AlB_{12} particles, it is envisaged that such reaction with residual Ti in the alloy should imbalance the stoichiometry of AlB_{12} and improves its suitability for acting as potential nucleant for α-Al.

The presence of agglomerated intermetallic particles is the main shortcoming of higher boron level. As the boron concentration increases, not all the boride particles could act as nucleation sites for aluminum. Furthermore, the limited solubility of these particles in α-Al may lead to the formation of agglomerates within the matrix.

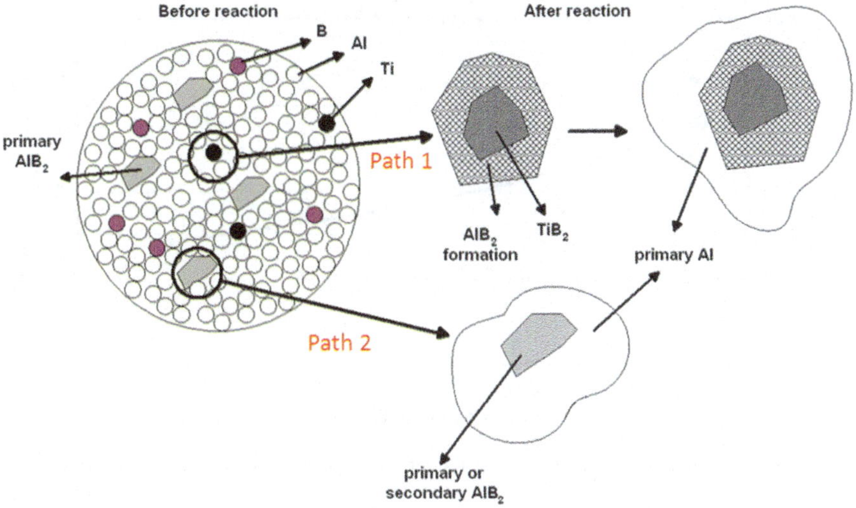

Fig. 6.38 Schematic sequence of the primary α-Al particles formation, either from reaction of Ti and B (path 1) or directly from AlB_2 (path 2) [25]

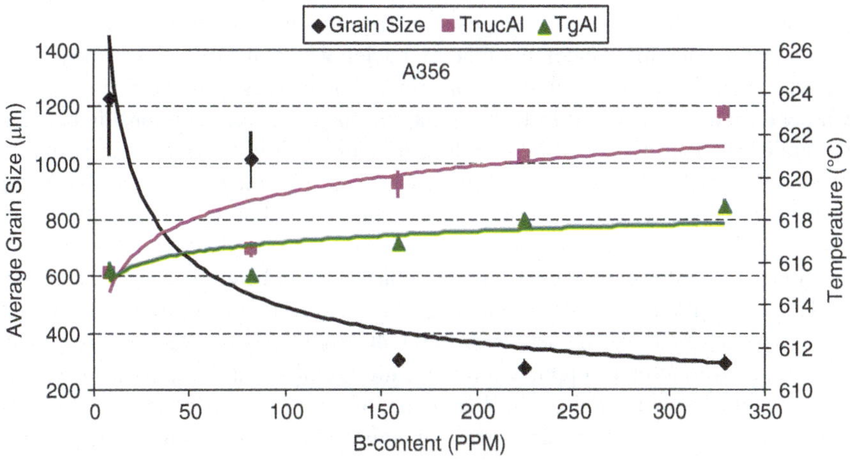

Fig. 6.39 The interrelationships between grain size, $T_{nuc_{Al}}$, and $T_{g_{Al}}$ in case of Al5B master alloy addition

Figure 6.39 shows the relation between thermal analysis data, grain size, and B-concentration. The grain size is slightly reduced by addition of 82 ppm boron and after that, there is a sharp change in grain size around 159 ppm B where the grain size reaches to about 300 μm. The ~3–4 fold decrease in grain size is a marked indication of boron potency.

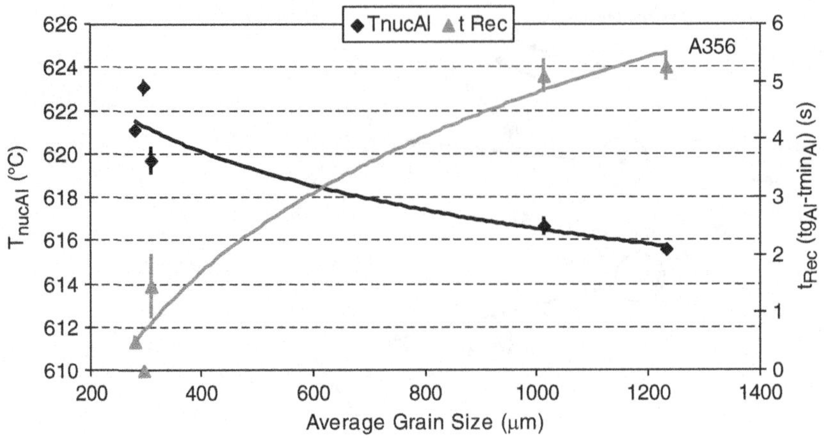

Fig. 6.40 Variation of grain size (% B-refiner) with primary α-Al nucleation temperature and recalescence time

There is a strong interrelationship between the nucleation temperature of the primary α-Al and the grain size. The graph indicates that the nucleation temperature rises up gradually to ~623 °C and the grain size reduced to ~300 μm and remained constant with further addition. The inverse trend for nucleation and growth temperatures with the grain size suggests an increase in nucleation temperature is equivalent to having greater number of nuclei which means lesser growth per nucleus even though the magnitude of growth has increased as the rise in growth temperature may suggest. It further emphasizes the direct interrelationship between the grain size and thermal history of the alloy during solidification. Comparing the results of boron addition, Fig. 6.39, with Al5Ti1B, Fig. 6.24, it is clear that boron addition not only increased the nucleation temperature, but its effect on the reduction of grain size is more pronounced. In addition, the standard deviation of grain size in the case of boron is much smaller which is an indication of refiner stability.

The correlation between grain size, nucleation temperature of α-Al, and recalescence time (t_{Rec}) are shown in Fig. 6.40. By sole boron addition, t_{Rec} value approaches zero after 159 ppm addition (grain size of about 300 μm). The absence of prolonged growth time at the beginning of solidification coupled with large number of nucleants are the precursors for equiaxed morphology to be the predominant grain form.

Semi-Solid Processing

Structural Analysis

The optical micrographs in Fig. 6.41 show the effect of boron on the primary α-Al phase. With increasing the percentage of boron, the structure is not only refined but also the degree of globularity of α-Al particles has increased.

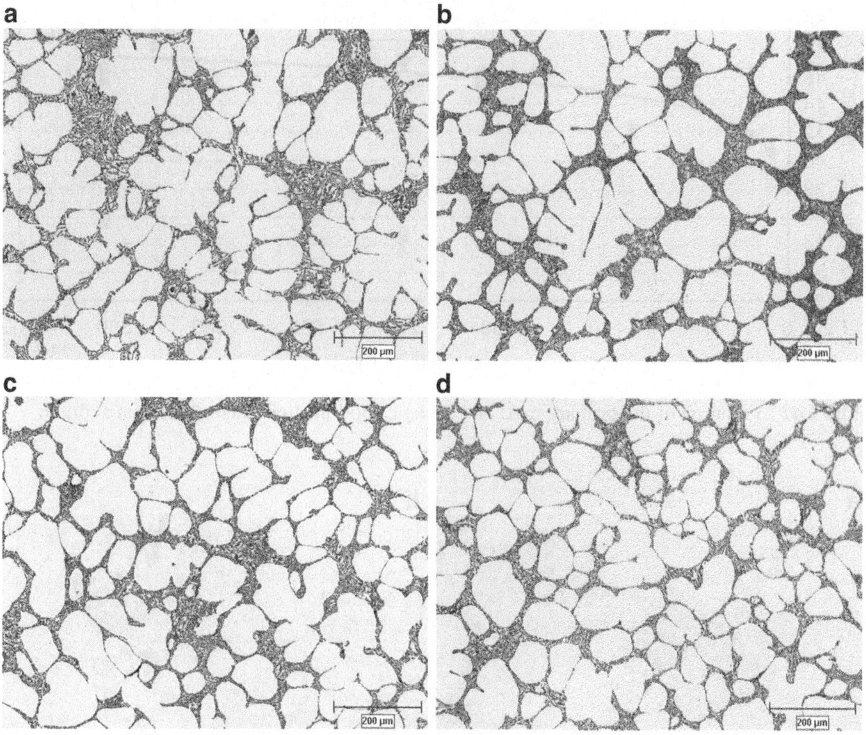

Fig. 6.41 Optical micrographs to show the effects of B-refiner addition in the SEED process: (**a**) without addition, (**b**) 82 ppm, (**c**) 159 ppm, (**d**) 225 ppm

Thermal analysis confirmed that by boron addition, nucleation temperature of α-Al particles increases significantly. By increasing the nucleation temperature, the solidification temperature range of the alloy system also increases. Figure 6.42 demonstrates the variations of the two distinct solidification range of primary α-Al (ΔT_{α}) and eutectic (ΔT_{eut}) for the A356 alloy with boron. By boron addition, ΔT_{α} increases by nearly 8 °C with 200 ppm boron and remains almost constant with higher B addition. However, such conclusion is not fully supported by ThermCalc prediction shown in Figs. 6.32 and 6.33 simply because the refiner has aluminum and AlB_2 particles already while ThermoCalc considers pure boron and aluminum and then their interaction based on the temperature changes. Since refiner particles are stable and have almost no solubility at the casting temperature, therefore they only act as nucleant without interfering on the thermodynamic of the mixing of Al-Si-B as is the basis of the ThermoCalc prediction.

The drawback of boron addition is the formation and agglomeration of boron-based intermetallics with increasing boron content beyond 300 ppm as shown in Figs. 6.43 and 6.44. Microprobe analysis confirmed these particles are AlB_2. In the case of master alloy containing AlB_{12} particles (Al4B, waffle form), it is understandable to find agglomerated AlB_{12} particles and these agglomerations are evident beyond 230 ppm (Fig. 6.43b).

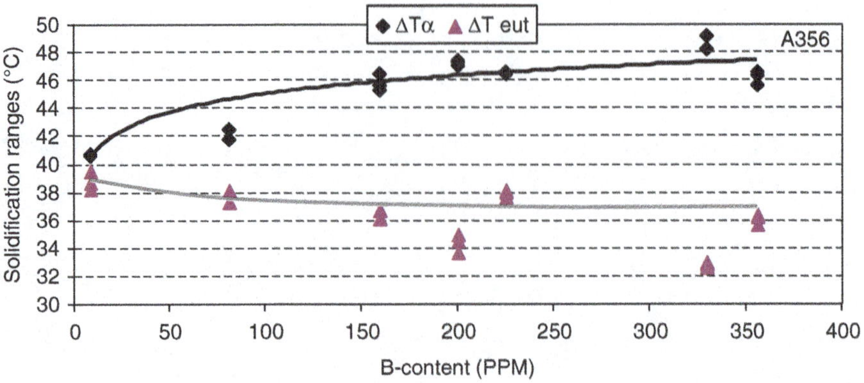

Fig. 6.42 Variation of the primary α-Al and eutectic solidification ranges by boron addition

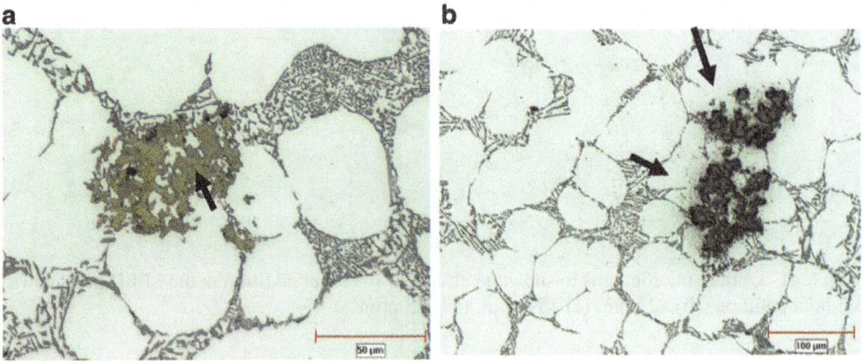

Fig. 6.43 Optical micrographs showing the agglomeration of boride particles; (**a**) addition through Al5B master alloy, sample with 329 ppm B, (**b**) through Al4B master alloy, sample with 803 ppm B

ThermoCalc calculation indicates that in Al7Si1Mg, boron solubility limits are 0.027 % and 0.014 % at 700 °C and 650 °C, respectively, and at eutectic reaction, this value further decreases to less than 0.0018 % (Fig. 6.32).

As stated previously, one of the key parameter for a refiner agent is its disregistry with the matrix and the smaller the disregistry, the better is the refinement. According to the Appendix B, the measured disregistry between AlB_{12} and the α-Al matrix is 151 % and therefore it is unlikely for AlB_{12} particles to act as an effective nucleant in the system. So, the question may arise as why and how the improved refinement efficiency of this master alloy around 200 ppm B could have resulted. It is believed that the refinement efficiency initiates from the following sources;

- The presence of AlB_2 particles within the master alloy
- Dissolved boron in the alloy
- Transformation of AlB_{12} to AlB_2

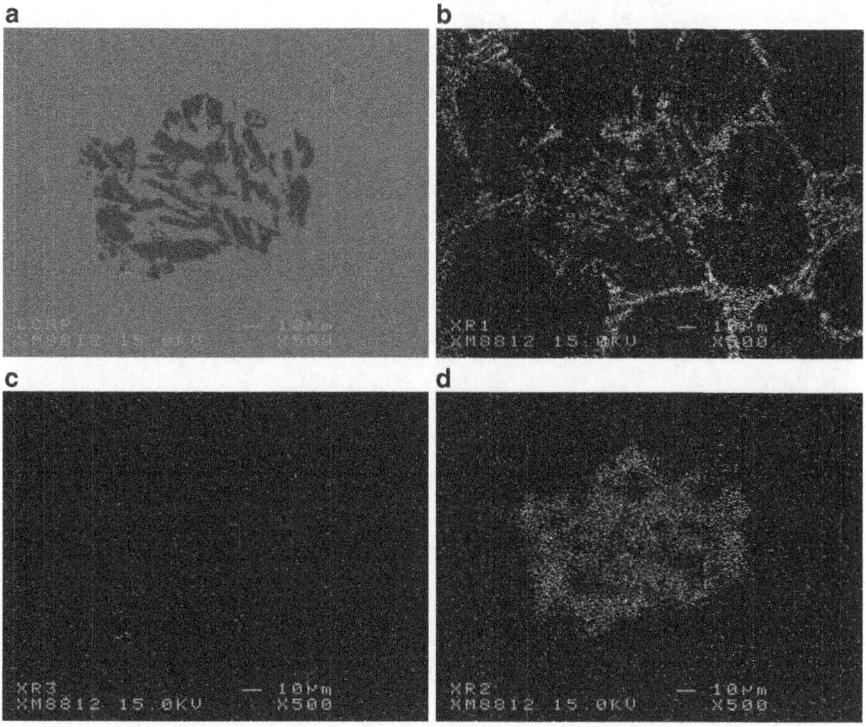

Fig. 6.44 (**a**) Backscattered electron micrographs showing the agglomeration of boride particles and X-ray maps for: (**b**) Si, (**c**) Ti, and (**d**) B map (sample with 329 ppm B)

By examining the master alloy structure more closely, it is clear that there are AlB_2 particles in the matrix hidden among the AlB_{12} agglomerates as shown in Fig. 6.45a where another master alloy with mainly AlB_2 particles is shown to highlight this point (Al5B rod form, Fig. 6.45b). It is believed that these particles and smaller ones which could not be resolved by optical microscopy play fundamental role on improving the refining efficiency of the master alloy (the ultimate purpose of this master alloy is to eliminate transitional elements from the pure aluminum alloy [33]).

Figure 6.46 shows the solubility limit of B in the melt at different temperatures. At 700 °C, there is about 0.05 % boron in the solution and this amount is quite enough to react with aluminum and form indigenous nucleants. Further, as shown in Fig. 6.18, the majority of the boride particles in this master alloy (Al4B, waffle form) are AlB_{12}. ThemoCalc calculation shows that AlB_{12} partially transforms to AlB_2 at 1289 °C and moreover is only stable at B concentrations in excess of ~3 % (not shown in the graph). Therefore, it can be concluded that it is almost unlikely that AlB_{12} could transform to AlB_2 within processing window (note that every predictive software calculates the changes based on their logic and database and therefore the result may not be identical).

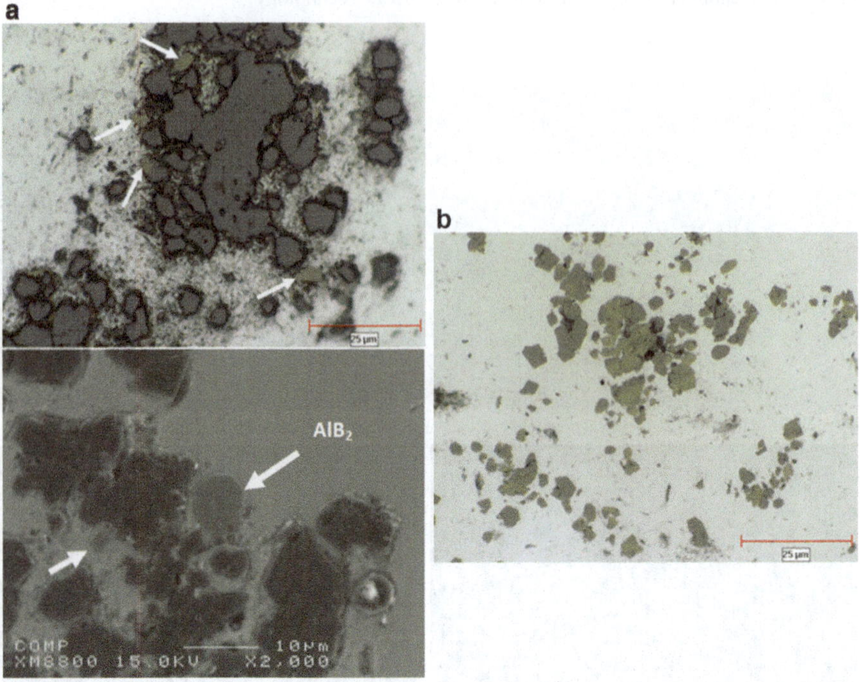

Fig. 6.45 AlB_2 particles in different master alloys: (**a**) Al4B, waffle form (*arrows* show the AlB_2 particles), (**b**) Al5B, rod form [25]

Fig. 6.46 Thermocalc calculation of isopleth for Al7Si1Mg versus B

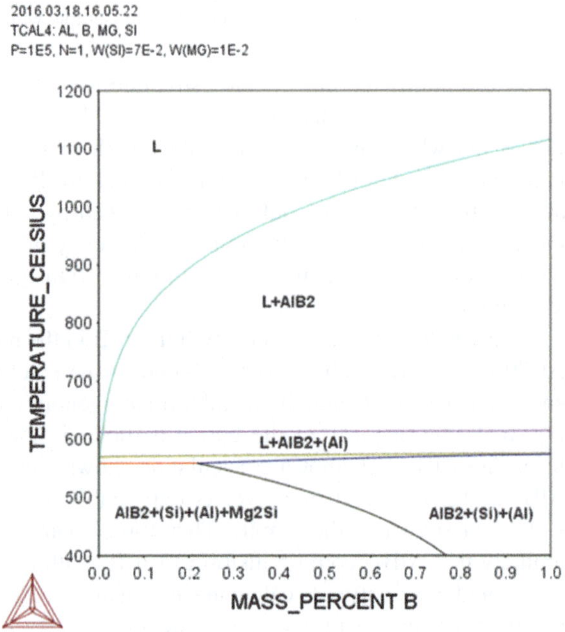

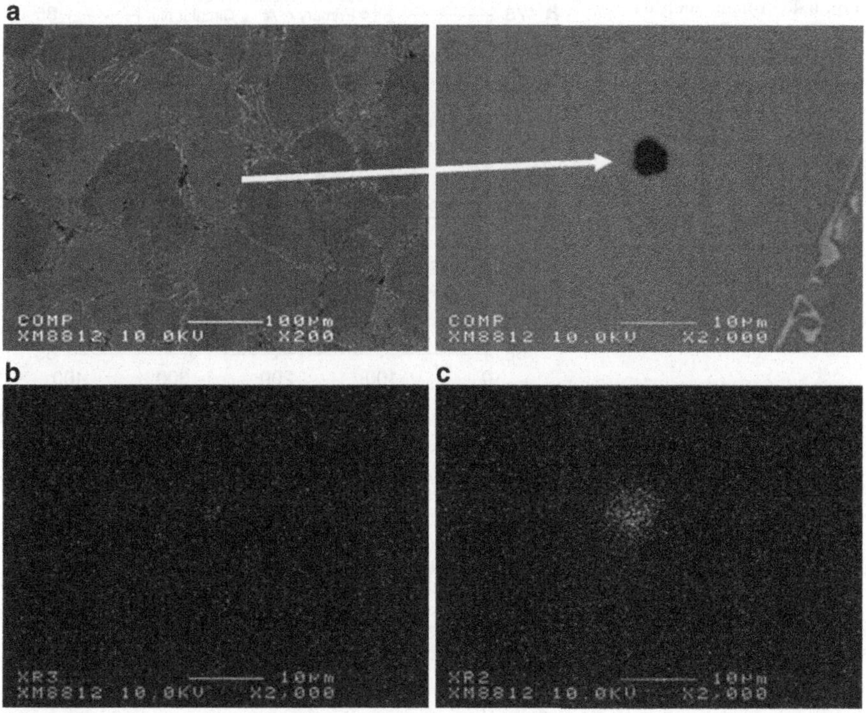

Fig. 6.47 (**a**) Backscattered electron micrographs of a nucleant in the sample with 803 ppm B (through Al4B master alloy), and the respective X-ray maps of: (**b**) Ti, (**c**) B [37]

In addition, there is another possibility for this master alloy to be an efficient refiner and that is the dissolved Ti in the alloy and its high affinity for B to partially transform the AlB_{12}/AlB_2 particles to complex compound of AlBTi, suitable for acting as nucleation site for the primary α-Al.

As stated earlier, identification of nucleants in the as-cast billets is challenging as the detection probability depends on the polished surface and whether it intersects the nucleant or not. Figure 6.47 shows a nucleant which has been characterized by the backscattered electron micrograph and X-ray maps of the particle. Due to the size of this particle, quantitative microprobe analysis was impossible; however, X-ray maps of the nucleant confirm its AlB stoichiometry and knowing the inability of AlB_{12} to act as a nucleant, it is true to conclude this particle being AlB_2.

Image Analysis

In order to fully characterize the microstructure of semi-solid slugs, it is necessary to quantify the percentage, size, and morphology of primary particles. For this purpose, image analysis was employed and Figs. 6.48 and 6.49 illustrate the results.

Fig. 6.48 Image analysis results for A356 with B addition: (a) primary α-Al percentage, average circular diameter, (b) number density of α-Al particles, percentage of α-Al particles having aspect ratio >2, (c) area/perimeter

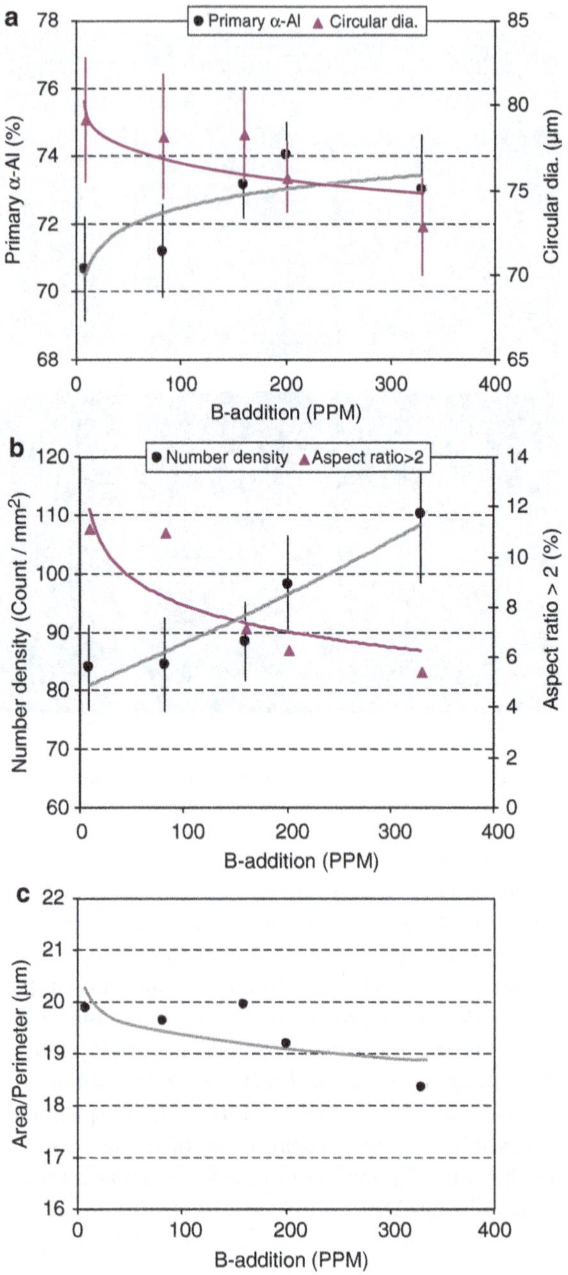

Increasing of primary α-Al percentage is noticeable since not only numerous nucleation sites are introduced to the system but also liquidus line is shifted to higher temperatures. In contrast to Al5Ti1B addition, boron proved to be more efficient and this potential is visible through more globule size reduction and sharp

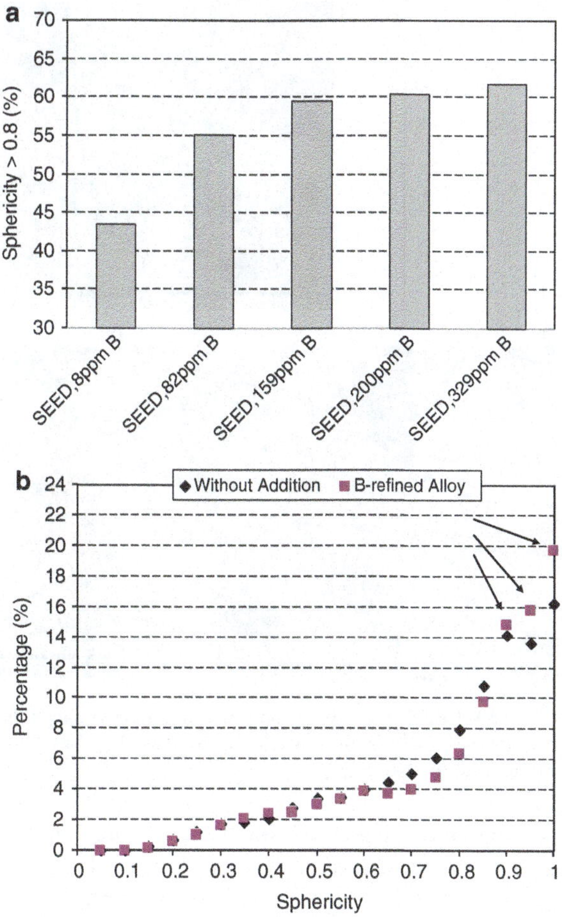

Fig. 6.49 (**a**) Variation of sphericity with boron addition (Al5B), (**b**) data sorting without B and with 225 ppm B

increase in the number of α-Al particles per unit area, number density. The reduction of A/P and percentage of α-Al particles having aspect ratio >2 do also confirm the formation of smaller and rounder primary particles.

Figure 6.49a shows the percentage of the particles with sphericity greater than 0.8. The greater the sphericity number, the more globular particles are in the structure. Figure 6.49b compares statistical significance of sphericity. With refining, data is more concentrated on larger values of sphericity, i.e., near 1. As it will be shown in rheological experiments, this image analysis parameter is a clear indication of the flowability of slurries.

For better understanding of refiner capabilities, a comparison is made by electron backscatter diffraction (EBSD) mapping between the unrefined and refined alloys by AlTiB and Al4B master alloys (Fig. 6.50). As evident, the untreated slurry is a mixture of dendrites and globules created during SSM processing. By addition of TiB_2 particles (through AlTiB master alloy), the number of individual globules

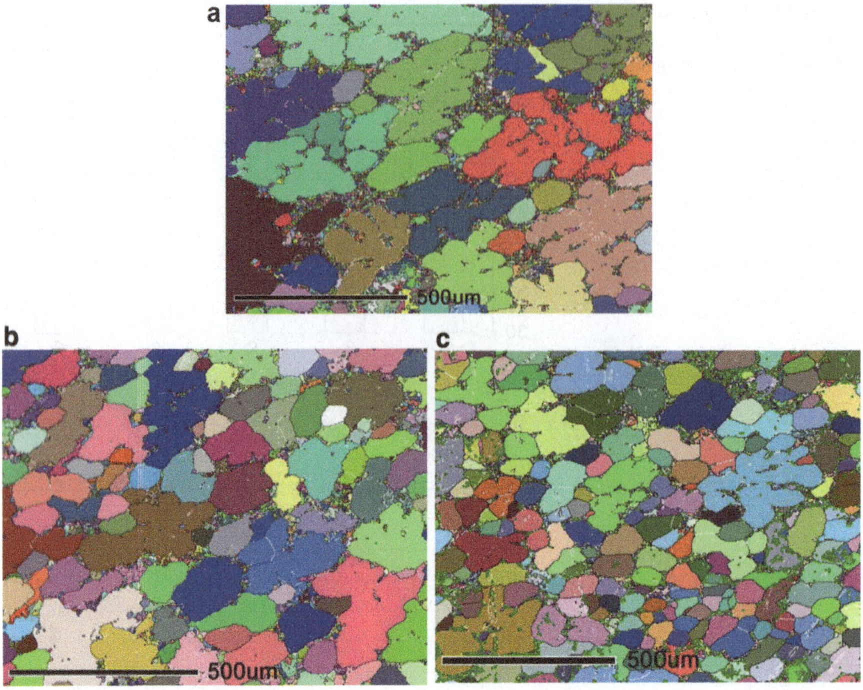

Fig. 6.50 EBSD maps of the untreated and treated alloys, *black lines* denote high angle grain boundaries with misorientations greater than 10°; (**a**) without addition, (**b**) 1300 ppm Ti, 210 ppm B, (**c**) 226 ppm B

per unit area is increased. Improved distribution of globules and the reduction in globule size is achieved by boron addition. However, some technical points should be considered; for example, grain refined billet is more sensitive to movement (due to lower viscosity) which makes it more difficult to handle from the billet preparation unit to die casting machine. As an additional remark, agglomeration of primary α-Al dendrites facilitates billet handling during the casting process.

6.2.1.4 Alloy Chemistry and Grain Refining: Effect of Titanium in Solution

The chemistry of molten alloy plays an important role on grain refining. Small addition of alloying elements reduces the grain size of the as-solidified structures to some degree depending on the alloy efficiency. The fact that solute elements in aluminum alloys have beneficial effects in reducing the grain size and improving mechanical properties of as-cast products goes back to Kissling and Wallace's work in 1963 [23]. It is generally believed that small addition of alloying elements interferes with grain growth during solidification by providing conditions to initiate greater number of nuclei to form and to promote the formation of finer grains. The solutes form an enriched boundary layer ahead of solidification front in which the actual temperature is lower than that of solidification temperature, constitutional undercooling [38]. The constitutional undercooling as given in (6.1) may be interpreted in terms of "Growth Restriction Factor," GRF. That is when the solidification is controlled by solutal not thermal diffusion and the degree of partitioning $\left(C_l^* - C_s^*\right)$ may be approximated to $C_0(1-k)$ when the solutal undercooling, ΔT_s, is small, i.e., $\Delta T_s \ll (T_m - T_l)$ where T_m is the melting temperature of pure aluminum and T_l the liquidus temperature for the alloy of C_0 composition [39]. The presence of solute leads to growth restriction as defined by GRF [40, 41] in (6.2). The constitutional undercooling is therefore expressed in terms of growth restriction factor in (6.3).

$$\frac{G_l}{R} = -\frac{m_l C_0(1-k)}{D_l k} \tag{6.1}$$

$$\mathrm{GRF} = m_l C_0(k-1) \tag{6.2}$$

$$\frac{G_l}{R} = \frac{GRF}{D_l k} \tag{6.3}$$

where: G_l = Temperature gradient in the liquid (K/m), R = Growth rate (m/s), C_0 = Initial alloy concentration (wt%), m_l = Liquidus slope (dT_l/dC) (K/wt%), k = Distribution or partition coefficient, C_l^* *and* C_S^* = Equilibrium solute concentrations of the liquid and solid at the interface (wt%), and D_l = Diffusion coefficient in the liquid (m^2/s).

It is therefore concluded that the greater the degree of constitutional undercooling, the more effective the solute would be in restricting the grain growth as GRF is inversely proportional to the growth rate. It was suggested that the growth restriction factor in an alloy should be the sum of GRFs for each alloying element (assuming that there is no interaction between the solutes) [1, 42]:

$$\mathrm{GRF}_{\mathrm{total}} = \sum_{\mathrm{i}} m_{l,i}\left(C_{0,i} - C*_{l,i}\right) \tag{6.4}$$

Table 6.6 shows the values of the parameters needed for the calculation of GRF numbers, (6.2), for the five important elements usually added to Al alloys.

Table 6.6 Data required for calculating the growth restriction factor, GRF [1, 41–43]

Element	k	m_l	$m_l\,(k-1)$	Max C_0 (wt%)	System
Ti[a]	7.8, 9	33.3, 30.7	220, 245.6	0.15	Peritectic
Si	0.11	−6.6	5.9	12.6	Eutectic
Mg	0.51	−6.2	3.0	34.0	Eutectic
Fe	0.02	−3	2.9	1.8	Eutectic
Mn	0.94	−1.6	0.1	1.9	Eutectic

[a]Depends on the reference

Table 6.7 Chemical analysis of the melts (wt %)

	Si	Mg	Fe	Mn	Cu	Ti	B	Al
356 without Ti	6.5–7.1	0.35–0.4	0.07–0.08	0.002–0.012	0.001	Max 0.0058	Nil	Bal.
356 with Ti						0.1–0.13		

The largest GRF value belongs to Ti having the highest growth restriction effect on aluminum. However, for Al alloys where Si concentration is appreciable, such as hypoeutectic Al–Si foundry alloys, the high Si content gives an overall higher GRF number. For instance, Al alloy with 7%Si and 0.1%Ti has GRF values of 41.3 and ~22 for Si and Ti, respectively.

Effect of Titanium in Solution

The effect of Ti as refiner was studied previously but as dissolved alloying element is revisited in this section and it will be shown that it could restrict the grains and globules size both in conventional and semi-solid casting. The chemical compositions of the alloys tested to highlight the effect of minor Ti addition as a routine practice in foundry and casting industry are listed in Table 6.7.

Conventional Casting

Thermal Analysis

Figure 6.51 illustrates the effect of dissolved Ti on the early stages of solidification for two different Ti concentrations in 356 alloys. The solidification events, nucleation and growth of the primary particles, have been shifted to higher temperature with increasing titanium addition. This is also predictable by thermodynamic calculations, Fig. 6.21b, to confirm that the liquidus temperature slightly increases with Ti addition.

Figure 6.52 shows nucleation temperature ($T_{\mathrm{nuc}_{\mathrm{Al}}}$) and recalescence value of the primary α-Al particles. The concept of increasing the liquidus temperature due to Ti addition is apparent where $T_{\mathrm{nuc}_{\mathrm{Al}}}$ is increased by 2–3 °C to about 617 °C. The predicted rise as shown by ThermoCalc calculation is lower, Fig. 6.21. It has to be emphasized that such discrepancies between the predicted and experimental results are mainly attributed to ThermoCalc calculation being under equilibrium condition while the current study is under nonequilibrium condition including higher cooling rate and the presence of other alloying or trace elements.

Commercial 356 alloy has lower recalescence value which means there is less growth of the nuclei with smaller recalescence. Bearing in mind that this is a localized region near the tip of thermocouple and with extending this logic across the bulk liquid, it is true to assume growth restriction is effective and the formation of a refined structure should be expected.

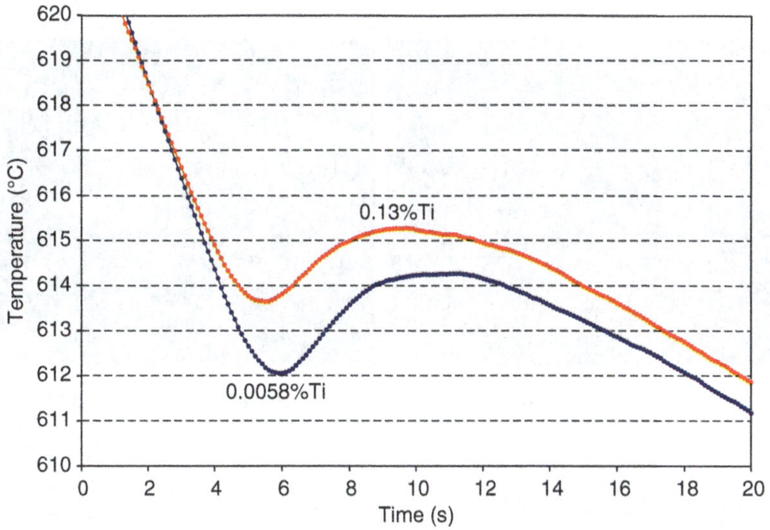

Fig. 6.51 Initial segment of typical cooling curves for a selection of samples (central thermocouple) of 356 alloys with different Ti levels [44]

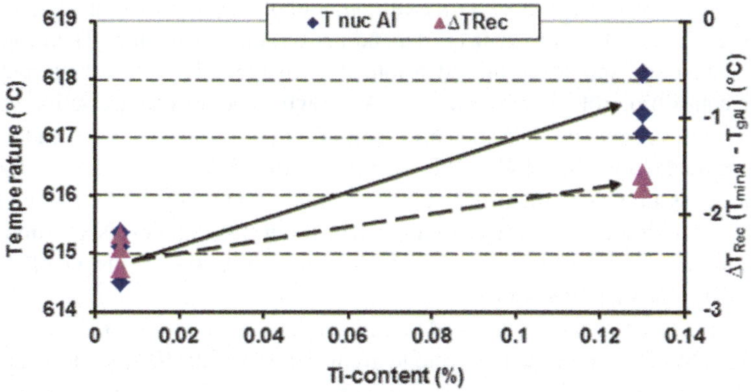

Fig. 6.52 The effect of dissolved Ti on the nucleation temperature and recalescence value (ΔT_{Rec}) of the primary α-Al particles [44]

Structural Analysis

The polarized light images in Fig. 6.53 show the microstructural differences in conventional casting between the 0.0058 % and 0.13 % Ti-containing alloys. The alloy with 0.13 % Ti has finer dendritic structure, and the grain size appears to have been reduced considerably. It may be argued that such refinement is due to more effective nucleation, not the growth restriction capability of Ti. In order to clarify

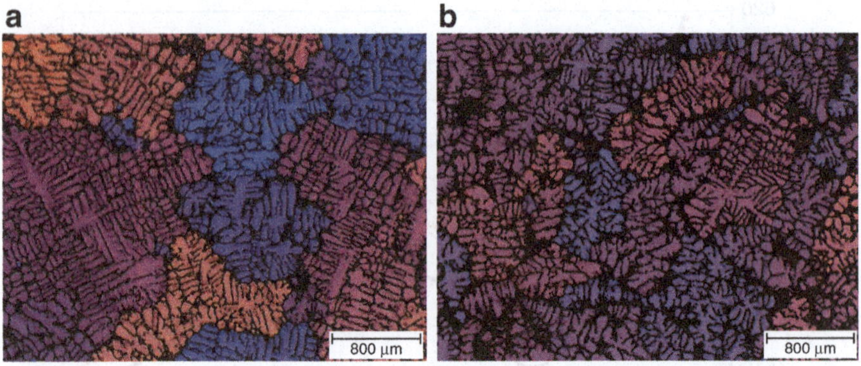

Fig. 6.53 Effect of Ti in solution on the microstructure: (**a**) with 0.0058%Ti, (**b**) 0.13%Ti

this issue, it is helpful to define the two concepts of "grain refining" and "grain growth restriction." The former is dependent primarily on the presence and then the size, population, and morphology of nucleants, i.e., independent Ti-based compounds, while the latter is about the dissolved Ti in the α-Al matrix and its diffusion across solid–liquid interface. It is believed at this level of Ti concentration, it is rather unlikely and remote to have independent Ti-based particles, since as shown in Fig. 6.21, the formation of Ti-based compounds, the potential nucleants, only begins at Ti% > 0.11 or > 0.17 depending on the chemistry. Therefore, the concerning issue should be the diffusion of Ti at the solidification interface and its rate controlling role on the primary α-Al growth. The growth restriction concept is thus more acceptable than the formation of Ti-based nucleants to render more effective nucleation, especially at the early stages of solidification. However, it may be true to assume that such diffusion of Ti may eventually lead, in a localized volume, to the formation of Ti-based nucleants, but this is a far-distant time event and surely not at the beginning of solidification, i.e., the diffusion of Ti is a prerequisite for such mechanism.

Figure 6.54 shows the effect of Ti on the grain size of 356 alloys. For the alloy with 0.0058%Ti in solution, the average grain size is about 1100 μm and reduces to around 850 μm when Ti concentration is increased to 0.13 %, a grain size reduction of almost 22 %. Furthermore, the value of standard deviation is much greater for the alloy with lower Ti content which is a clear indication of wider distribution of grain size in this alloy. For higher Ti content, the smaller value of standard deviation suggests a more uniform grain size, and consequently the properties are expected to be much more uniform.

It is further clear from Fig. 6.54 that the grain size and nucleation temperature of the primary α-Al particles vary inversely with increasing Ti in solution; the average α-Al grain size decreased with increasing nucleation temperature. The rise in nucleation temperature simply extends the nucleation period and generates more nuclei.

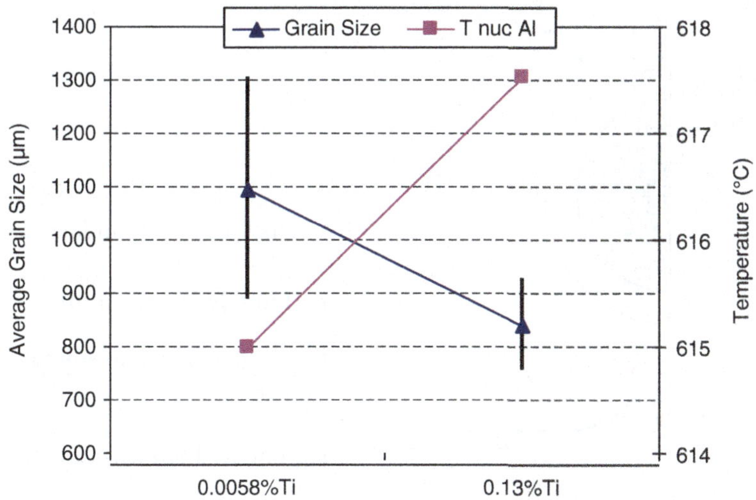

Fig. 6.54 Relation between grain size and T_{nucAl} in different Ti content [44]

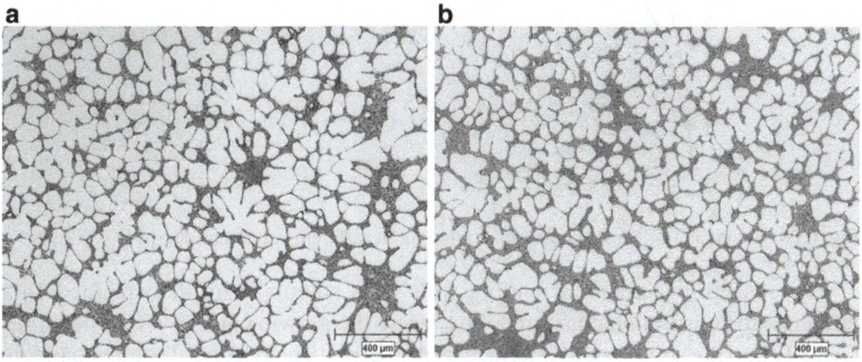

Fig. 6.55 Optical micrographs in the SEED process: (**a**) 0.0058%Ti, (**b**) 0.13%Ti [44]

Semi-Solid Processing

Structural Analysis

The optical micrographs in Fig. 6.55 show the typical microstructure of the billets. The primary α-Al particles are almost fully globular irrespective of the Ti content. Although the microstructure of the alloy containing 0.13 % Ti appears to have been slightly refined, the effect of dissolved Ti on the microstructure is better realized if it is characterized quantitatively. Therefore, both quantitative metallography and rheological tests were carried out on different Ti-containing alloys to highlight the effect of Ti on grain growth restriction.

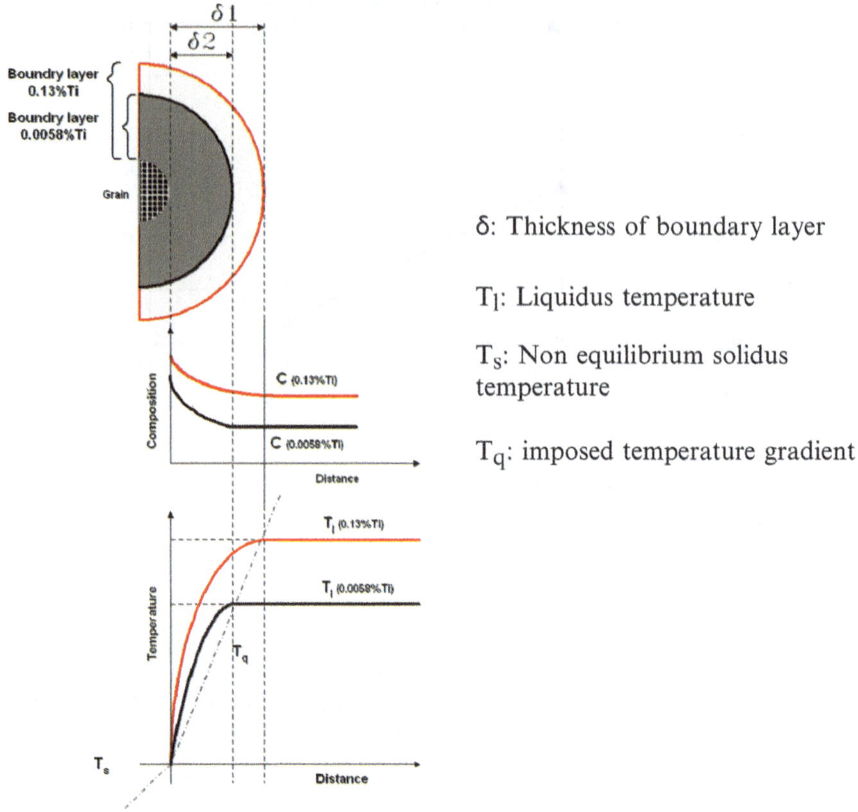

Fig. 6.56 Schematic of constitutional undercooling in different solute buildup [44]

The concept of structural refining with Ti in solution may be explained with reference to constitutional undercooling where the growing primary α-Al dendrites reject Ti into the solid–liquid interface as they grow. The rejected Ti atoms build up ahead of the solidification interface and form a Ti-enriched boundary layer. This is graphically illustrated in Fig. 6.56 where the concentration of Ti at the boundary layer varies for different Ti concentrations due to differences in the diffusion fluxes for different Ti contents. The value of diffusion flux (J), $J = -DA\frac{\partial c}{\partial x}$, is greater for 0.13%Ti than that of 0.0058, since $\frac{\partial c}{\partial x}$ is greater for 0.13%Ti. Therefore, the degree of constitutional undercooling is much greater for 0.13 % Ti than that of 0.0058 % Ti. Equally, the boundary layer thickness is greater for higher Ti content alloy as schematically illustrated in Fig. 6.56. As pointed out by Kurz and Fisher [45], the equivalent boundary layer thickness, (δ_c), is inversely proportional to growth rate, (R), for an alloy solidified at different growth rates:

$$\delta_C = \frac{2D}{R} \tag{6.5}$$

where D is the diffusion coefficient for dissolved titanium. Since higher growth rates are equivalent to lesser diffusion and rejection of solute element into the boundary layer, i.e., lower $\frac{\partial c}{\partial t}$, it may be acceptable to consider higher concentration of solute element within the boundary layer is equivalent to that of lower growth velocity or greater thickness of this layer.

Furthermore, as the concentration of Ti increases at the interface during later stages of solidification, it may reach the level where new nucleants of Al_3Ti could form within the interface layer. The presence of such particles encourages the formation of new α-Al nuclei within the interface. The newly formed α-Al particles are expected to reject Ti into the newly formed interfaces as they grow. The repetition of such mechanism ensures the formation of finer and more equiaxed and spherical particles as observed in 0.13%Ti added 356 alloy.

The results of quantitative metallography are presented in Fig. 6.57. The mean globule size "equivalent circular diameter" appears to have decreased with increasing Ti content. The rise in the number of α-Al particles per unit area (mm^2), number density, coupled with the reduction of area to perimeter ratio, A/P, indicate microstructural refinement leading to the formation of more isolated and finer particles in

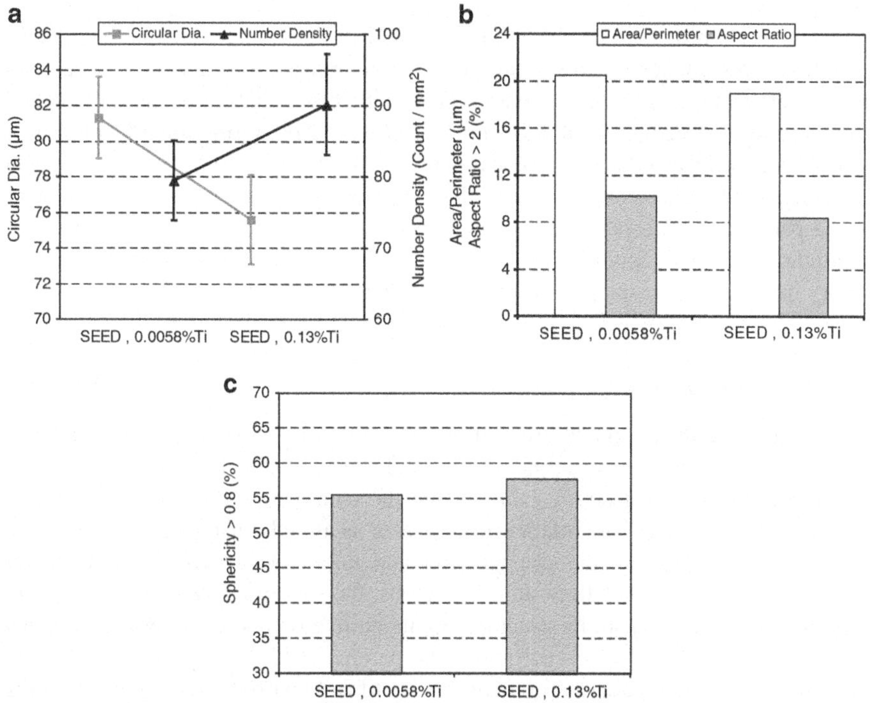

Fig. 6.57 Selected image analysis results: (**a**) circular diameter and number density of α-Al particles, (**b**) area/perimeter and percentage of particles having aspect ratio >2, (**c**) the percentage of particles having sphericity >0.8 [44]

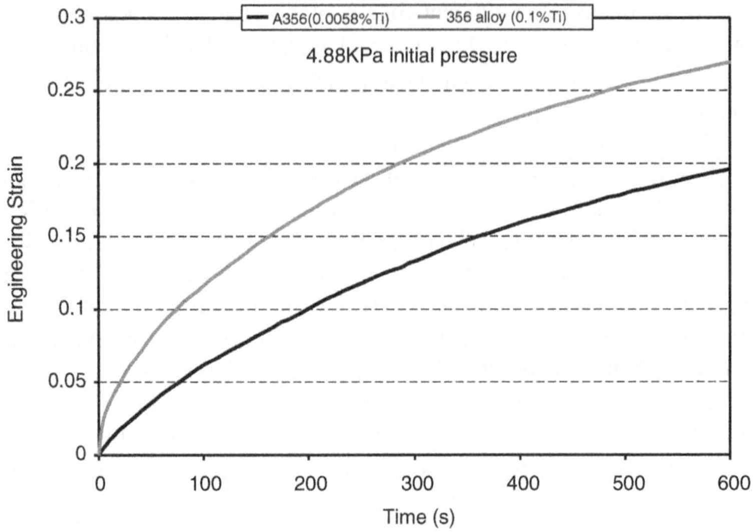

Fig. 6.58 Strain–time graphs for different Ti in solution [44]

the Ti-content billets. With the addition of growth restricting elements such as Ti, dendrite growth tends to slow down in direction opposite to heat flow approaching those of the lateral growth. The overall growth therefore tends to be more radial and the resulting structure should be more globular. The formation of spheroidal particles is critical on the rheological behavior of the slurries as discussed later in this chapter. This is clearly shown by the reduction in the percentage of the primary α-Al particles with aspect ratio greater than 2. The formation of more spherical particles is further supported by the rise in the percentage of particles with sphericity number greater than 0.8 with increasing Ti content (Fig. 6.57c).

Strain–Time Graphs

The changes in the billets deformability due to the primary **α**-Al particles size and distribution are clearly detectable on the graphs in Fig. 6.58. The alloy with almost no Ti shows the lower strain values while the commercial 356 billet deforms at much higher range due to smaller globule size as already discussed. As shown in Sect. 4.2, the billets with more globular and finer microstructure yield greater engineering strain, better flow. In other words, the superior flow of the alloy with higher Ti content is due to its structure having more globular and finer grains which enables it to flow better. Furthermore, the parameters from Fig. 6.57, especially average circular diameter and number density of α-Al particles, clearly support the conclusion that the billet with higher dissolved Ti has more globules with smaller size.

6.2.2 *Modification*

Silicon is one of the most significant alloying elements incorporated in Al alloys. Its addition is to improve castability, fluidity, reduce shrinkage, and to render superior mechanical properties. The morphology of silicon however plays an important role on the mechanical properties of finished products. Therefore, the common practice is to modify the as-cast flake or acicular silicon morphology employing special heat treatment or addition of certain modifiers. As a result, the silicon morphology changes to fibrous form with enhanced mechanical properties of the as-cast parts.

For conventional casting, Si modification has been investigated widely by many researchers but in the SSM processing, few published data could be found as stated earlier in this chapter (e.g., [20–22]). The purpose of this contribution is to report the effects of Sr addition as a modifier to A356 Al–Si alloy for both conventional and SSM casting methods.

6.2.2.1 Master Alloys

Al10%Sr

The optical and backscattered electron micrographs together with the associated X-ray maps show the microstructure of the commercially produced master alloy (Fig. 6.59). The structure of Al10Sr master alloy consists of blocky Sr-rich particles, distributed within Al matrix. The average particle size is about 16 ± 8 μm and microprobe analysis results suggest the Al_4Sr stoichiometry.

6.2.2.2 Modification in Al7%Si and A356 Alloy

The modification treatment of A356 and Al7Si alloys was carried out by addition of Al10Sr rod master alloy. The base chemical compositions are listed in Table 6.8.

Conventional Casting

Thermal Analysis

Figure 6.60 shows typical cooling curves and first derivatives for Al7Si and A356 alloys. Although both alloys solidify over solidification range but there are differences if examined closely. For Al7Si alloys, solidification basically consists of two major reactions: formation of primary α-Al dendrites and subsequent formation of eutectic phases (first and second peaks in the first derivative, respectively).

The addition of Mg and other alloying elements, intentionally or as trace elements, has resulted in pre- or post-eutectic reactions and changed the critical points or temperatures as shown in Fig. 6.60. For example by Mg addition to the

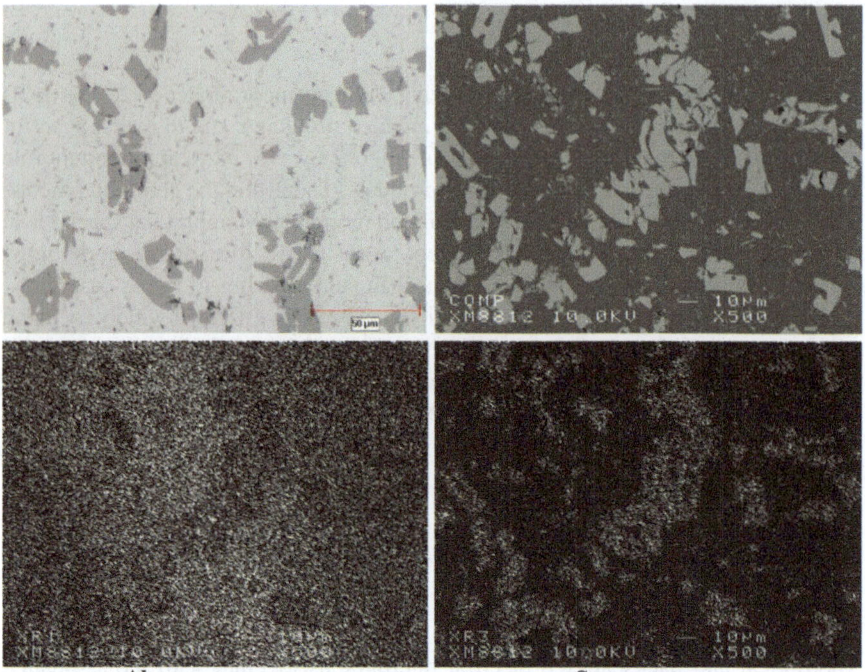

Al x-ray map Sr x-ray map

Fig. 6.59 Typical optical image and backscattered electron micrographs with X-ray maps of Al10Sr master alloy, cross section. rod form [46]

Table 6.8 Chemical analysis of the base alloys (wt. %)

	Si	Mg	Fe	Mn	Cu	Ti	B	P	Sr	Al
Al7Si	7.0–7.3	Nil	Max 0.09	Nil	Nil	Nil	Nil	Max 0.0003	Nil	Bal.
A356	6.62–6.81	0.36–0.4	Max 0.08	Max 0.003	Nil	Max 0.0058	Nil	Max 0.0003	Nil	Bal.

alloy, the main eutectic reaction shifted to a lower temperature and formed over a longer period. Normally, the solidification termination of pure binary Al–Si alloys is sharp and distinctive while this was not the case for Al7Si alloys used in this study. This is mainly due to the trace elements such as Fe which was part of the alloy used in this study, confirmed by optical emission spectroscopy (OES) to contain 0.03–0.09%Fe. As reported by Backerud et al. [27], alloys containing trace amount of iron have smooth solidification termination.

By magnesium addition, eutectic compounds do not form at a constant temperature. This concept is well depicted by performing thermodynamic calculation using FactSage software [48]. An isopleth diagram, pseudo-binary phase diagram, was constructed for Si addition as shown in Fig. 6.61. By addition of Mg to the hypoeutectic Al–Si alloys, not only solidification terminated at lower temperature,

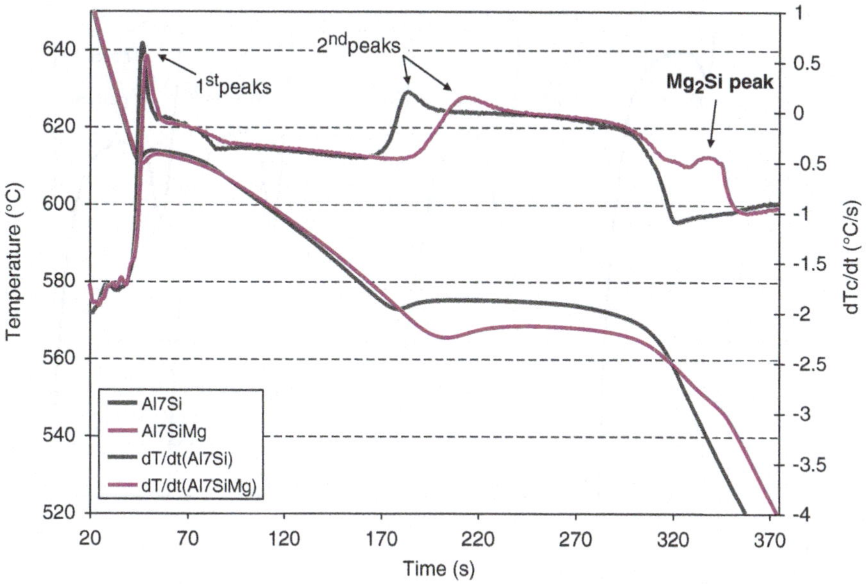

Fig. 6.60 Comparing cooling curves of Al7Si and Al7Si0.35Mg alloys [47] (reprinted by permission of Taylor & Francis Ltd)

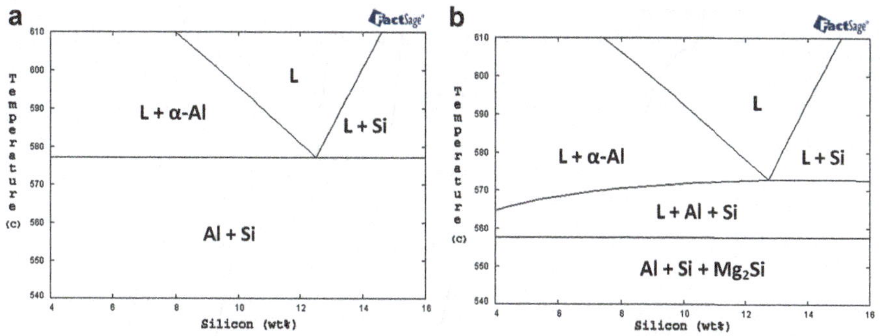

Fig. 6.61 FactSage calculation of phase diagram for: (**a**) Al7Si and (**b**) isopleth of Al7Si1Mg [47] (reprinted by permission of Taylor & Francis Ltd)

but also the eutectic reaction is no longer isothermal, as seen for high purity Al–Si alloys, and forms over a temperature range.

Figure 6.62 shows the eutectic segment of the cooling curves. For both alloys, the addition of small percentage of strontium (about 50 ppm) has depressed the eutectic plateau by 3–6 °C. After the initial sharp drop in the eutectic temperature, it shifts up slightly by further Sr addition.

Figure 6.63 illustrates the changes in the cooling curve parameters due to the Sr addition. The eutectic nucleation temperature, $T_{\mathrm{nuc_{eut}}}$, decreased to its minimum value at about 50–100 ppm Sr and increased slightly before becoming constant at Sr concentrations of about 300 ppm and more. A similar trend is observed for the

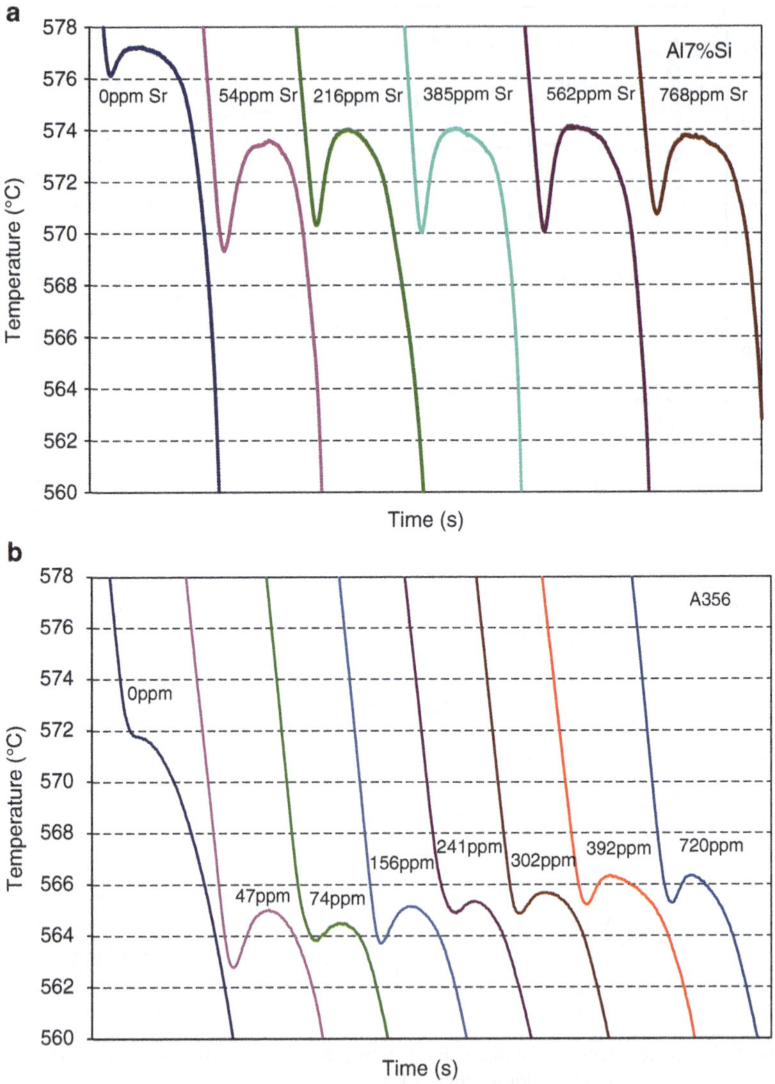

Fig. 6.62 The eutectic segment of the cooling curves taken from the thermocouples at the center of graphite cups with the addition of Sr, (**a**) Al7Si, (**b**) A356 [46]

eutectic maximum temperature of both alloys, $T_{\text{max}_{\text{eut}}}$. Addition of Sr depresses the eutectic temperature parameters, $T_{\text{max}_{\text{eut}}}$ and $T_{\text{nuc}_{\text{eut}}}$. Basically, the higher the difference of $T_{\text{max}_{\text{eut}}}$ or $T_{\text{nuc}_{\text{eut}}}$ before and after Sr addition, the stronger is the silicon modification. Therefore, the optimum addition seems to be around 50–100 ppm.

Eutectic recalescence (Fig. 6.63c) also increased to its maximum value at about 50–100 ppm Sr and decreased slightly with further addition. Reaching the maximum value is indication of greater barrier for the eutectic nucleation due to the

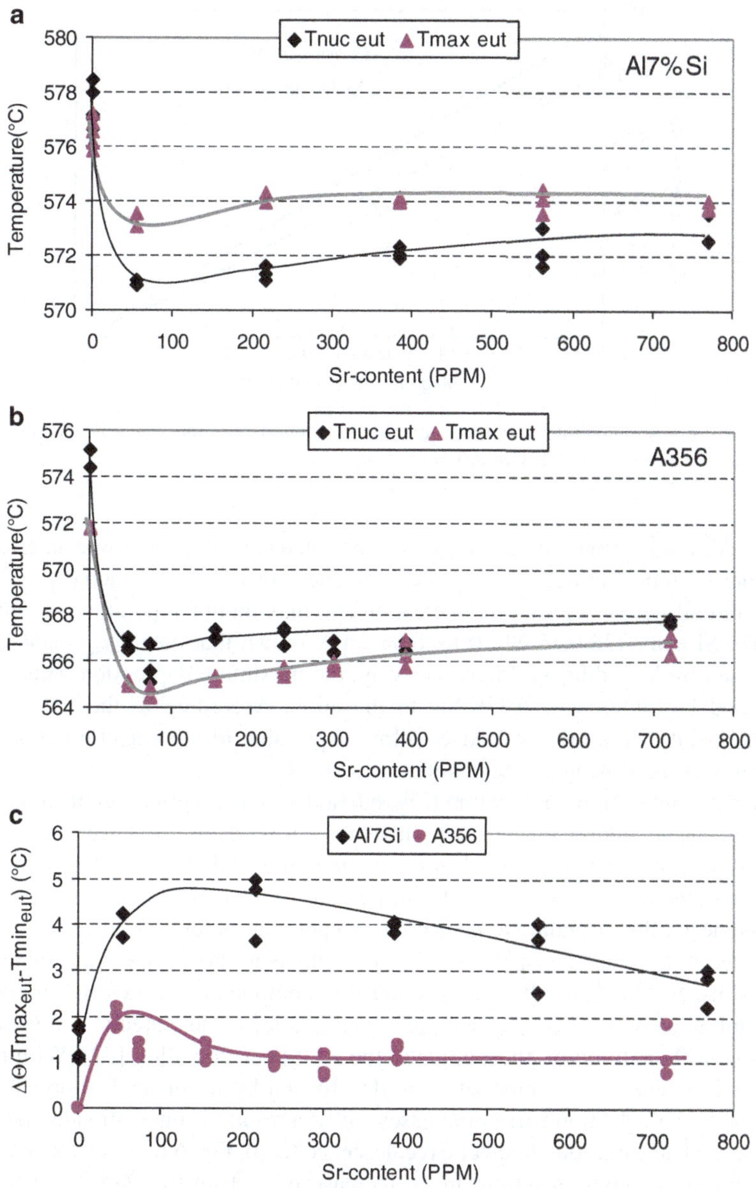

Fig. 6.63 Effect of Sr addition on various thermal analysis parameters

Sr addition. The reduction in eutectic temperature and the fact that the alloy liquidus remained unchanged have caused the primary α-Al solidification range to increase. It is therefore anticipated to have an increase in the percentage of primary α-Al phase as will be discussed later.

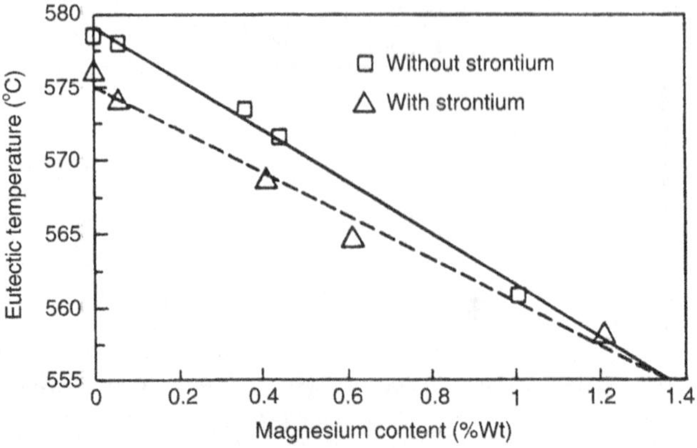

Fig. 6.64 Depression of eutectic temperature as a function of Mg content of Al7%Si alloys [49]. (Reprinted by permission of Taylor & Francis Ltd)

In addition to strontium, alloying and trace elements may also have an effect on the eutectic temperature. An example is magnesium which intentionally added to industrial alloys. By comparing nucleation and maximum temperatures of eutectic in Al7%Si and Al7Si0.35 Mg (Fig. 6.63), it is shown that both $T_{\mathrm{nuc}_{\mathrm{eut}}}$ and $T_{\mathrm{max}_{\mathrm{eut}}}$ decreased by Mg addition. In another way, the maximum depression values were increased by addition of 0.35%Mg to the alloy. According to the Joenoes and Gruzleski [49], in a modified Al–Si alloy, depression of the eutectic temperature is more with increasing the Mg content (Fig. 6.64).

Eutectic nucleation temperature (Fig. 6.63a) has a direct effect on the number of potential nuclei found in the melt. In other words, there are fewer barriers for nucleation with increasing nucleation temperature and thus greater number of isolated eutectic Si particles may form, i.e., this is the case for flake morphology. The highest value for eutectic nucleation temperature before modification may be attributed to the trace phosphorus and iron contents in the alloy used in this study. As confirmed [50, 51], AlP particles and iron compounds are favorite nucleation sites for silicon. It was reported that both liquidus and eutectic temperatures increase with addition of phosphorus to the Al–Si eutectic alloy [52]. With modifier, such potential nucleation sites are deactivated by fewer nucleation sites and consequently nucleation barrier increases and as a result the nucleation temperature decreases. The variation of eutectic recalescence ($\Delta\Theta$), Fig. 6.63c, shows that small Sr addition (47 ppm) resulting in an increase of $\Delta\Theta$ up to ~2–4 °C. This is a nucleation barrier equivalent to having a reduction in the number of active nucleation sites which forces Si to branch out to form a three-dimensional interconnected morphology, fibrous, to keep up with the growing aluminum phase.

Furthermore, from growth point of view, Sr was reported to poison the re-entrant edge growth steps in silicon crystals [53, 54], thus deactivating the "TPRE" (Twin Plane Re-entrant Edge) growth. Consequently, due to interruption in the effective growth mechanism for Si, high degree of $\Delta\Theta$ is needed to drive an alternative growth mechanism for the development of the fibrous silicon structure which is "Impurity Induced Twinning—IIT" mechanism [53].

In conclusion, Sr addition affects both nucleation and growth behavior of eutectic Si to transform silicon morphology from flake to fibrous. Furthermore, the overall results of thermal analysis for this study are in line with the previous reports [55–57].

Structural Analysis

The morphology of the primary α-Al phase appears to be somewhat different to that of nonmodified alloy as presented in Fig. 6.65. If the micrographs are examined closely, the eutectic regions within the modified alloy appear to have boxed-in the primary α-Al phase, while this is not the case for nonmodified structure. In other words, eutectic regions within modified alloy are continuous and interconnected while there are isolated individual patches of eutectic regions for the nonmodified alloy. Such distinct difference may be attributed to the fibrous morphology of Si and the fact that each individual Si nucleus grows through branching and expanding three-dimensionally rather than as an isolated flake for nonmodified alloy.

The concept of "boxing-in" of the primary α-Al particles by the eutectic mixture in the modified alloy is well illustrated in Fig. 6.66 along with the changes related to the silicon morphology. In the nonmodified samples, silicon plates up to 120 μm long exist in the Al matrix with the boundaries of primary α-Al dendrites being not clear. It may be true to say that in nonmodified alloy, the remaining liquid does not solidify as a duplex mixture of Al and Si, but there is an overshoot of primary α-Al dendrites, due to difficulty of nucleation of Si, to take with them the Al content of eutectic mixture. Employing electron backscattered diffraction (EBSD), Nogita and Dahle [58] reported that in non-modified alloys there is not any difference in the

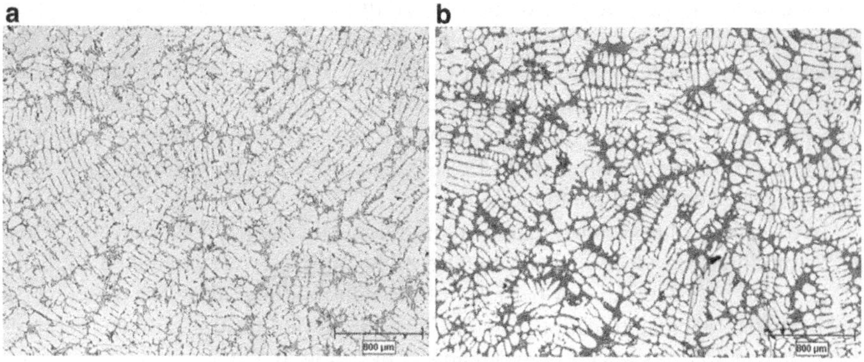

Fig. 6.65 Optical micrographs showing the structural changes in the primary α-Al particles due to Sr addition, A356; (**a**) without Sr, and (**b**) 241 ppm Sr [46]

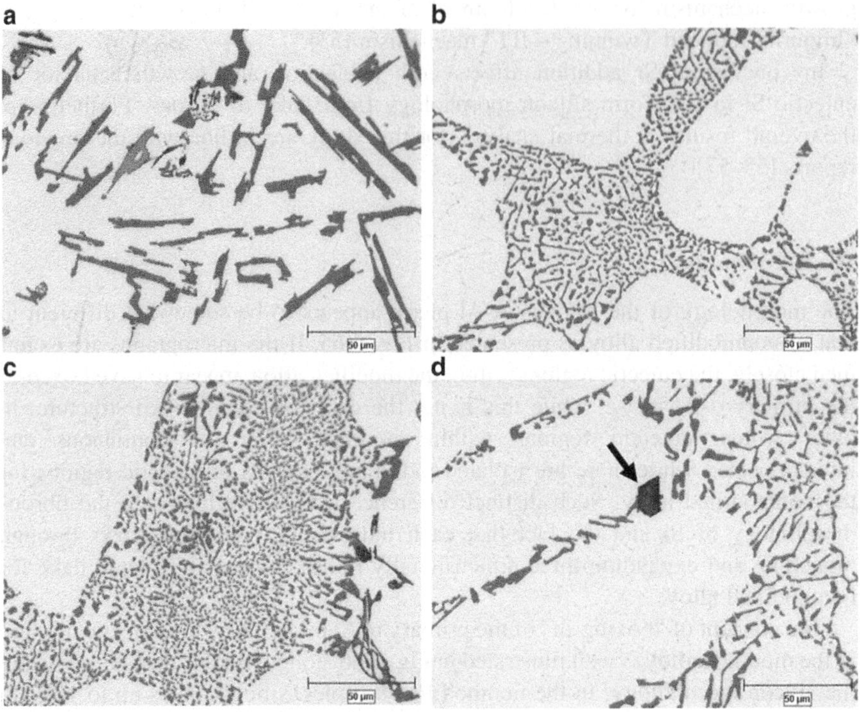

Fig. 6.66 Optical micrographs showing the effect of Sr addition on the eutectic morphology in the conventional process, A356; (**a**) without Sr, (**b**) 47 ppm Sr, (**c**) 156 ppm Sr, and (**d**) 392 ppm Sr (*arrow* showing Sr-based intermetallic compound)

crystallographic orientation between the primary α-Al phase and the eutectic aluminum at the vicinity of the eutectic pools.

With modification however, the eutectic melt solidifies as a duplex mixture due to the changes in the silicon morphology. Addition of ~50 ppm Sr is sufficient to transform the flake Si to fibrous morphology and with further addition of Sr, the structure becomes fully modified and eventually overmodified. Overmodification can easily be detected by the coarsening of silicon particles and increasing their interparticle spacing and the segregation of Sr-based compounds (Fig. 6.66d). Electron probe microanalysis (EPMA) has confirmed Al_2Si_2Sr stoichiometry for the segregated Sr-based compounds.

In order to further characterize the morphological changes of the eutectic Si due to Sr modification, deep etched graphite mold samples were examined by SEM (Fig. 6.67). The flake silicon particles of the non-modified alloy are quite coarse even at such low magnification, Fig. 6.67a. With Sr modification, silicon morphology changes to fibrous with seaweed appearance, Fig. 6.67b.

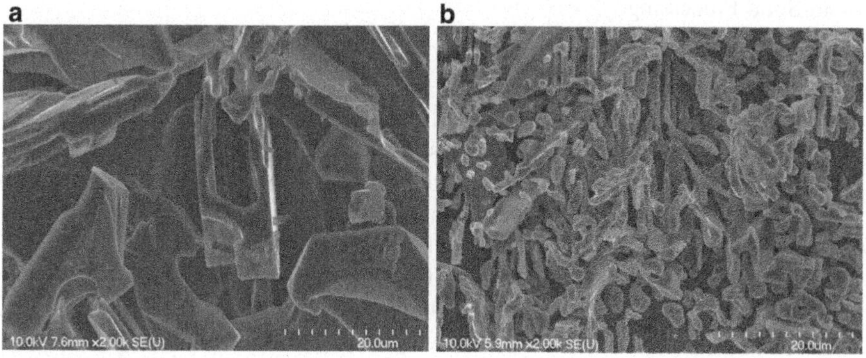

Fig. 6.67 SEM secondary electron (SE) micrographs of the deep etched (10%HF) graphite mold samples; (**a**) without Sr [46], (**b**) 156 ppm Sr

Image Analysis

The effect of Sr modification on the microstructure of A356 alloy is quantitatively presented in Fig. 6.68. No data is provided for the primary α-Al percentage in the non-modified samples, Fig. 6.68a, which is due to the difficulties for the image analysis system to differentiate between the primary and eutectic α-Al phases, i.e., it was impossible to measure the primary phase alone.

The percentage of the primary α-Al as well as eutectic silicon increases slightly with Sr addition, Fig. 6.68a, and b. This is attributed to the Sr effect on the Al–Si phase diagram. As mentioned in the thermal analysis section, the addition of modifier to the Al–Si alloys shifts the eutectic point to the higher silicon contents and lower temperatures to expand the mushy zone. The question may arise as how greater amounts of both the primary α-Al and eutectic silicon could form with Sr modification. If a simple lever rule calculation is carried out, it would confirm such conclusion when eutectic point is depressed down and to the right with Sr addition.

The trends observed for the eutectic silicon number density, number of silicon particles per unit area—Fig. 6.68c, its average circular diameter, and area-to-perimeter ratio, Fig. 6.68d, confirm refining and reducing the size of eutectic silicon. This is further supported by the percentage of silicon particles having aspect ratios greater than 2, Fig. 6.68c, where its percentage decreases with Sr addition up to 100 ppm and as for the other parameters remains almost constant with more Sr. The quantitative data supports microstructural observations where flake silicon transforms to fibrous morphology with Sr (as noted earlier, the calculation of circular diameter in the software is based on the area not the shape of the silicon particles and this results in inaccurate measurement. However, error bars of data become smaller when the structure is modified which is an indication of more circular morphology in 2D).

Semi-Solid Processing

Structural Analysis

The optical micrographs in Figs. 6.69 and 6.70 show the effect of Sr addition on the morphology of the primary α-Al particles. The Sr-treated structure appears to have more globular characteristics with isolated primary α-Al particles dispersed in a refined eutectic matrix. The Si morphology, Fig. 6.70, however is fully fibrous as expected with modification treatment but denser and finer than those of the modified conventionally air-cooled casting of Fig. 6.66. This is, mainly due to the higher cooling rate of the quenched billets, and to lesser degree to the SSM process itself. Stirring not only leads to the formation of globule structure in the slurry but also

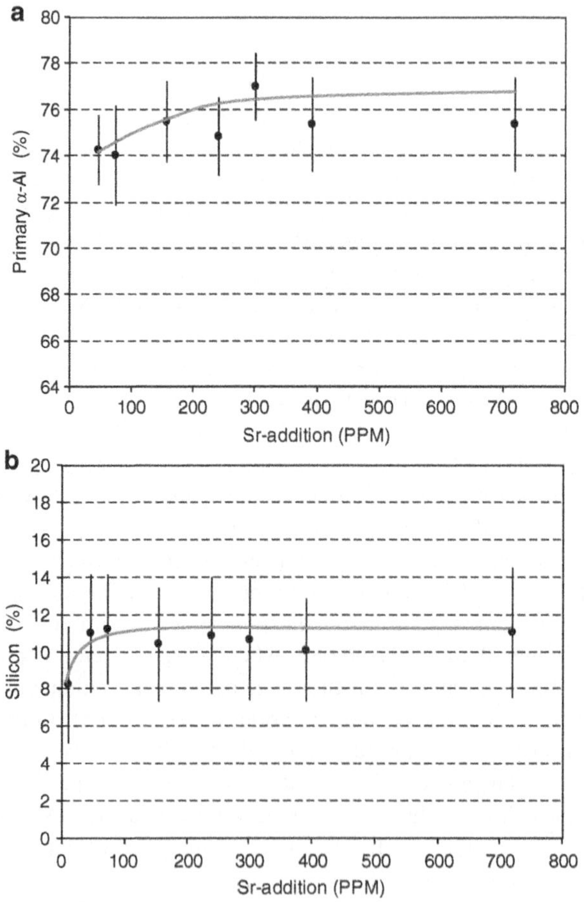

Fig. 6.68 Quantitative metallographic results, A356 alloy; (**a**) α-Al percentage, (**b**) Si percentage, (**c**) Si number density and Si particles percentage having aspect ratio greater than 2, (**d**) Si circular diameter and area/perimeter ratio (reproduced from [46])

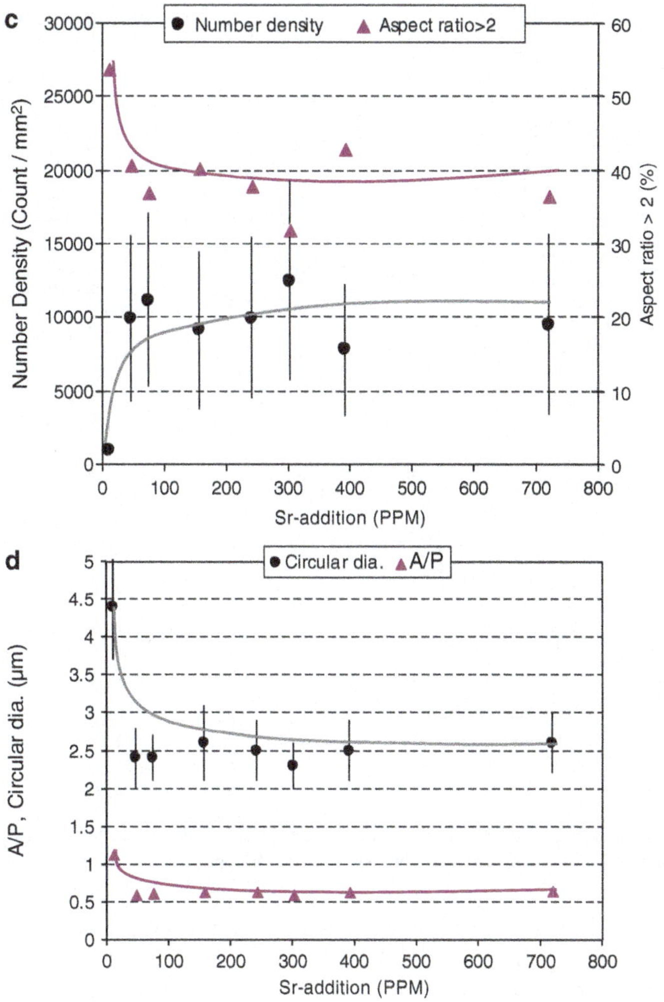

Fig. 6.68 (continued)

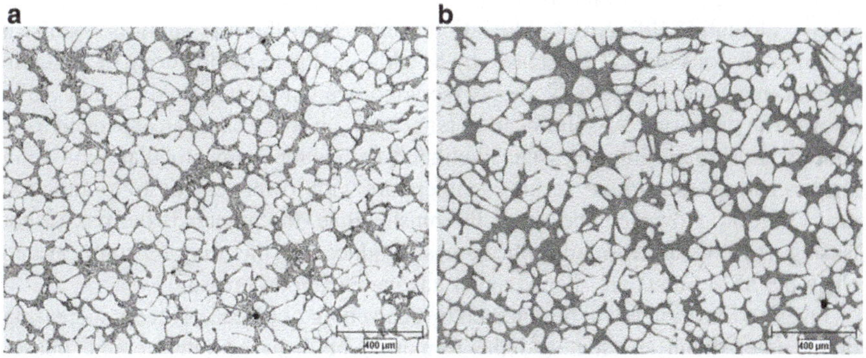

Fig. 6.69 Optical micrographs of the SEED billet quenched at ~598 °C: (**a**) non-modified, and (**b**) modified with 156 ppm Sr

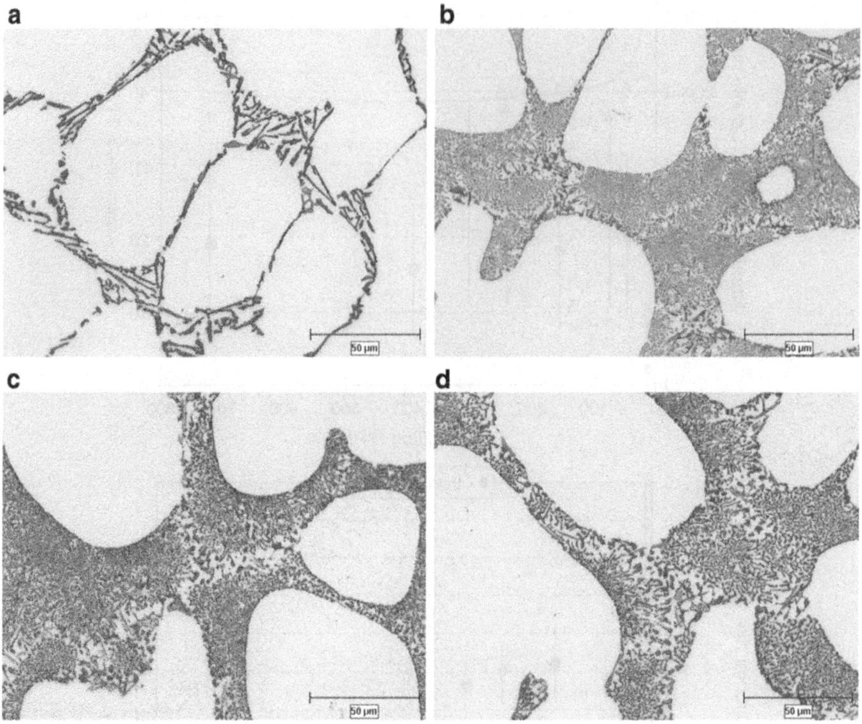

Fig. 6.70 Optical micrographs showing the effect of Sr addition in the SEED process: (**a**) non-modified, (**b**) 47 ppm Sr, (**c**) 156 ppm Sr, and (**d**) 392 ppm Sr

imparts forced convection onto the remaining liquid resulting in better fluid flow and smaller silicon particles (Sect. 3.3.3).

Comparing the microstructure with diecast part (injected billets) reveals similar globule size; however, the primary phase in die-cast parts is partially deformed by the applied pressure. The magnified region shows a Sr-modified structure comparable to the quenched billet (Fig. 6.71).

By calculating different solidification ranges from thermal analysis results, it becomes obvious that ΔT_{α} increases by Sr addition due to the lowering of the main eutectic reaction. Figure 6.72 shows the solidification range of Al7Si and A356 alloys. The primary α-Al solidification range (ΔT_{α}) is a key since an expanded α-Al solidification range improves the likelihood of having more globular semi-solid slurries and gives a wider working temperature with lower sensitivity to temperature variation. However, the larger the mushy zone, the greater is the risk of hot tearing and porosity formation in the cast product, but this may be compensated when the SSM billets are shaped under adequate external pressure.

Overmodification could be detected by coarsening of eutectic silicon, increasing of the intersilicon spacing, and the formation of Sr-based intermetallic phases within the eutectic pools. It is noticeable from Fig. 6.73 that Sr compounds may

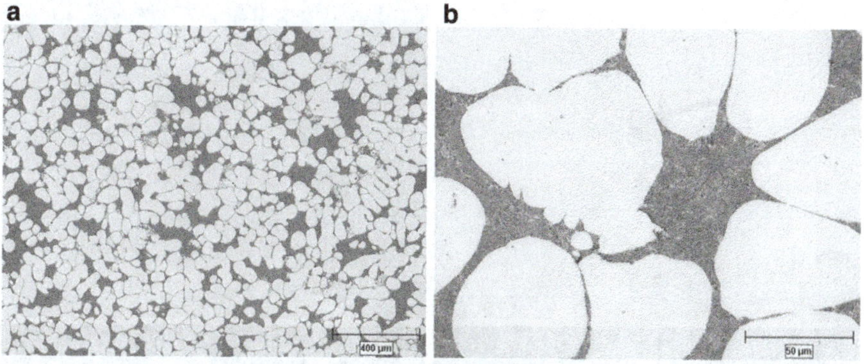

Fig. 6.71 HPDC structure of the Sr-modified A356 alloy, with 110 ppm Sr

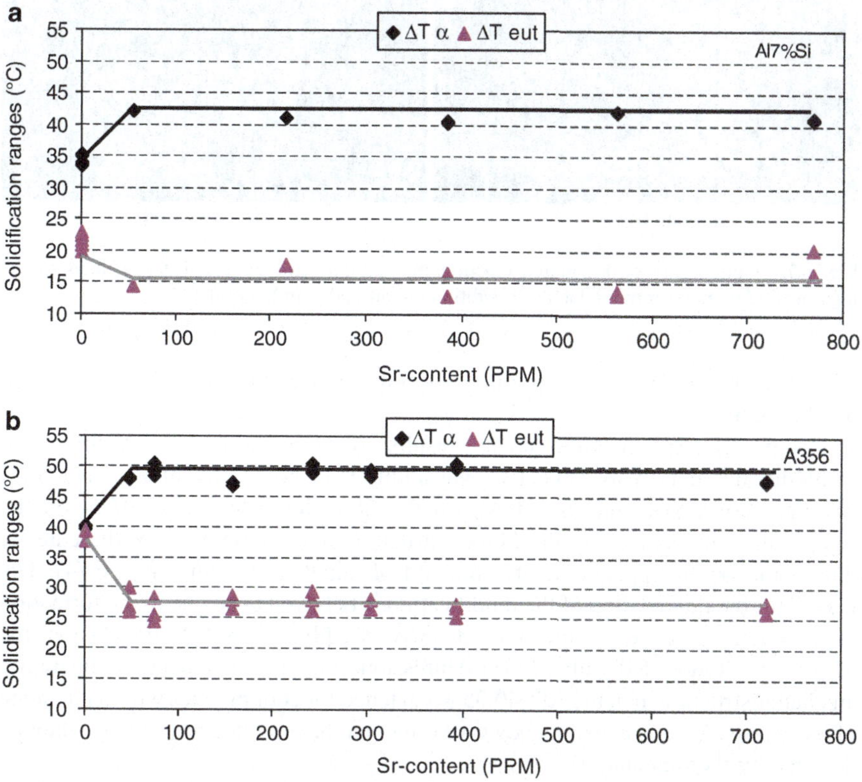

Fig. 6.72 Effect of Sr on the solidification range of various phases; (**a**) Al7Si, (**b**) A356

segregate in two distinct morphologies of cuboidal and flake/plate (or needle). The microprobe analysis of the Sr-based compounds suggests Al_2Si_2Sr stoichiometry (with no chemical variation) in confirmation of the results reported before [27, 28]. These particles were found to form in castings with Sr additions in excess

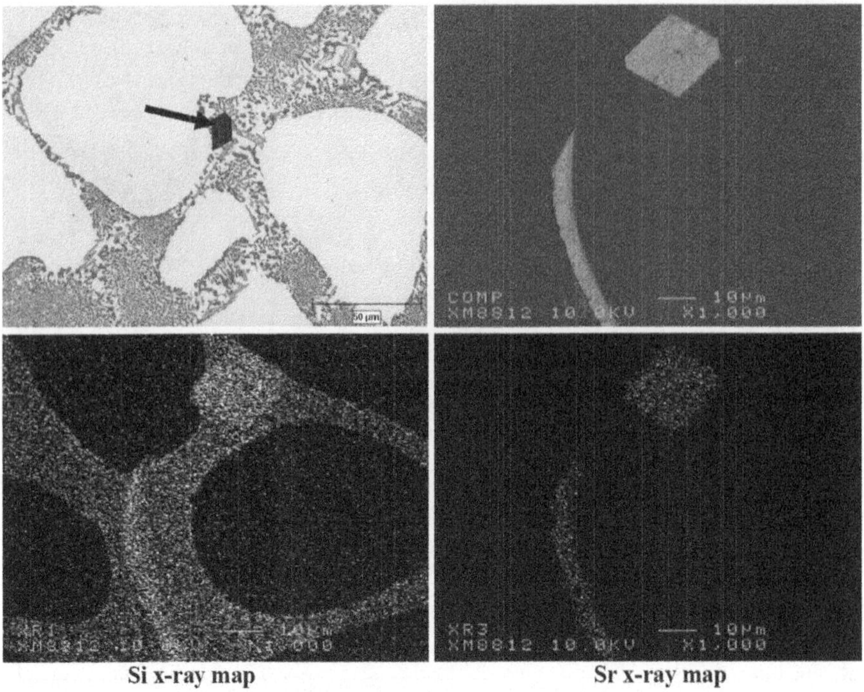

Fig. 6.73 Optical and backscattered electron micrographs and respected X-ray maps of the intermetallic particles formed due to Sr addition (about 700 ppm Sr) [46]

of 390 ppm. Therefore, the optimum concentration of Sr appears to be between 50–200 ppm.

Comparing the metallographic results of microstructural evolution of Sr-modified billets with respect to Mg addition in both conventional and SSM castings shows that addition of Mg to the alloy has a negative effect on the modification by coarsening the fibrous structure and deterioration of the eutectic microstructure by appearance of some sort of lamellar structure (Fig. 6.74). The reason for the degradation of the modification is believed to be due to the formation of intermetallic phases of the type $Mg_2SrAl_4Si_3$ [49, 59, 60]. In conclusion, the Mg-free melt modified with a lower depression value for the eutectic temperature and better structure than the Al7Si0.35 Mg. Hence for cooling rates within the range of these tests, Mg has a distinct negative effect on the modification which cannot be detected by thermal analysis.

The SEM analysis of the non-modified deep-etched specimens, Fig. 6.75a, suggests the plate like silicon crystals have grown out of individual nucleation sites, while the modified structure consists of highly branched silicon fibers, but with much lower number of isolated fiber. The interlamellar/interrod spacing appears to have decreased in the rheocast billets when Figs. 6.67 and 6.75 are compared. Such refinement is partly due to the quenching of SSM slurries from the

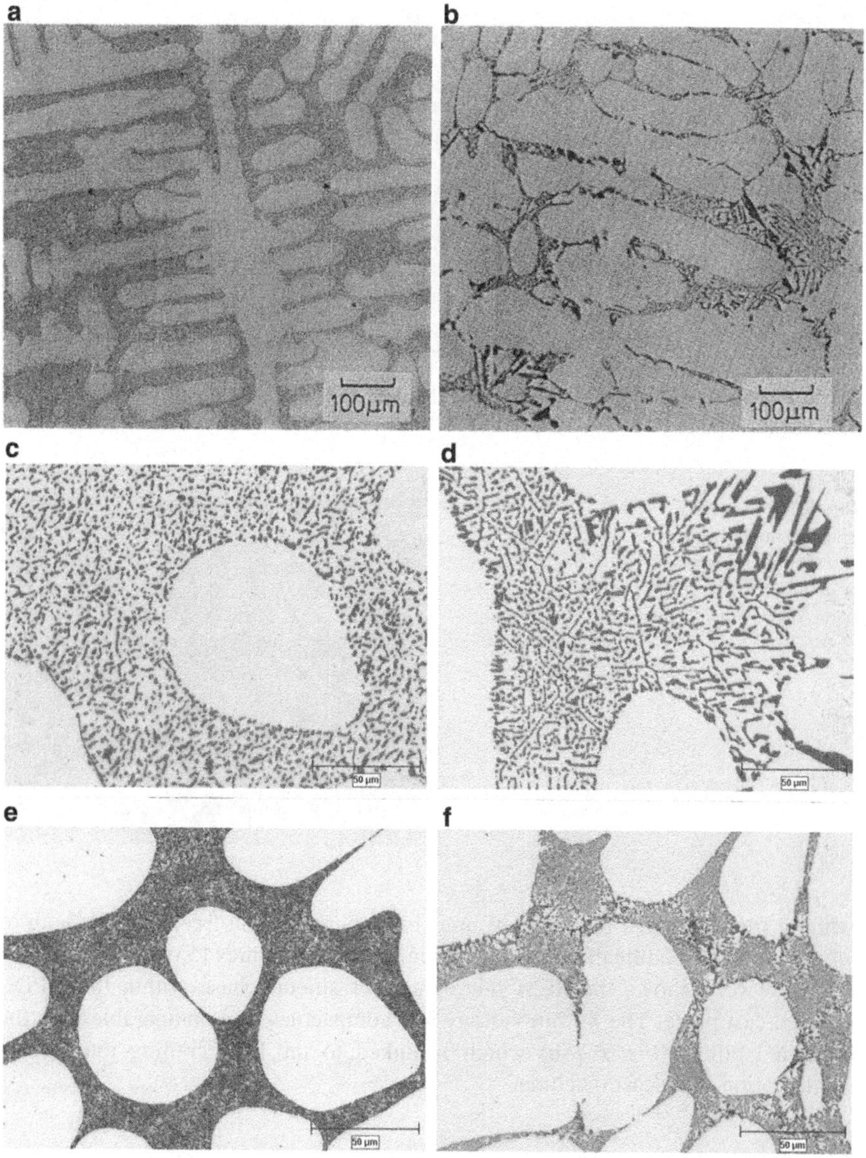

Fig. 6.74 Comparison of Sr-modified microstructures with varying Mg content: (**a**) 0.0%Mg, 300 ppm Sr, Al7%Si (**b**) 1.21%Mg, 250 ppm Sr, Al7%Si [49] (reprinted by permission of Taylor & Francis Ltd), (**c**) graphite cup, without Mg, 54ppmSr, (**d**) graphite cup, 0.35 Mg, 47ppmSr, (**e**) quenched billet, without Mg, 54ppmSr, (**f**) quenched billet, 0.35 Mg, 47ppmSr

mushy zone and partly due to the fluid flow and thermal uniformity within the SSM mold. This means that the branching of Si fiber is not hindered and it can branch in any direction, since there is not any directional heat flow within the mold. From the modification mechanism point of view, with the presence of modifier, both the

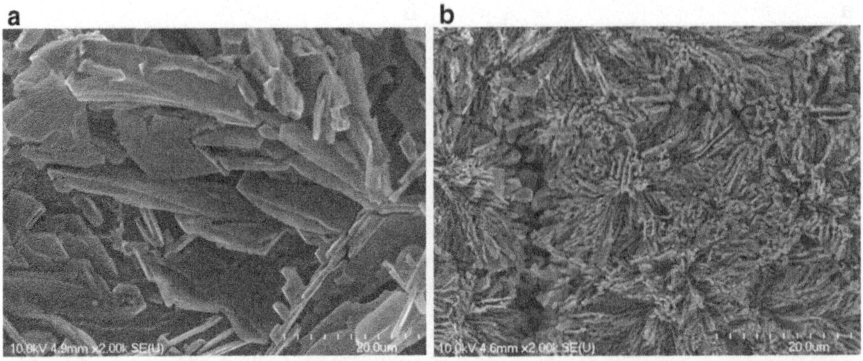

Fig. 6.75 Scanning electron micrographs of the deep etched (10%HF) SEED billets to show morphological changes of eutectic Si with Sr addition; (**a**) without Sr [46], (**b**) 156 ppm Sr

Fig. 6.76 Scanning electron micrograph showing the fibrous structure in HPDC SSM part, A356 with 110 ppm Sr

twining frequency and the angle of branching increase with freezing rate, both of which promote modification and formation of finer structures [53].

Figure 6.76 shows the SEM micrograph of silicon phase within the HPDC injected cast piece. The Si morphology and compactness are comparable with the quenched billets (Fig. 6.75b) which is linked to the high cooling rate during solidification in diecast machine.

Image Analysis

The percentage of the primary α-Al particles increases slightly with strontium (Fig. 6.77a). With addition of Sr, the eutectic line moves downward and as a result, the eutectic point shifting to the right, thus expanding the α-Al freezing range and increasing the fraction of primary α-Al. This is of great importance in the SSM processing since a wider solidification range renders more flexibility in SSM processing especially when the commercialization of the process is intended.

The average primary α-Al particle size, equivalent circular area diameter, increases slightly with Sr addition while the number density remains approximately constant (Fig. 6.77a, b). This is in line with the increase in the primary α-Al percentage and since number density remains constant, it is expected that the α-Al particle size to increase. The area-to-perimeter ratio is measured and showed an increase with Sr addition in support of the findings for the average circular diameter [46]. In order to further highlight the globularization effect with Sr

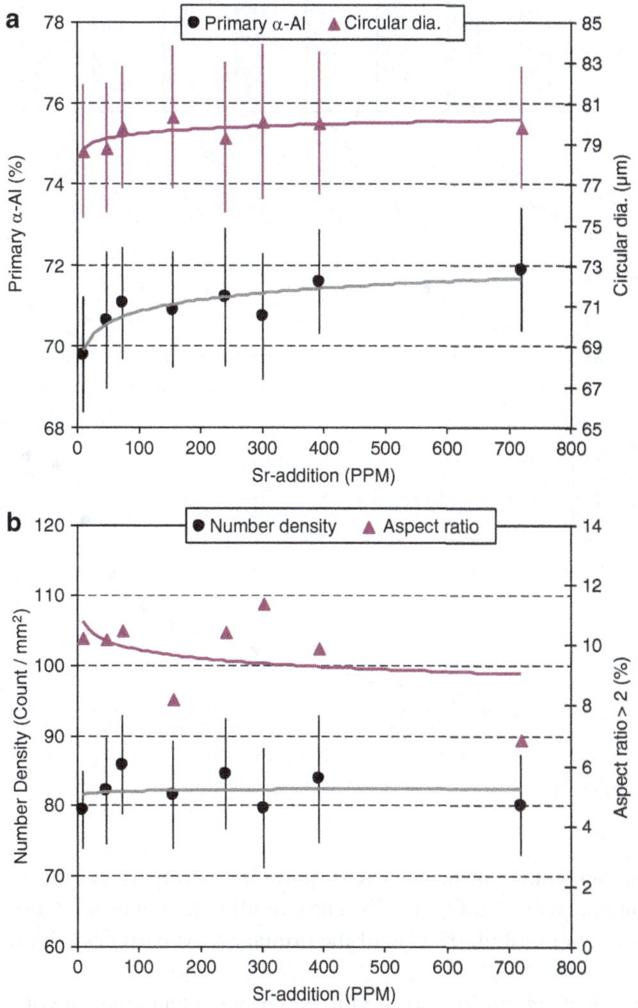

Fig. 6.77 Image analysis results for A356 with Sr addition: (**a**) primary α-Al percentage, average circular diameter, (**b**) number density of α-Al particles, percentage of α-Al particles having aspect ratio >2, (**c**) the percentage of particles having sphericity >0.8, (**d**) data sorting with and without Sr (reproduced from [46])

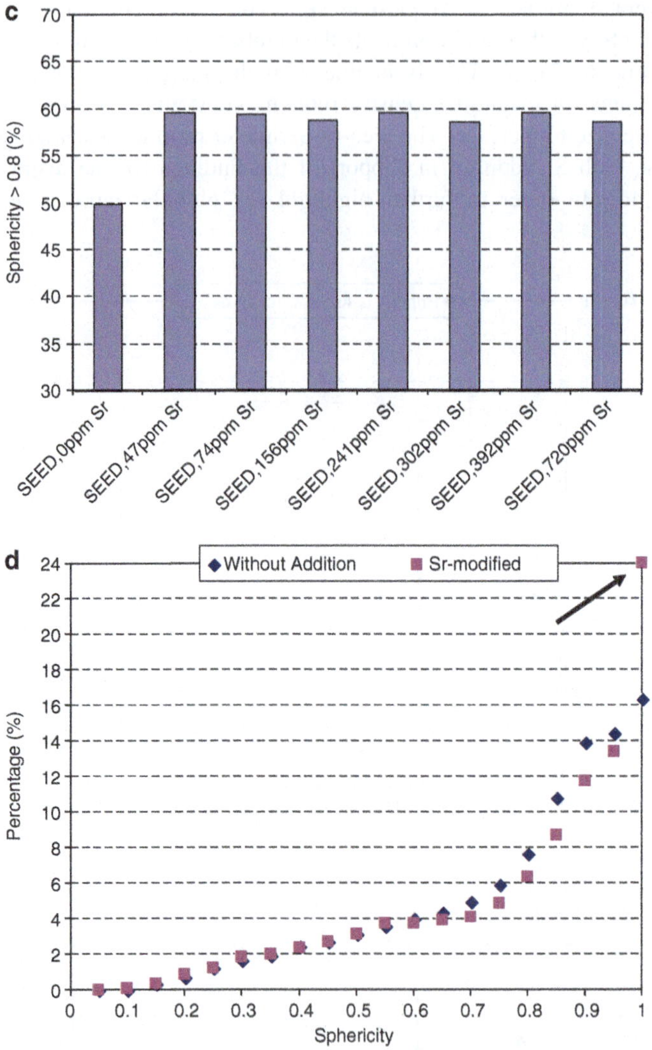

Fig. 6.77 (continued)

addition, the percentage of primary α-Al particles having aspect ratio above 2 was measured and presented in Fig. 6.77b. The overall reducing trend in this figure is an indication of better globularization of the primary α-Al particles with increasing Sr addition.

The concept of sphericity was used to better understand the role of modification on the morphology of the primary α-Al particles. Figure 6.77c shows the percentage of α-Al particles with sphericity greater than 0.8. With modification, the percentage of particles having sphericity values closer to one increased with increasing Sr.

Since modification is entirely attributed to the alteration of silicon morphology, the question may arise if it has any contribution toward primary α-Al morphology as well. It is believed that any variation in the surface tension of the alloy plays an important role on the globularization of α-Al particles. Previously, it was reported that Sr and Na decrease the surface tension of the melt [61, 62]. By decreasing the surface tension, wettability of primary α-Al particles by the remaining liquid is improved. This is equivalent to having more contact between the remaining liquid and primary α-Al. Decreasing the surface tension makes the fluid flow around the α-Al particles easier and consequently renders enhanced shaving and trimming of these particles and therefore causes the formation of more spherical and rounder particles.

This is even more pronounced during stirring, where thermal and solutal homogenization take place and with consideration of better fluidity due to lower surface tension and better wettability, it may be true to conclude that the flow around particles could smooth off primary α-Al particles better and thus rounder particles obtain.

6.2.3 *Combined Effect of Refiner–Modifier*

Grain refining and modification offers substantial benefits in casting processes. Finer grain ensures better mechanical properties, improved machinability, better feeding while with modification silicon flakes change to fibrous resulting in better properties especially ductility. Therefore, it is reasonable to use combined refiner and modification treatments to take advantage of both as it has been the norm for the last three decades. However, each treatment has its own characteristics and with combination, considerable changes may occur in the percentages, shapes, and sizes of both primary α-Al and eutectic Si.

6.2.3.1 Addition of Ti, B, and Sr to the A356 Alloy

The objective is to study the simultaneous addition of refiner and modifier to the alloy. For this purpose, Al5Ti1B and Al10Sr master alloys were added to the melt to raise Ti, B, and Sr levels in the alloy. The optimum additional level was deduced from the information in Sects. 6.2.1 and 6.2.2 and the final compositions are listed in Table 6.9.

Table 6.9 Chemical analysis for before and after combined treatment (wt. %)

	Si	Mg	Fe	Mn	Cu	Ti	B	P	Sr	Al
Base alloy	6.68	0.4	0.07	0.003	0.001	0.0058	0.0001	0.0003	0.000	Bal.
Treated alloy	6.68	0.39	0.08	0.003	0.001	0.058	0.0098	0.0003	0.014	Bal.

Conventional Casting

Thermal Analysis

Figure 6.78 reveals the nucleation and growth segments of typical cooling curves for primary α-Al formation and eutectic reaction of both untreated and treated alloys. By combined treatment, the nucleation and growth temperatures of primary α-Al shifted up while the eutectic reaction has been suppressed to lower temperatures.

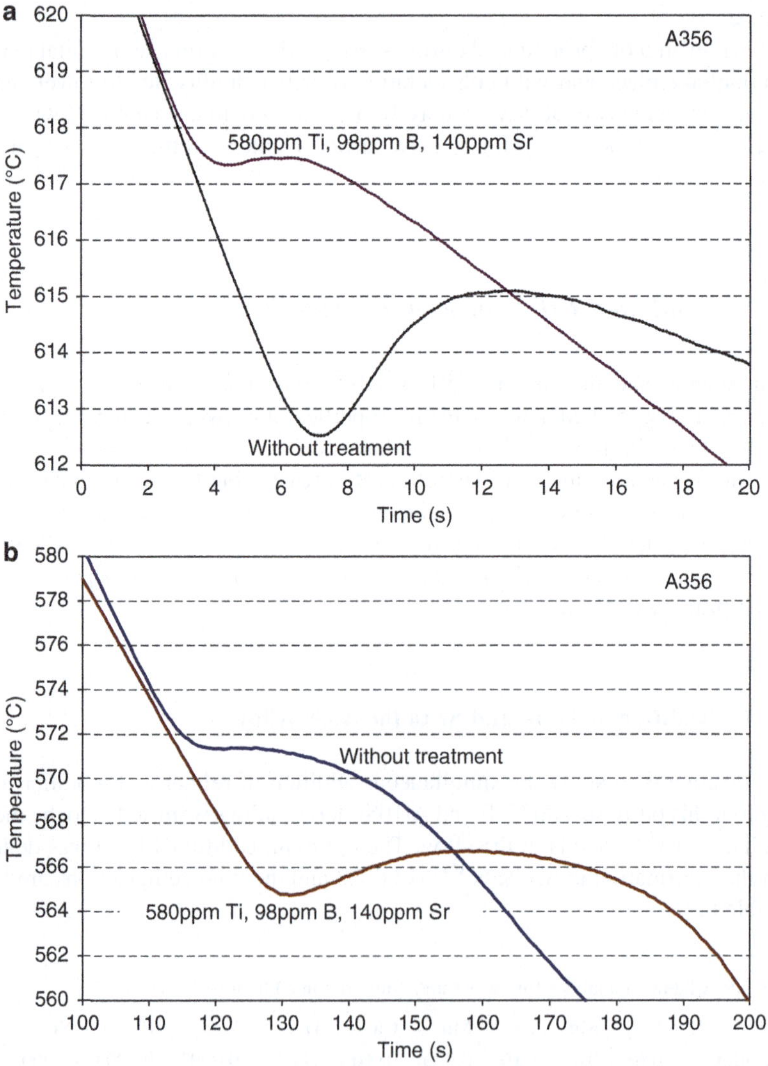

Fig. 6.78 Cooling curves before and after additions: (**a**) beginning of solidification, (**b**) eutectic reaction zone [63]

Small addition of Al5Ti1B shifted the curves up and recalescence decreased. Furthermore, both nucleation and growth temperatures of α-Al particles increased but the temperature rise is greater for $T_{nuc_{Al}}$ as clearly seen in Fig. 6.79a. This means that there are more nuclei with lower chance of growth at any one time during nucleation and early growth of primary phase.

The changes in $T_{nuc_{eut}}$, $T_{max_{eut}}$, $\Delta\Theta$, and ΔT_{α} due to combined treatment are shown in Fig. 6.79b. The changes in eutectic thermal analysis parameters are the same for sole and combined treatments which may confirm the independent function of refiner and modifier during solidification. In other words, the strong affinity of B for Ti to form TiB_2 nucleants impedes any reaction between Sr and B to form SrB_6 compounds and thus cannot inactivate Sr as a modifier. The eutectic nucleation temperature and eutectic maximum temperature decrease with Sr addition (Fig. 6.79b). The eutectic recalescence, $\Delta\Theta$, also increased by ~2 °C with 140 ppm Sr. Also depression of the eutectic temperature causes larger α-Al solidification range and consequently more α-Al formation.

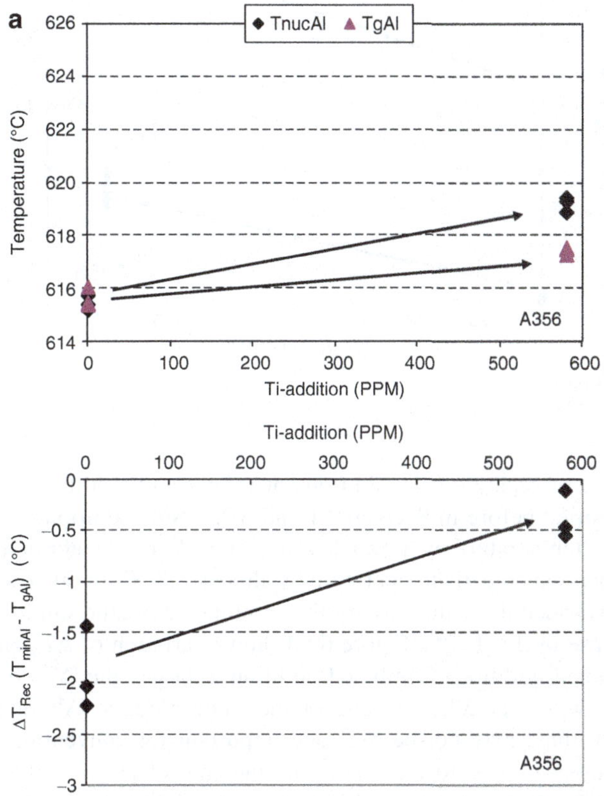

Fig. 6.79 Variation of: (**a**) nucleation and growth temperatures of α-Al particles and ΔT_{Rec}, (**b**) nucleation and maximum temperature of eutectic, $\Delta\Theta$ and ΔT_{α} [63]

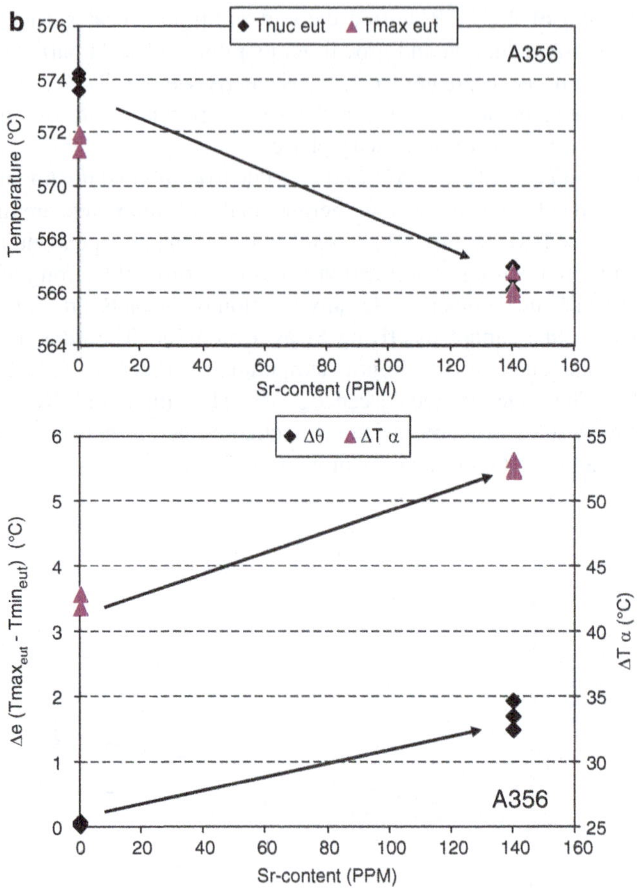

Fig. 6.79 (continued)

The effects of separate sole additions of refiner and modifier to A356 Al–Si alloy are reported before in Sects. 6.2.1 and 6.2.2. Sole addition of refiner lifted up nucleation temperature by as much as 4–5 °C, while sole addition of Sr-based modifier depressed eutectic temperature by almost 7–8 °C. In other words, if both additives performed the same way for the combined case, the solidification range should increase by 11–13 °C. Figure 6.80 shows variation of ΔT_{α} and ΔT_{eut} due to the combined addition. Combined addition enlarges the ΔT_{α} by more than ~12 °C and depresses ΔT_{eut} by almost the same amount. Although widening solidification range may not be regarded important for conventional casting or even be a drawback due to increasing the incident of porosity, but surely it is quite crucial for SSM processes. Furthermore, development of wider freezing range slows down directional solidification which is again a positive point in SSM processing.

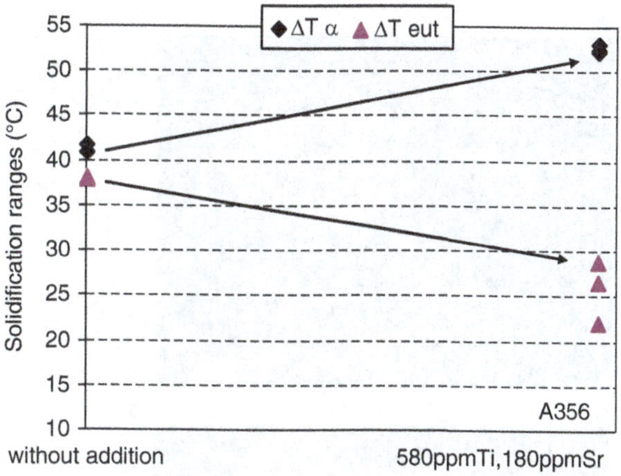

Fig. 6.80 Effect of combined addition on the solidification range

Structural Analysis

The optical micrographs in Fig. 6.81 depict microstructural changes due to the combined treatment. By refiner addition α-Al nucleation temperature increased due to more potent and effective nucleants in the bulk liquid. When there is a high density of nucleants in the system, the mean free path between the nuclei becomes smaller and thus grain growth is restricted. Furthermore, the heat flow becomes multidirectional due to high density of nucleants. Such solidification conditions should eventually lead to the formation of somewhat spherical primary phase particles.

The correlation between the average grain size, the α-Al nucleation temperature, and the recalescence time is demonstrated in Fig. 6.82. With the combined addition, the grain size was reduced more efficiently than when the sole refiner was added, the final grain size being ~520 μm (a 56 % reduction against 33 % with refiner by itself). It seems that combined effect gives better refinement than the addition of sole refiner, shown in Fig. 6.25. Such a conclusion may be reached if the recalescence values for the combined (−0.5 to −0.1) and the sole treatment (−1.0 to −0.5) are compared. It is well established that the smaller the magnitude of recalescence, the more effective is the grain refiner. Interestingly, by measuring of the liquidus undercooling time, t_{Rec}, it was observed that this time decreased by refiner addition. A small value of t_{Rec} signifies that the nuclei do not have a long nonsteady growth period at the beginning of solidification. In other words, longer values of t_{Rec} are related to more nonsteady growth opportunity and consequently larger dendrite formation. This is in line with the micrographs shown in Fig. 6.81.

Semi-Solid Processing

Structural Analysis

The optical micrographs in Fig. 6.83 show the effect of combined treatment. Inoculation is obvious through globule size reduction and more primary particles,

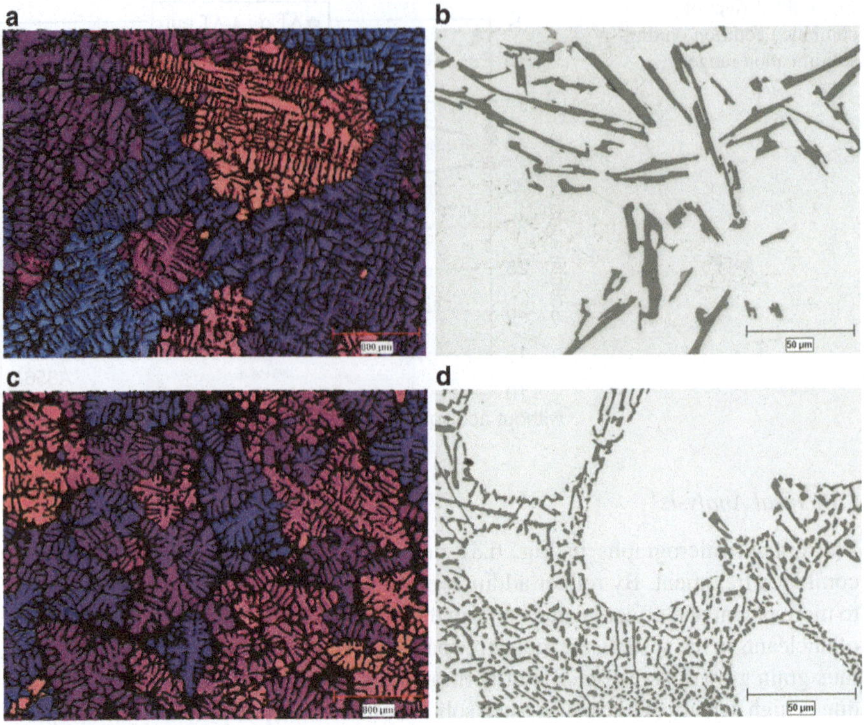

Fig. 6.81 Micrographs revealing the impact of combined treatment: (**a, b**) Without addition, (**c, d**) 580 ppm Ti, 98 ppm B, 140 ppm Sr

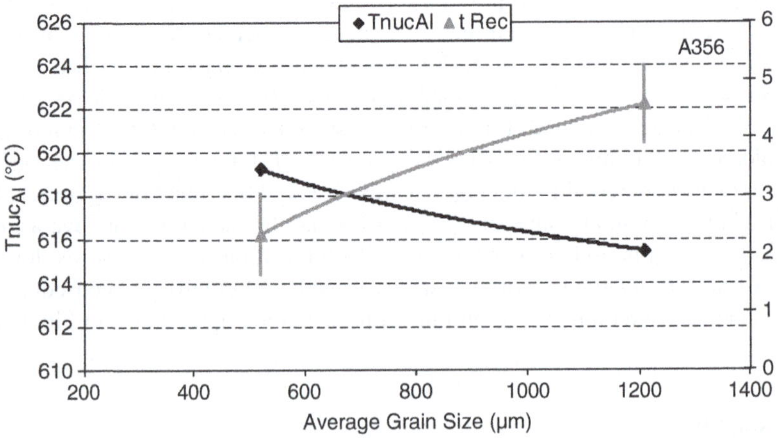

Fig. 6.82 Relation between grain size and thermal analysis parameters [63]

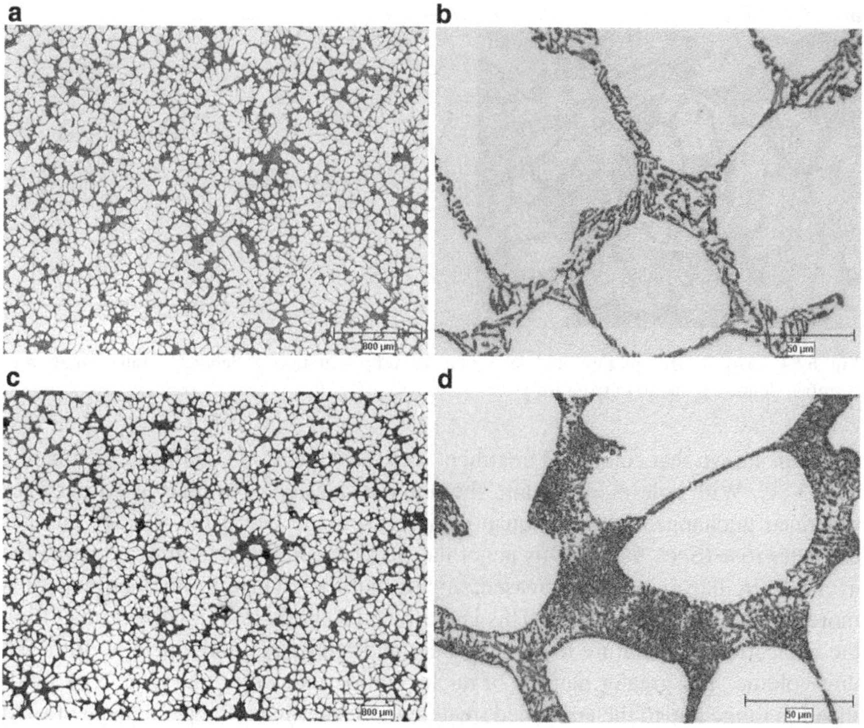

Fig. 6.83 Micrographs of SSM structures: (**a, b**) Without addition, (**c, d**) 580 ppm Ti, 98 ppm B, 140 ppm Sr

while modification is visible at higher magnification by morphological changing of silicon particles from lamellar (flake) to fibrous.

By stirring and forced flow, convection, the initial dendrites obliged either to break up by fragmentation or root remelting mechanism (Sect. 3.2.1) or even plastically bend with the final shape of a "cramped dendrite." Figure 6.84 confirms that the individual globules are indeed part of one single cramped dendrite and only appeared as globules due to metallographic sectioning. In order to determine whether the globules are independent grains or segments of a larger cramped α-Al phase, polarized light microscopy is a suitable tool.

Image Analysis

As discussed in the previous sections, the scale of eutectic structure is quite fine in the SSM billets and optical quantitative metallography was not viable for the eutectic silicon. Therefore, the results as presented in Fig. 6.85 are solely concentrated on the primary α-Al phase.

The primary α-Al percentage appears to have increased slightly which is mainly due to the higher number of effective nuclei from the refining agent (Fig. 6.85a). It

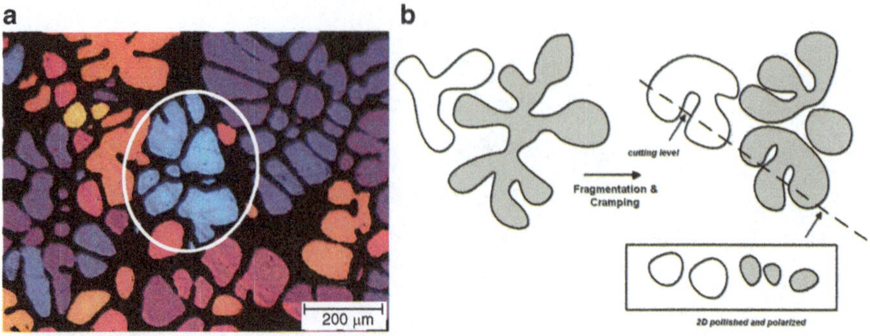

Fig. 6.84 (**a**) Polarized picture, and (**b**) schematic representation of dendrites transformation to cramped dendrites due to stirring [63]

has been shown that combined treatment increased the α-Al solidification range by 11–13 °C. With sole modification, the number density of primary α-Al particles remained unchanged while its quantity increased and hence the average circular diameter rose (Sect. 6.2.2). This is not the case for combined treatment, where the average circular diameter decreased slightly but the change in number density is more pronounced (Fig. 6.85b). This is attributed to grain refining effect in shifting the nucleation temperature to higher values and rendering more nuclei to form per unit volume. The greater number of nuclei compensates the enlargement of solidification range due to the combined treatment. Furthermore, the percentage of α-Al particles with aspect ratio >2 decreases. Such finding coupled with the increasing in sphericity values as shown in Fig. 6.85c is an indication of improving globularity.

6.2.3.2 Addition of B and Sr to the A356 Alloy

The optimum additional levels of boron and strontium were reported in Sects. 6.2.1.3 and 6.2.2, respectively. The selected chemical composition are therefore based on those findings and given in Table 6.10 (master alloys Al5B and Al10Sr).

Conventional Casting

Thermal Analysis

The combined addition appears to be effective in reducing the recalescence in the bulk liquid and lowering the eutectic reaction due to its strontium content (Fig. 6.86). The resulting thermal data extracted from further analysis of the cooling curves are plotted in Fig. 6.87. In the untreated alloy, the recalescence is about 2 °C and nucleation and growth temperatures of the primary phase are ~615–616 °C while with the combined treatment, both temperatures increased and recalescence

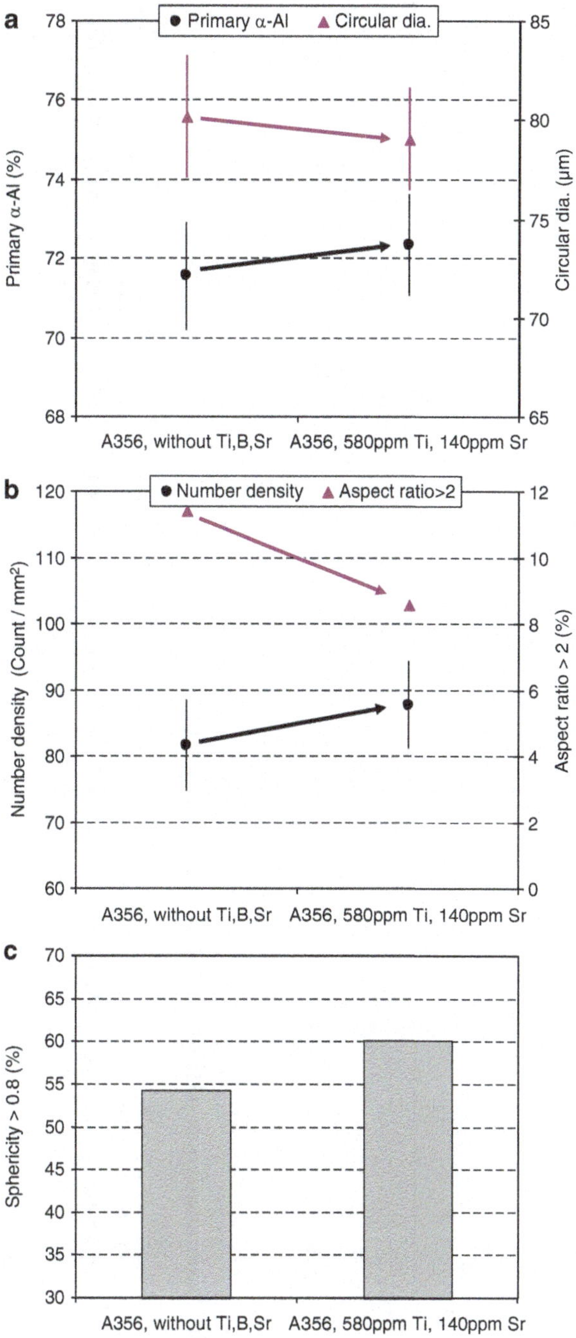

Fig. 6.85 Image analysis results; (**a**) primary α-Al percentage and circular diameter of α-Al particles, (**b**) number density and percentage of α-Al particles having aspect ratio >2, and (**c**) percentage of particles having sphericity >0.8 (reproduced from [63])

Table 6.10 Chemical analysis of A356 alloy before and after treatment (wt. %)

	Si	Mg	Fe	Mn	Cu	Ti	B	P	Sr	Al
Base alloy	6.66	0.4	0.07	0.002	0.001	0.0057	0.0001	0.0003	0.000	Bal.
Treated alloy	6.6–6.8	0.38–0.4	Max 0.07	Max 0.003	Max 0.001	Max 0.004	0.02	0.0003	0.01–0.014	Bal.

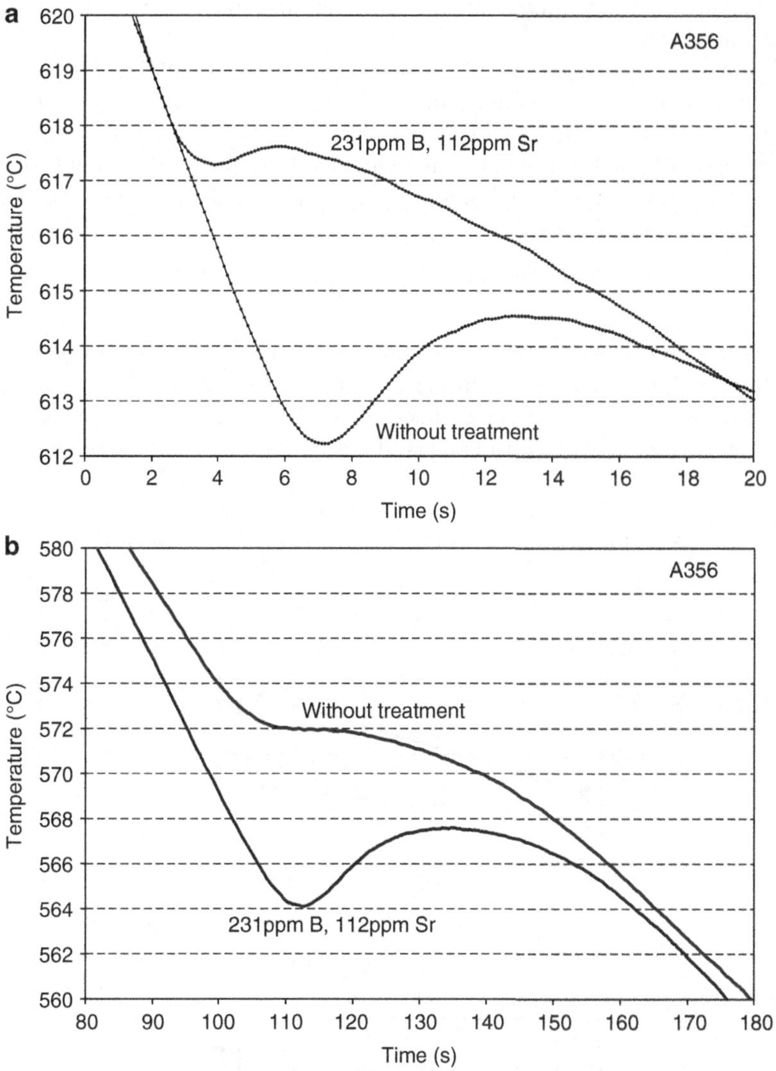

Fig. 6.86 (**a**) Commencement of solidification with combined B-Sr treatment, (**b**) eutectic reactions [64]

decreased. In the eutectic reaction zone, Sr modification causes lower eutectic nucleation temperature and the eutectic recalescence also increased.

Comparing the results with the sole boron addition indicates that increasing of nucleation and growth temperatures of primary α-Al particles is slightly greater in the B-treated alloy and the recalescence for combined treatment could not reach zero. This is an indication for a bit larger grain size in B-Sr-treated alloys as will be described later. From modification perspective, Sr decreases the nucleation and maximum temperatures of the eutectic and thus increasing the α-Al solidification range.

Figure 6.88 shows the effect of combined addition on the solidification range. As confirmed in Sect. 6.2.1.3, boron is the most effective refiner and changes $T_{nuc_{Al}}$ by as much as ~8 °C. On the other hand, Sr reacts with B and as a result, the overall percentages of Sr and B to act as refiner/modifier decrease and thus could not be as effective as those of sole addition. Sr and B react and form SrB_6 compounds according to following reaction:

$$Sr + 6B \rightarrow SrB_6 \tag{6.5}$$

This reaction shows that each strontium atom could react with six boron atoms and form SrB_6 compound. According to Appendix B and the disregistry values, both SrB_6 and TiB_2 could act as nucleant however boron consumption in the compound is one of the key parameters. It means that AlB_2 consumes less amount

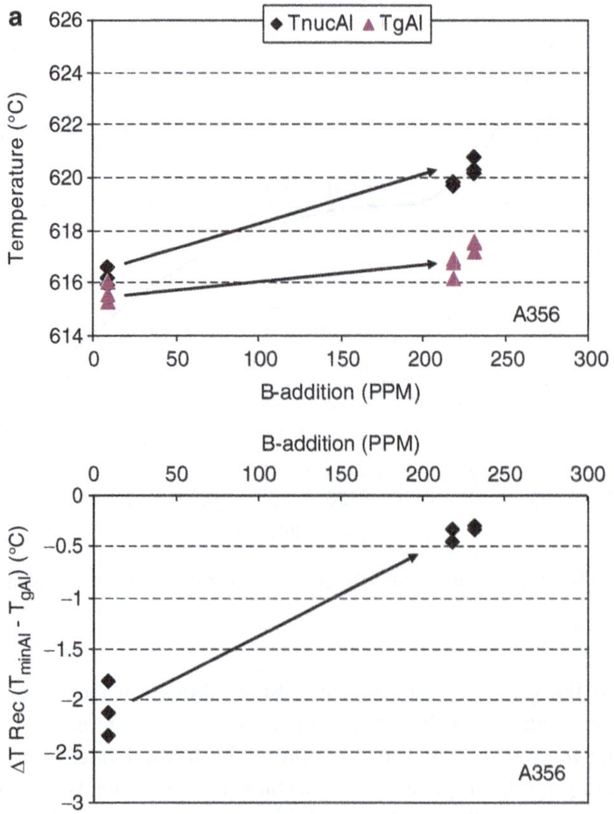

Fig. 6.87 Variation of: (**a**) $T_{nuc_{Al}}$, $T_{g_{Al}}$, and ΔT_{Rec} due to the refiner, (**b**) $T_{nuc_{eut}}$, $T_{max_{eut}}$, $\Delta\theta$, and ΔT_{α} due to modifier addition [64]

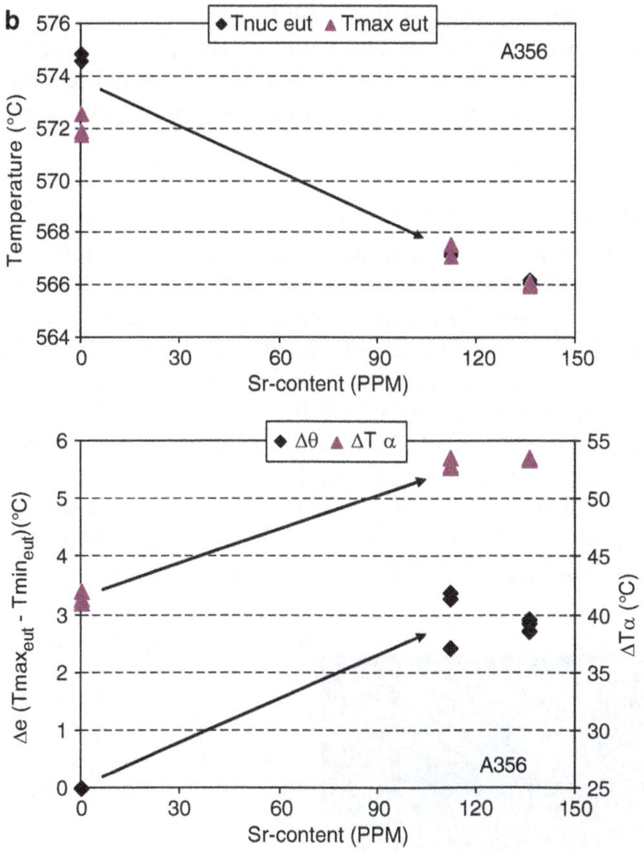

Fig. 6.87 (continued)

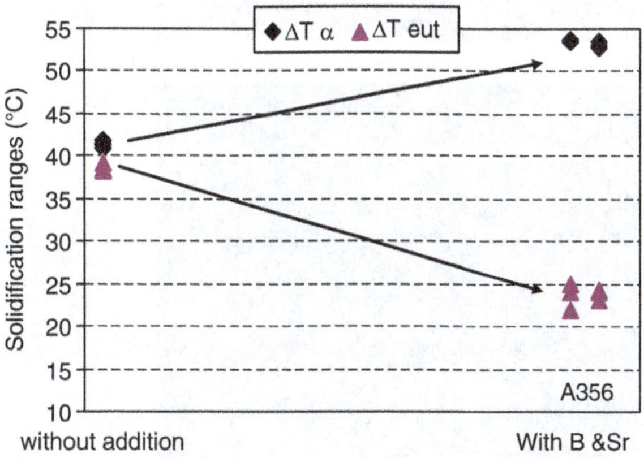

Fig. 6.88 Effect of combined addition to the solidification range [64]

of boron comparing to SrB_6 and as a result by considering a constant amount of boron, the density of nucleant particles is much higher in the case of AlB_2 as lower number of boron atoms are associated with this compound. Therefore by greater number of potent nucleants, the probability of having lower grain size is increased.

Structural Analysis

The microstructural transformation due to the combined addition of B and Sr is shown in Fig. 6.89. The microstructure obviously refines with the formation of smaller equiaxed grains. This is attributed to the large number of nucleants dispersed within the bulk liquid to initiate nucleation and coupled with multidirectional growth to end up with an equiaxed structure. By formation of SrB_6 compounds and decreasing of the free strontium atoms in the liquid, the remaining Sr atoms are not sufficient enough in altering the flake morphology of the eutectic silicon to fibrous, Fig. 6.89c, d.

The correlation between the microstructural parameters of grain size, $T_{nuc_{Al}}$, and t_{Rec} is shown in Fig. 6.90 where a reduction of the grain size from ~1200 μm to ~400 μm is accompanied by increasing the nucleation temperature of α-Al particles

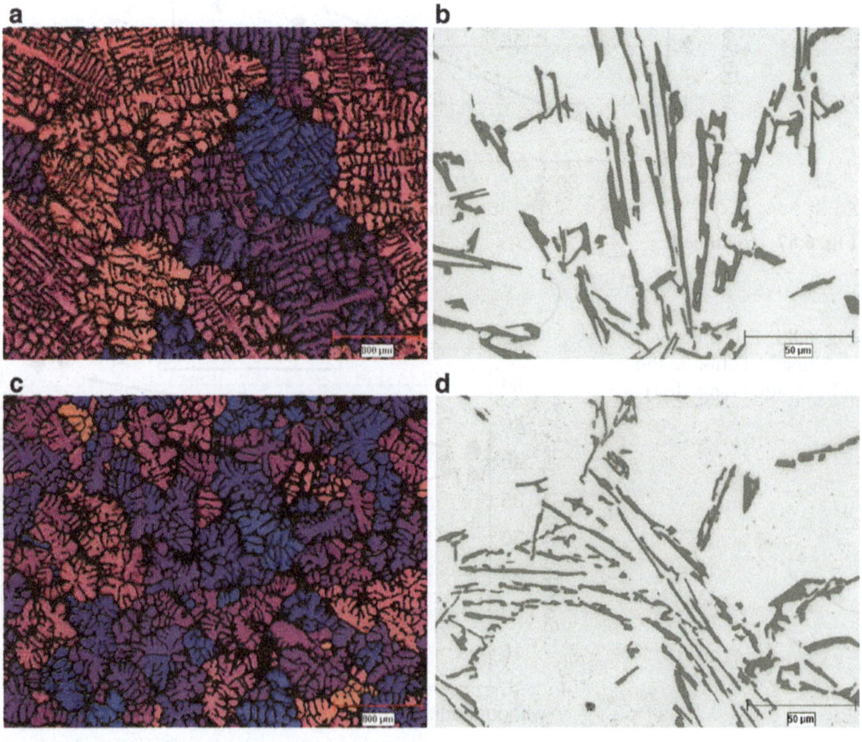

Fig. 6.89 Micrographs representing the effect of simultaneous melt treatment: (**a, b**) Without addition, (**c, d**) 231 ppm B, 112 ppm Sr

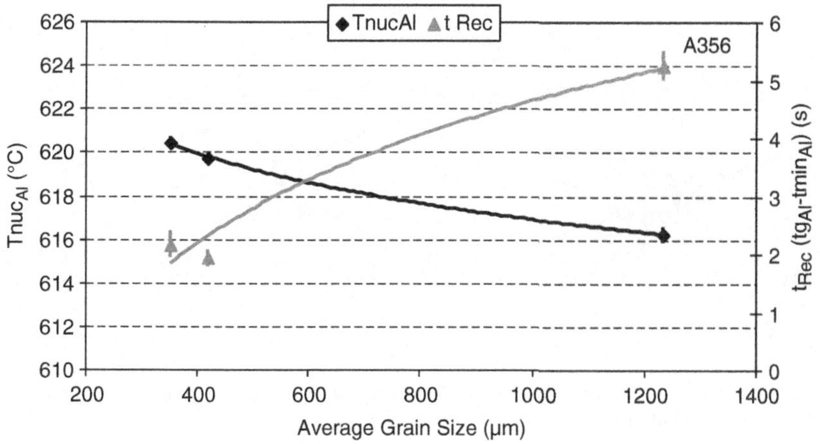

Fig. 6.90 The correlation of t_{Rec} and nucleation temperature on the grain size of B-Sr-treated alloys

by 3 °C. The combined addition of Sr and B has caused smaller grain size value comparing to that of Ti and B addition (Fig. 6.82) but it could not reach to the values of sole boron addition (Fig. 6.40). In case of sole boron addition, the grain size reached to a minimum value of ~300 μm while co-addition of Sr-B resulted in the formation of SrB_6 particles and thus consuming more boron atoms comparing to AlB_2 particles, the nucleant responsible for refinement in sole B-treated melt, the minimum value of grain size is greater than the sole boron addition. t_{Rec} is a parameter where its reduction is an indication of lower nonsteady growth rate of the primary particles at the initial stages of solidification. However, comparing the combined Sr-B results with the sole addition of boron confirms that this parameter could not approach zero for combined addition; it is not as effective as sole boron addition.

There is a strong affinity between Sr and B and therefore these elements immediately react and form SrB_6. Most of the boron atoms which should be utilized for refining purposes are consumed to form this compound and as a result, refining becomes less effective. Also Sr is removed from the melt and hence there is less strontium available for modification. The result is partial refining and modification. Figure 6.91 shows boron-based intermetallics within eutectic area in different additional levels.

Semi-Solid Processing

Structural Analysis

The optical micrographs in Fig. 6.92 show the effect of combined melt treatment with B and Sr. The addition of inoculant means having more nucleants in the system and by considering of higher $T_{nuc_{Al}}$ which was shown in Fig. 6.87a, higher

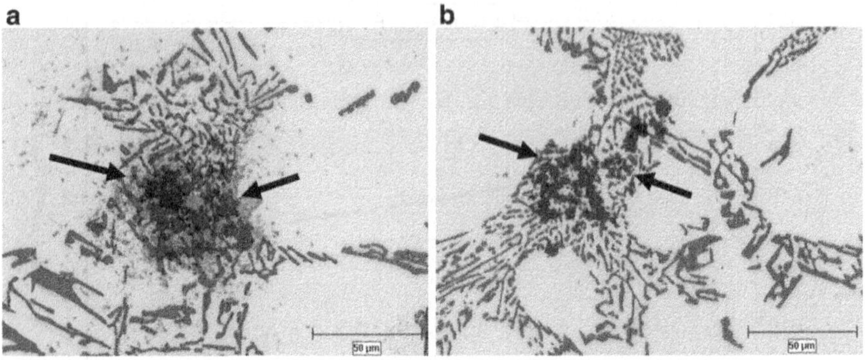

Fig. 6.91 Intermetallics formation and agglomeration due to: (**a**) 231 ppm B, 112 ppm Sr, (**b**) 218 ppm B, 136 ppm Sr [64]

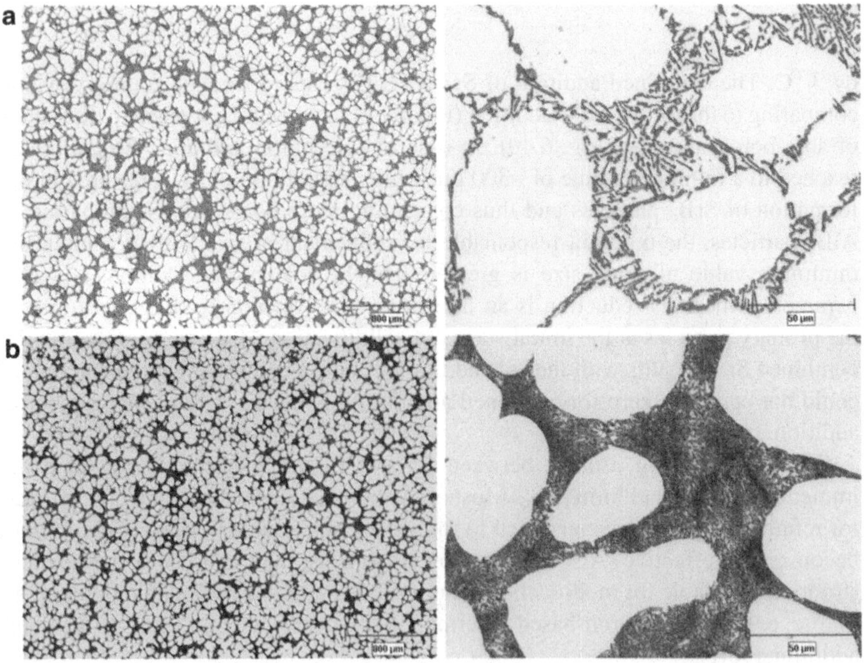

Fig. 6.92 Optical micrographs showing the effect of combined treatment: (**a**) Without addition, (**b**) 218 ppm B, 136 ppm Sr [64]

percentage of α-Al particles is expected. Sr addition increased the percentage of primary phase by depressing the eutectic temperature to lower values, and widening the solidification range. Such qualitative remarks will be supported when the quantitative results are discussed later.

The effect of modification is recognizable by the formation of rounder globules plus morphological evolution of silicon flakes to fibrous (Fig. 6.92). The evolution of eutectic silicon is not homogenous and a mixture of flake/fibrous structure is dominant. However, if Fig. 6.92 is compared with that of Fig. 6.89, the question may arise as why there is a change in the silicon morphology in the SSM cast billets. It is necessary to consider two distinct effects. By quenching the SSM billets, the silicon flake size decreases due to the higher cooling rate. From a kinetics point of view, the reaction of Sr and B in this system is time dependent and in conventional air cooled samples, there is enough time for SrB_6 formation prior to the main eutectic reaction and therefore the remaining Sr is inadequate to cause morphological changes of silicon flakes. In quenched billets, the story is different as the billets were quenched from 598 °C and this would result in shorter reaction time for Sr and B (SrB_6) and therefore the unreacted Sr in the system is greater.

Combined addition of Sr and B has disadvantages of intermetallic formation in the eutectic regions due to their high affinity. Figure 6.93 shows typical agglomeration in the billets. Electron microprobe analysis of these particles confirmed SrB_6 stoichiometry.

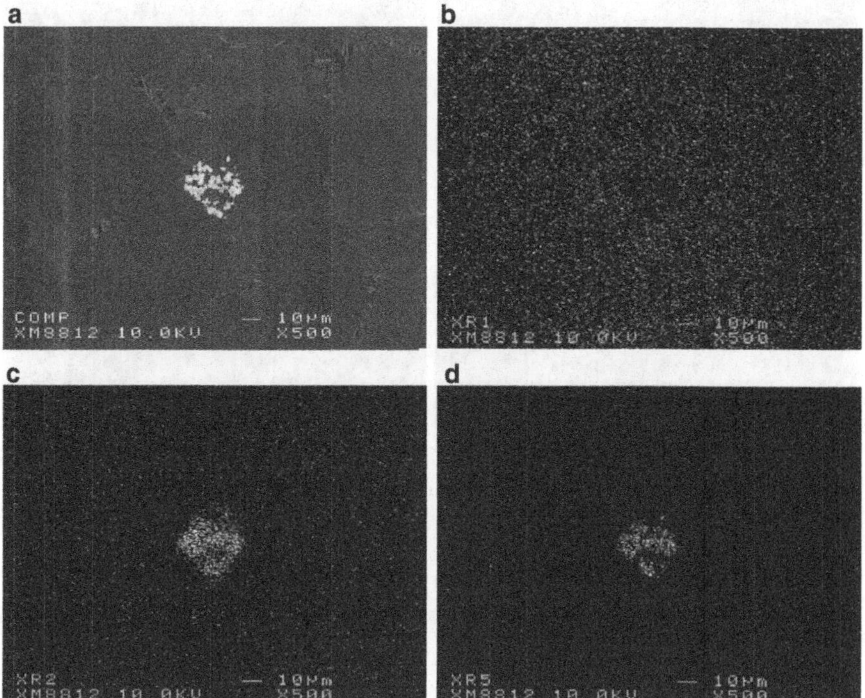

Fig. 6.93 (**a**) Backscattered electron micrograph of agglomerations, and X-ray maps for: (**b**) Al, (**c**) B, and (**d**) Sr (SEED sample with 231 ppm B, 112 ppm Sr) [64]

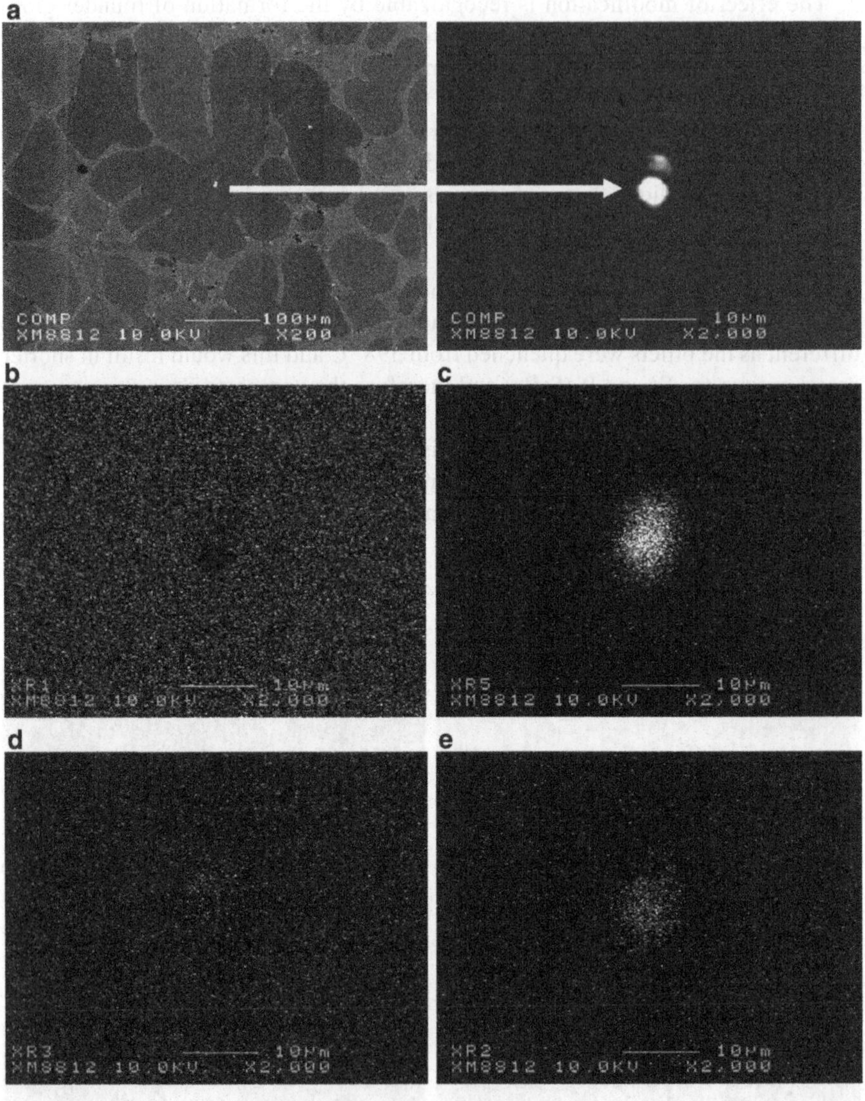

Fig. 6.94 (**a**) Backscattered micrographs of a nucleus in the sample with 231 ppm B, and X-ray maps of (**b**) Al, (**c**) Sr, (**d**) Ti, and (**e**) B [64]

The nucleation sites for the primary α-Al particles were searched in the as-quenched billets. The backscattered electron micrograph and corresponding X-ray maps in Fig. 6.94 confirms that the nucleant is a particle consisting of Sr, B, and Ti elements.

Figure 6.95 shows the SEM (secondary electron) micrographs of selected deep etched samples. Comparing to the conventional cast specimens, finer and fibrous structure is more obvious in the combined treated samples. With modifier addition,

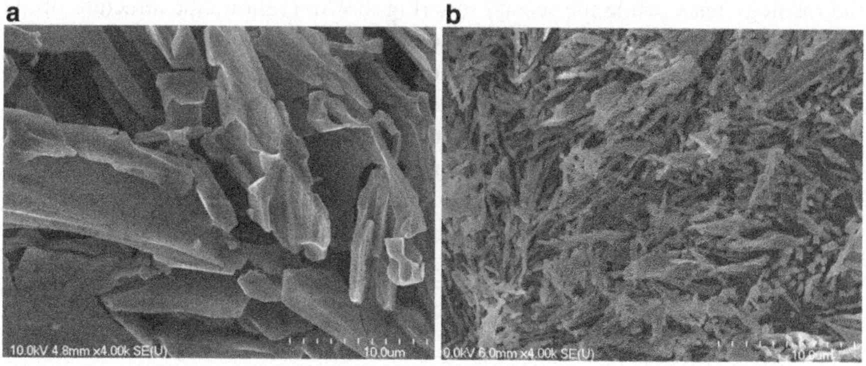

Fig. 6.95 SEM secondary electron (SE) micrographs, deep etched in 10%HF, (**a**) without addition, (**b**) 218 ppm B, 136 ppm Sr addition [64]

the flaky structure of silicon changes to fibrous and the seaweed-like structure is visible but its distribution is not uniform in the sample and some flakes are also visible.

Image Analysis

The results of image processing are shown in Fig. 6.96. The primary α-Al percentage increased which is mainly associated with the numerous nucleation sites as discussed before, i.e., shifting up the solidification temperature due to the B addition and the benefit of lowering the eutectic reaction due to the Sr addition. Globule size reduction was noticed through different parameters such as decreasing of circular diameter, the percentage of α-Al particles having aspect ratio >2, and the area-to-perimeter ratio and increasing of particles number density. Also, the percentage of particles having sphericity greater than 0.8 increases, meaning higher degree of globularity.

6.2.4 Rheological Characteristics

6.2.4.1 Rheological Studies of A356 Alloy/Addition of Ti, B, and Sr

A limited number of rheological tests were performed with the optimum additional level of refiner and/or modifier using Al5Ti1B and Al10Sr master alloys. The optimum rates were obtained from the previous sections in this chapter and the final chemical compositions are listed in Table 6.11.

The optical micrographs in Fig. 6.97 show the microstructural evolution of the untreated and treated billets before and after deformation (the entire specimens were quenched in water from 598 ± 2 °C). The micrographs in the first row (Fig. 6.97a–c) show the structure of the billets at the end of casting and just before the rheology tests, while the second row (Fig. 6.97d–f) shows the structure of the same billets after isothermal reheating and compression tests (10 min at 598 ± 2 °C) which obviously show some degree of coarsening due to isothermal holding. This coarsening is expected, since the billets were held isothermally under pressure which not only induced better contact between particles, but also improved wetting between particle and liquid. In other words, the diffusion and transport of constituent elements as the prerequisite for coarsening were improved. The disappearance of finer particles in the deformed billets is a clear indication of Ostwald ripening as the dominant mechanism for coarsening.

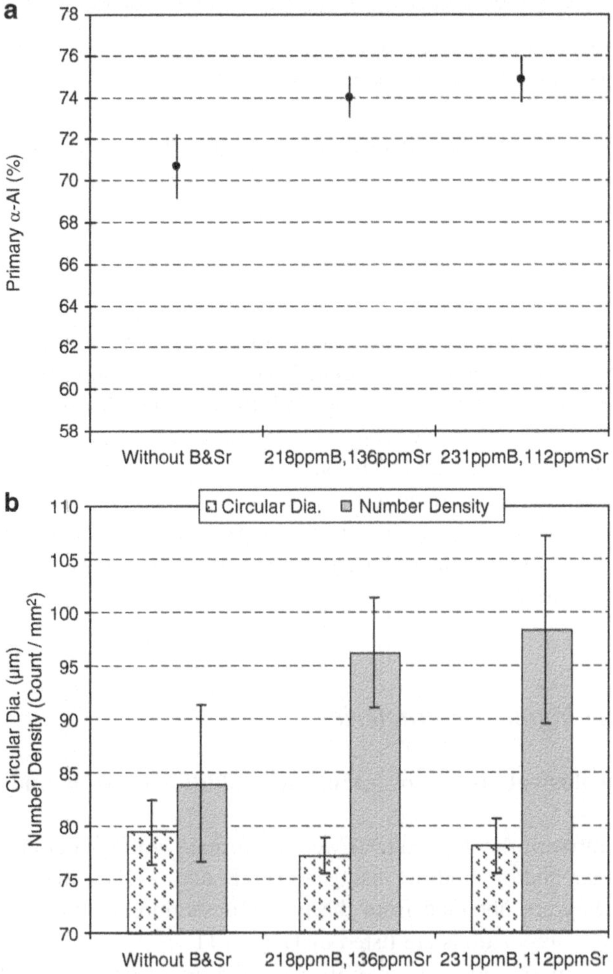

Fig. 6.96 Image analysis parameters for B-Sr-treated A356: (**a**) primary α-Al percentage (**b**) circular dia. and number density of α-Al particles, (**c**) A/P ratio and percentage of α-Al particles having aspect ratio >2, (**d**) % of particles having sphericity >0.8 [64]

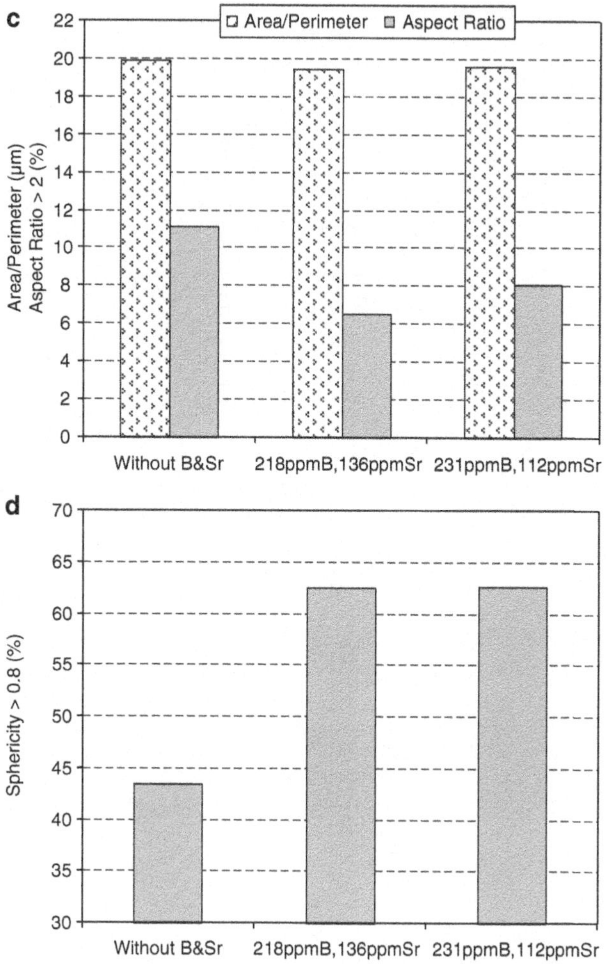

Fig. 6.96 (continued)

Strain–Time Graphs

Figure 6.98a shows a typical deformation–time curve. The graph typically has three deformation stages [65];

1. Stage-I, unsteady state, where the billet flows almost without any resistance to applied pressure. This behavior is attributed to the easy movement of the primary α-Al particles within the residual liquid and without appreciable collision. The extent of this region is dependent on the particle size and applied pressure where it is greater at smaller particle size or higher applied pressure.
2. Stage-II, unsteady state, where there is some degree of resistance to flow due to the collision of solid particles and formation of α-Al agglomerates. The agglomerated chunks are the resisting constituents to billet flow.

Table 6.11 Chemical analysis of the melts (wt. %)

	Si	Mg	Fe	Mn	Cu	Ti	B	Sr	Al
Original	6.5–6.75	0.36–0.4	0.07–0.08	0.002–0.003	0.001	0.0058	Nil	Nil	Bal.
Refined						0.06–0.07	0.010–0.014	Nil	
Modified						0.0058	Nil	0.017–0.018	
Combined						0.06–0.07	0.010–0.014	0.015–0.018	

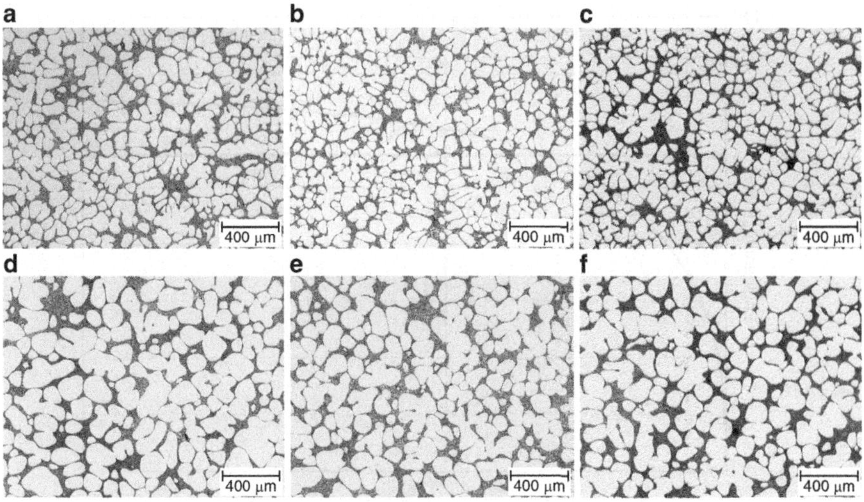

Fig. 6.97 Microstructural evolution of quenched samples ((**a**) to (**c**)): (**a**) untreated alloy, (**b**) grain refined with 620ppmTi, 110 ppm B, (**c**) combined treatment, 610ppmTi, 100 ppm B, 160 ppm Sr, and after isothermal pressing ((**d**) to (**f**)): (**d**) sample "(**a**)," (**e**) sample "(**b**)," (**f**) sample "(**c**)" [65]

Both stages I and II are regarded as nonsteady states, since the slope of the graph varies with time, i.e., there is mainly agglomeration. That's why the first two regions are classified as unsteady state segment in Fig. 6.98a.

3. Stage-III, quasi steady state, where the billet deforms steadily with a constant slope. It is believed the solid–liquid mixture, "the mush," has reached a state where it deforms as a single phase. The processes of agglomeration and deagglomeration establish a quasi-steady-state equilibrium condition in the mush at this stage of deformation.

Figure 6.98b shows the effect of various melt treatments on the deformation behavior of the SSM billets. The change in the billets deformability due to α-Al particles size and distribution is clearly visible on the graphs. The untreated alloy (base alloy) shows the lowest strain values while the grain refined and combined treated alloys deform at much higher range due to smaller globule size as indicated in the previous section. It has to be emphasized that the flow characteristics of the refined and combined treated billets are almost the same and only one deformation band is presented for both alloys.

For the shaped engineering components, the mechanical properties are not only dependent on the primary α-Al phase, but also on the morphology of silicon. It is believed that the combined treatment is the preferred choice for melt treatment resulting in better die filling and mechanical properties. The modified alloy lies somewhere between the refined/combined and the base alloys.

To better understanding the particle size effect during deformation, several tests were performed on the overrefined billets and typical strain–time graphs are given in Fig. 6.99. Overrefinement is defined as the addition of inoculant in excess of the

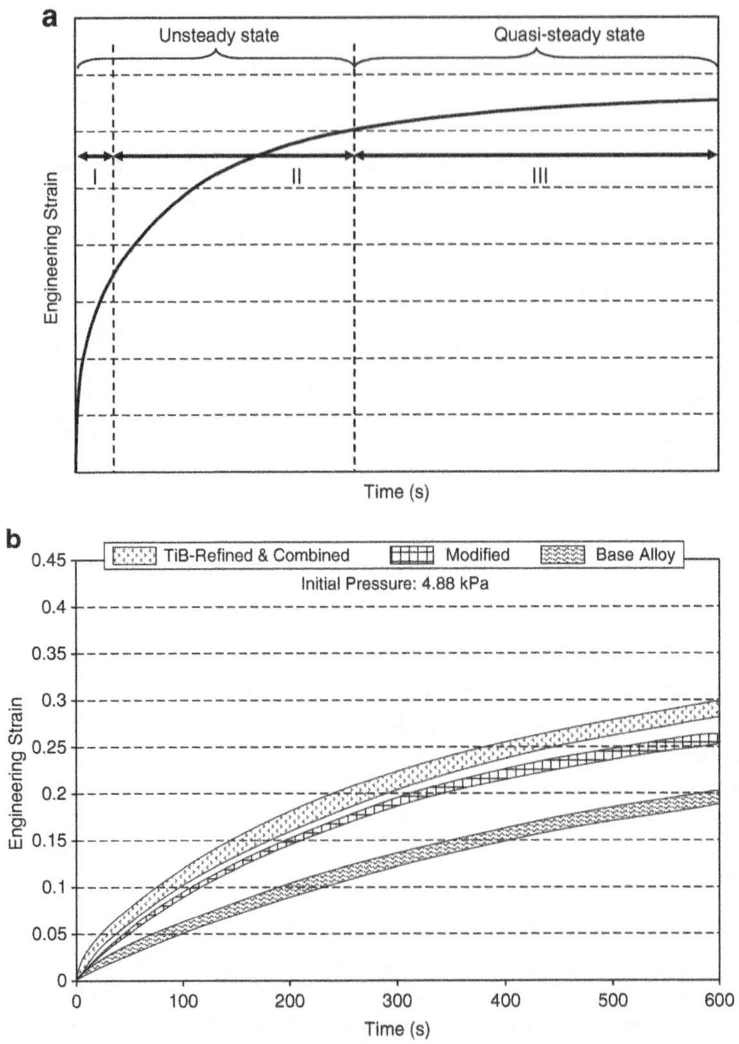

Fig. 6.98 (**a**) A typical graph for different deformation stages during parallel plate compression test, (**b**) strain–time graphs for melt treatment [65]

optimum percentages of ~0.06–0.08 % Ti, and ~0.01–0.02 % B, respectively. With over addition, the average circular diameter of the primary α-Al particles in the quenched billets was reduced by ~5 % and number density increased by ~20 % (Sect. 6.2.1.2). The lower resistance to flow for the overrefined billet shown in Fig. 6.99 is solely due to the smaller particle size of the primary α-Al phase. The deformation in the overrefined billet is even higher than that of the billet with optimum refiner addition throughout the compression test. However, the drawback of overrefined billets is the formation of Ti-based intermetallic compounds as

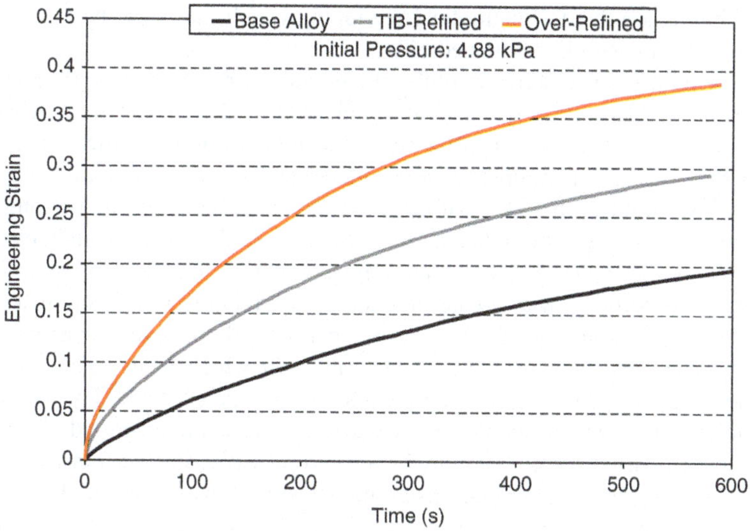

Fig. 6.99 Comparison of the typical strain–time graphs for the untreated, refined, and overrefined alloys [65]

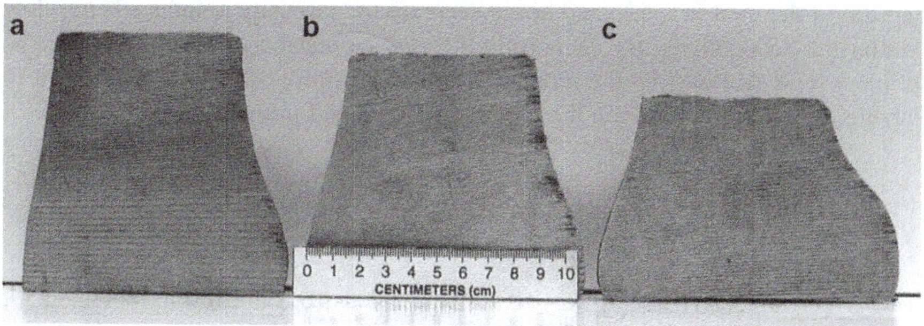

Fig. 6.100 Deformed billets under 4.88 kPa pressure; (**a**) base alloy, (**b**) refined, and (**c**) overrefined [65]

discussed earlier. It is important to note that the formation of Ti-based intermetallics in the overrefined condition did not impede the flow during rheological test. The main negative result of these intermetallics is on the mechanical properties of the finished product. Once the billet is shaped to a product, they may act as potential crack initiation sites.

The optical photomicrograph in Fig. 6.100 compares the deformability of the base alloy with that of the refined and overrefined billets to further confirm the strain–time graphs presented in Fig. 6.99. Before the tests, billets were of the same original height and diameter and the reduction of height shows the billet capability to flow under constant applied pressure. The slight width differences in the three treated billets are due to longitudinal sectioning which may have been somewhat off-center for the combined treated billet.

The better flow characteristics of the refined billets are attributed to the reduction of primary α-Al particle size and lower dendrite coherency point (DCP). For conventionally cast alloys, it is reported that refining and modification reduces the DCP to lower temperatures and thus induces less restriction to liquid mass flow [66]. For SSM alloys, the fact that large and extended dendrites are no longer the dominant morphology, the tendency to form a three-dimensional interconnected solid network is reduced. The refined alloy contains finer and more spherical primary α-Al particles with greater amount of solid–liquid interfacial area where the interparticle liquid film could assist solid particles gliding over one another. The finer the particles, the greater the interfacial area which tends to facilitate the particle slipping and thus easier particle glide should be expected. For particle-to-particle contact, first the smaller particle size increases the magnitude of contact area between particles and therefore the applied pressure is better transferred from particle to particle resulting in easier movement of particles and better flow. Second, the energy required to push smaller particles over each other is smaller than that required for larger ones.

For the case of Sr modification, it is proposed that the reductions in the surface tension of the remaining liquid and lower DCP are the factors that improve flow characteristic of the SSM billets. It has been reported that strontium reduces surface tension of the melt [61, 62], and hence decrease the viscosity to induce better flow within the SSM billets. By decreasing the surface tension, the wettability of the primary α-Al particles by the remaining liquid is improved. This is equivalent to having more contact between the remaining liquid and primary α-Al.

Viscosity

The graphs presented in Fig. 6.101 are calculated from the results obtained for height changes with time during the quasi-steady-state deformation, ~200–600 s after the beginning of each compression test (see Fig. 6.98a). The results were calculated on the assumption that the semi-solid billets behave as Newtonian fluids at this stage of deformation.

The viscosity is calculated through inverse slope of these graphs according to the equations explained in Sect. 4.2 and the results are listed in Table 6.12. The results are similar to viscosity of semi-solid alloys of Sn–15%Pb [67], A356 [68, 69], and Al–SiC composites [70] having globular primary particles at similar solid fraction and shear rate (see Fig. 4.16). The working temperature in these series of tests was 598 ± 2 °C which according to the thermal analysis and ThermoCalc calculations result in maximum 5 % variation of fraction solid. The 5 % solid fraction variation is normal in SSM processing and should not be a matter of concern in viscosity measurement. Besides, A356 alloy is a well-researched alloy and has low fraction solid sensitivity to minor temperature variation, which is due to its wide

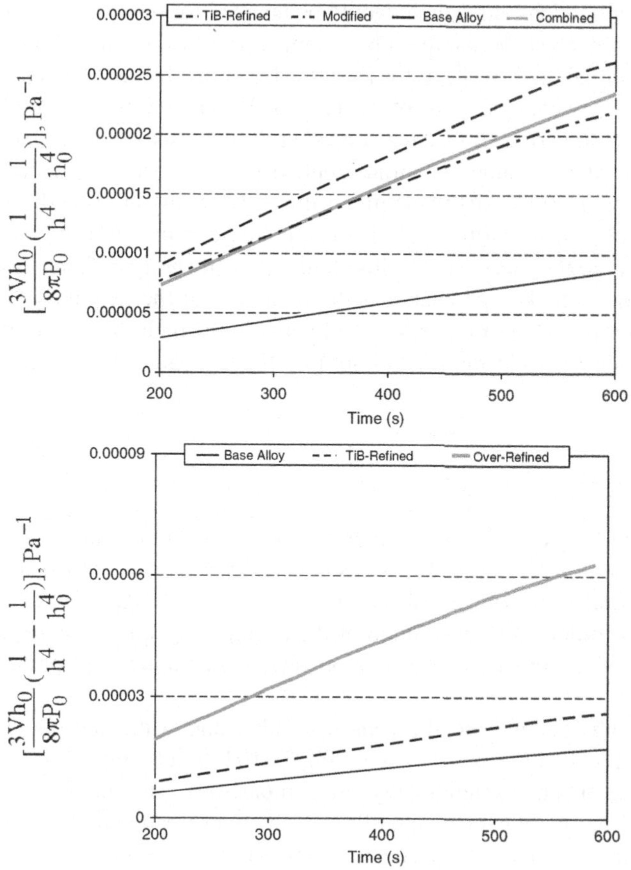

Fig. 6.101 Quasi-steady-state part of the graphs, where the left-hand side of equation 4.30 (Chap. 4) is plotted against time within 200–600 s after the beginning of each test to calculate the viscosity [65]

Table 6.12 Viscosity values with different treatments [65]

	Base Alloy	Ti-B Refined	Modified	Combined	Overrefinement
Log η (Pa.s)	8.0–8.1	7.38–7.4	7.39	7.3–7.4	7

solidification range. This is particularly true for the narrow semi-solid processing window used in this study, 598 ± 2 °C.

Most of the previous studies were carried out on small samples with $h \ll R$ (h is the height and R is the radius of billet). The fact that the current results are close to those reported for samples with $h \ll R$, suggests the irrelevancy of such criterion provided that all the measurements taken from regions with maximum radial deformation (see Sect. 4.2.2—sample size). This has an important implication in industrial applications since there would be no need for resizing the billet. Furthermore, it may suggest the parallel plate compression test on actual billets could be used as an online quality check for rheocast feedstock in cast shops [71].

In all cases where the size of primary α-Al phase was decreased by refiner addition or by lowering surface tension using Sr, the viscosity values were found to be lowered. At low values of initial applied pressure, the non-refined structures have shown higher viscosity than refined alloys. Normally in a specific alloy, higher percentage of primary particles leads to an increase in the viscosity value which is linked to the larger particles with less liquid percentage (which acts as lubricant). By inoculation, the key parameter is the reduction of the globule size and higher sphericity values. Therefore, prevailing factors in the melt treatment procedure are size and sphericity of primaries and not the slight rise in α-Al percentage.

Liquid Segregation

As described in Sect. 4.2, the billets were compressed uniaxially. The rapid initial vertical descent of the top plate causes lateral liquid movement in the billet while the solid particles are compacted and may collide to form agglomerated particles at the center of billets. The magnitude of the segregated liquid zone depends on the percentage of deformation, deformation rate, temperature, and particle size and morphology [70].

Micrographs taken from the center, middle, and wall sections of the billets deformed at 4.88 kPa are shown in Fig. 6.102 (micrographs taken at various locations are attached intentionally for comparison). The microstructures of the central zone are homogenous regardless of the melt treatment. The eutectic liquid (dark regions) appears to have accumulated mainly at regions near the billet wall. This was expected, since the applied pressure displaces the residual liquid laterally toward the walls. This phenomenon is intensified during the injection of semi-solid billets in high pressure die casting due to greater magnitude of applied pressure [72]. The formation of segregated liquid regions is due to the differences in the density and flowability of the residual liquid and the primary α-Al particles.

The untreated base alloy along with the modified, and refined samples show some degree of liquid segregation in the outer layers near the billet wall, while the combined and overrefined cases show the least segregation. The absence of liquid segregation for the latter samples may be attributed to better distribution of primary particles (α-Al), having finer and more spherical morphology in the billets. Fine α-Al solid particles move with liquid without excessive collision and bring about homogenous distribution of primary α-Al phase. The micrographs presented in Fig. 6.102 confirm such hypothesis and show almost no segregation of liquid for the billets with the combined treatment.

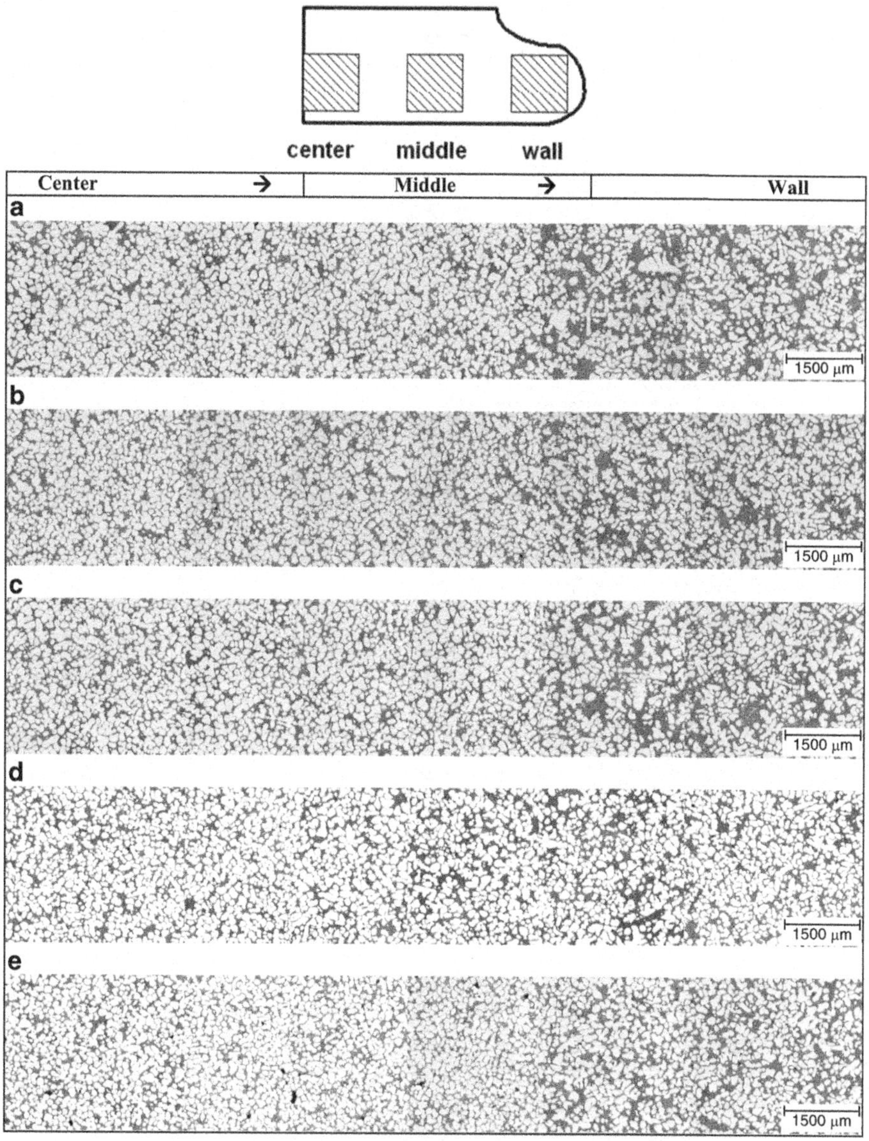

Fig. 6.102 Microstructure of deformed SSM billets from center to the wall, (**a**) untreated alloy, (**b**) refined, (**c**) modified, (**d**) combined, and (**e**) overrefined [65]

6.2.4.2 Rheological Studies of A356 Alloy/Addition of B and Sr

The optimum additional levels were measured in Sect. 6.2.3.2. The final chemical compositions are shown in Table 6.13.

Table 6.13 Chemical analysis of A356 alloy (wt. %)

	Si	Mg	Fe	Mn	Cu	Ti	B	Sr	Al
Original	6.5–6.75	0.36–0.4	Max 0.08	Max 0.003	Max 0.001	Max 0.0058	Nil	Nil	Bal.
Refined							0.02–0.03	Nil	
Combined							0.02–0.03	0.012–0.016	

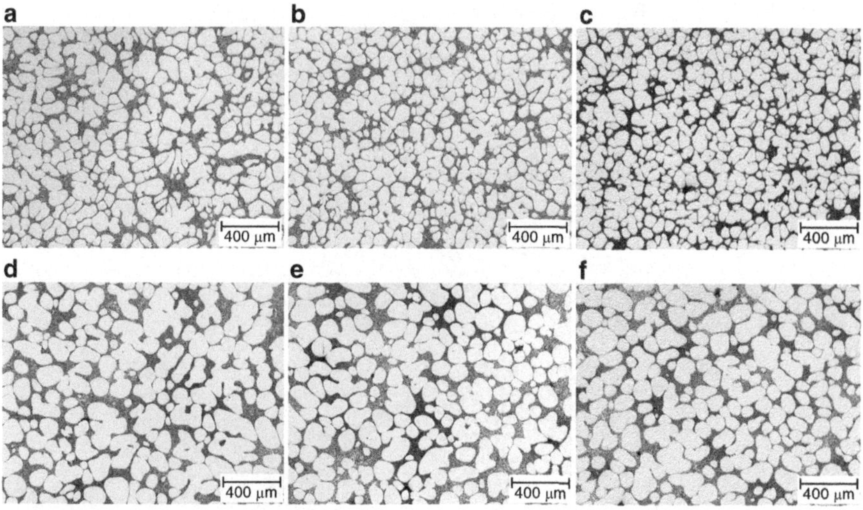

Fig. 6.103 Optical micrographs of quenched samples ((**a**) to (**c**)): (**a**) untreated alloy, (**b**) grain refined with 225 ppm B, (**c**) combined treatment, 230 ppm B, 112 ppm Sr, and after isothermal pressing ((**d**) to (**f**)): (**d**) sample "(**a**)," (**e**) sample "(**b**)," (**f**) sample "(**c**)" [64]

Figure 6.103 shows the structural evolution of untreated, B-treated, and B-Sr-treated alloy. As confirmed in Sect. 6.2.1.3, boron is an effective refiner (enhanced performance) and its addition results in ~5 % reduction in average circular diameter and ~18–23 % increase in number density. The smaller and more spherical primary α-Al particles with overall higher amount are obvious in the structure as seen in Fig. 6.103b.

By Sr and B combination, it is expected to have fibrous morphology of eutectic silicon but as discussed in Sect. 6.2.3.2, this evolution is not uniform and homogenous throughout the billets. It means that some parts are fully modified while others are partially modified as noticeable in Fig. 6.104b.

With isothermal holding, entrapment of liquid (eutectic) is evident which is normally seen in thixocasting. As well, α-Al particles coarsen and with increasing in size and coalescence of primary α-Al particles (i.e., pseudo-sintering the primary particles), liquid is encapsulated and finally form entrapped areas (later this liquid transforms to eutectic mixture during quenching process). This process of liquid entrapment may also be assisted by agglomeration mechanism related to the application of external forces as clearly shown in Fig. 6.105.

Strain–Time Graphs

The formability of the refined and combined treated slurries is shown in Fig. 6.106. The higher deformation of the B-treated alloy as an efficient refiner is clear in the strain–time graphs with the highest strain at any time belonging to the B-refined slurries. As discussed in Sect. 6.2.1.3, boron contained master alloy is capable of

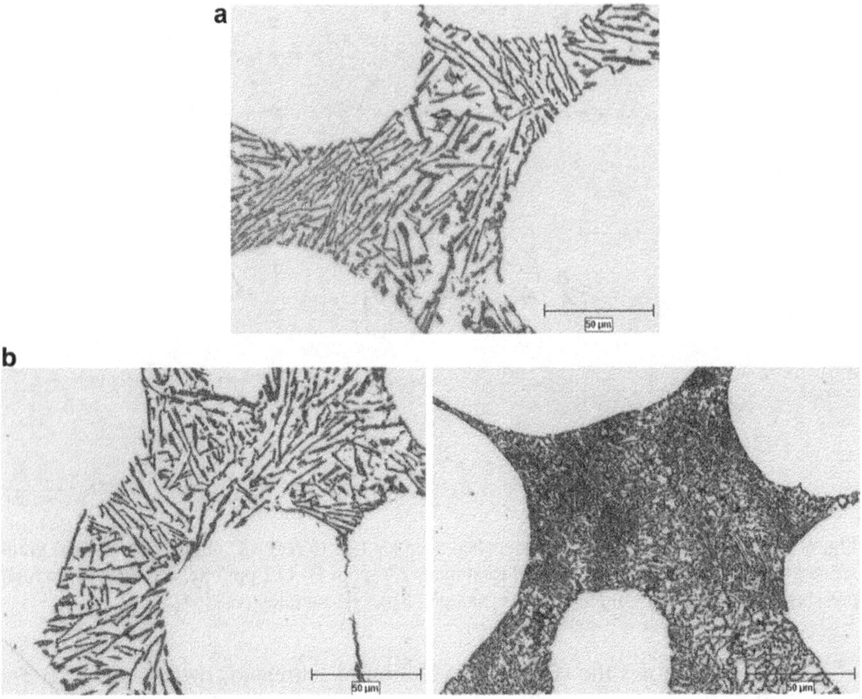

Fig. 6.104 Effect of Sr addition on the deformed billets: (**a**) 288 ppm B, (**b**) 313 ppm B and 160 ppm Sr [64]

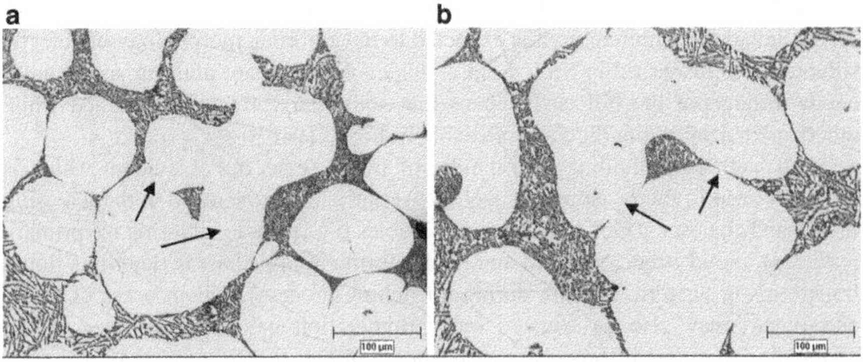

Fig. 6.105 Optical micrographs showing the entrapped liquid: (**a**) refined with 288 ppm B, (**b**) with 313 ppm B and 160 ppm Sr addition (arrows showing the sintering points) [64]

efficient refinement which is contributed to the presence of AlB_2 particles, and dissolved boron in the aluminum matrix. The untreated alloy has the lowest overall deformation and strain rate and highest deformation and strain rate is due to boron inoculation. The combined effect has the moderate value which confirms the loss of

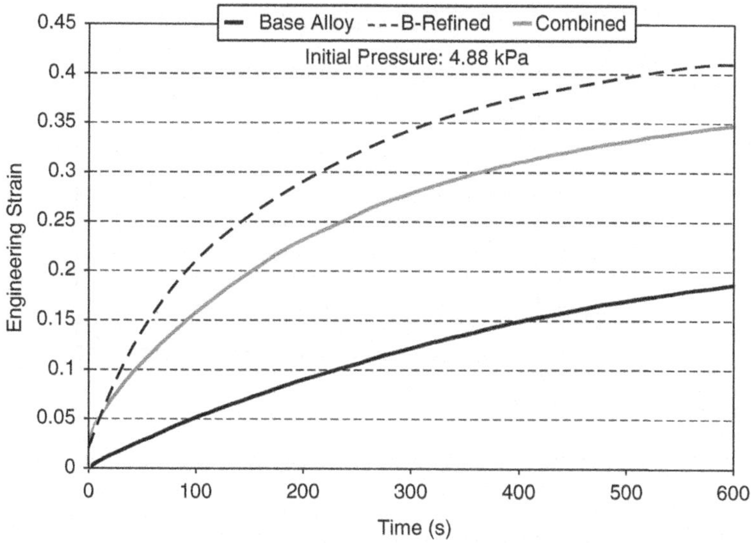

Fig. 6.106 Strain–time graphs for B and B-Sr additions [64]

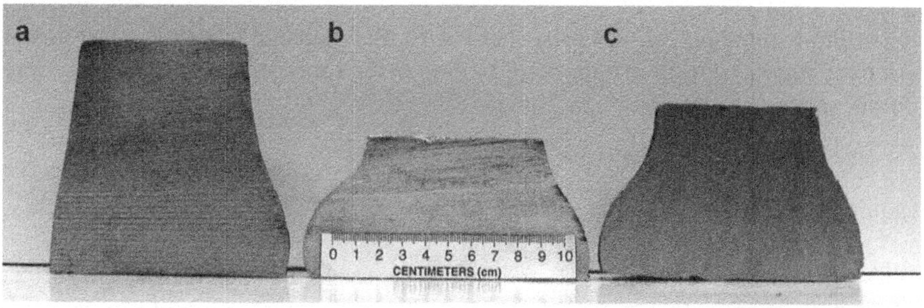

Fig. 6.107 Deformed billets under 4.88 kPa pressure (base alloy, B-refined, and combined from *left* to *right*) [64]

B due to the formation of SrB_6. The remaining B is therefore not adequate to effectively reduce particle size and thus resulting in less deformation. Another disadvantage of combined treatment is the nonhomogenous structure of eutectic silicon which leads to nonuniform mechanical properties. Better deformability of the boron-treated billet is also clear in Fig. 6.107.

Viscosity

As explained in Sect. 6.2.4, the viscosity of SSM billet was calculated from the quasi-steady-state region of the graphs in 6.106 (~200–600 s) (Fig. 6.108). The calculated viscosity values are given in Table 6.14. The lowest values of the

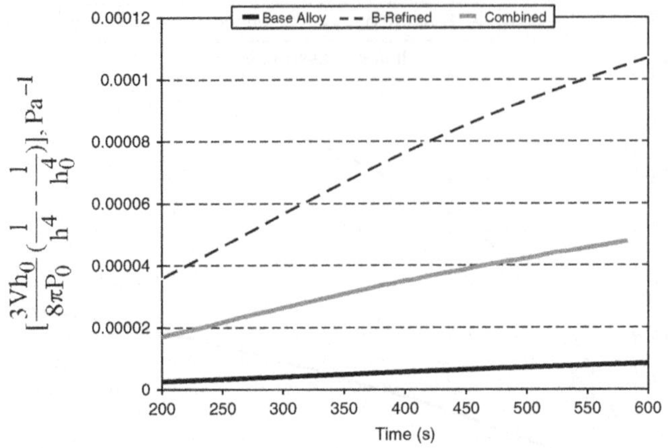

Fig. 6.108 Steady state part of the graphs used for viscosity calculation, equation 4.30 (Chap. 4)

Table 6.14 Viscosity values for refining and combined treatments

	Base Alloy	B-Refined	Combined
Log η (Pa.s)	8.0–8.1	6.6–6.7	7.1–7.3

logarithmic numbers of viscosity belong to the B-refined sample. The lower viscosity numbers result in better mold filling in the subsequent injection procedure during semisolid casting.

Liquid Segregation

With deformation, liquid moves laterally and causes liquid segregation. This effect is always a combination of liquid escaping due to the application of pressure and also tendency of liquid to migrate to lower billet extremes, "*elephant footing defect*" [73]. By comparing Fig. 6.102b which shows the best condition of Ti-B master alloy addition with Fig. 6.109b, it is concluded that boron refinement eliminates liquid segregation by improved distribution of primary particles and consequent enhanced flowability due to smaller particle size.

6.2.4.3 Rheological Studies of Commercial 356 Alloy/Addition of Ti, B, and Sr

Finding a correlation between the structural development and deformability of the alloy is the main concern in this chapter. In this section, the comparison is made for commercial alloy, with Ti content in the original ingots. The optimum additional level was measured in Sect. 6.2.1.4 and the final chemical compositions are listed in Table 6.15.

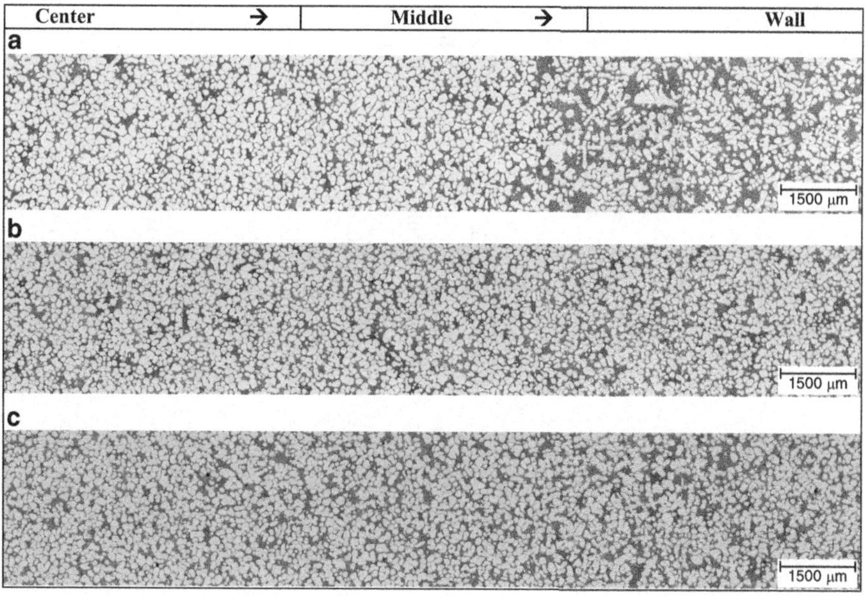

Fig. 6.109 Microstructure of deformed SSM billets from center to the wall; (**a**) original alloy, (**b**) refined, and (**c**) combined [64]

Due to the preexisted Ti in the alloy, the resulting melts have higher GRF (growth restriction factor) values (Sect. 6.2.1.4) and hence dendritic growth is restricted to some degree. Figure 6.110 shows the microstructural evolution due to the refining and combined treatments. With refining, the primary α-Al percentage and number density increase slightly. These augmentations are not as much as the laboratory prepared A356 alloy with trace level Ti content since the original titanium level in the commercial alloy could overshadow the refiner effect. Fig. 6.110d–f also show the usual coarsening effect of isothermal compression tests. It has been shown quite clearly that by isothermal holding at 598 ± 2 °C, the particles not only grow and become larger, but also contacts between the primary α-Al particles increase. The greater contact area is due to the compression force and diffusion at higher temperatures.

In the modified and combined cases, not only the percentage of the primary phase increases, but also the eutectic silicon morphology is changed. In the deformed micrographs, Fig. 6.111, the Sr addition results in morphological evolution of eutectic silicon from lamellar to fibrous structure. Also it seems that the primary α-Al particles are encapsulated by the eutectic regions.

Strain–Time Graphs

Effects of different treatments on the formability of billets are shown in Fig. 6.112. Higher strain or better flowability is related to the refined alloy treated by Al5Ti1B

Table 6.15 Chemical analysis of commercial 356 alloy (wt. %)

	Si	Mg	Fe	Mn	Cu	Ti	B	Sr	Al
Original	7.0–7.2	0.35–0.36	0.07–0.08	0.012	0.002	0.09–0.1	Nil	Nil	Bal.
Refined						0.12–0.13	0.006	Nil	
Modified						0.09–0.1	Nil	0.015–0.017	
Combined						0.12–0.13	0.006	0.011–0.012	

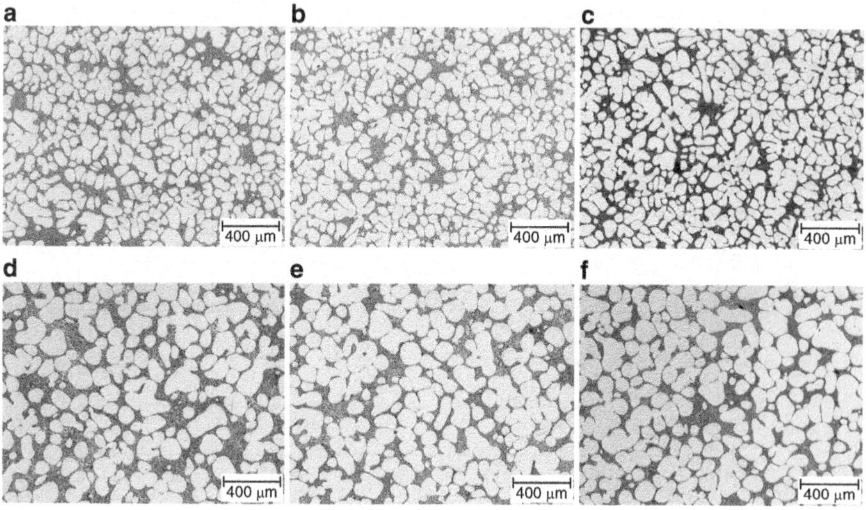

Fig. 6.110 Microstructural evolution of 356 quenched samples ((**a**) to (**c**)): (**a**) base alloy, (**b**) grain refined, (**c**) combined treatment and after pressing ((**d**) to (**f**)): (**d**) sample "(**a**)," (**e**) sample "(**b**)," (**f**) sample "(**c**)" [64]

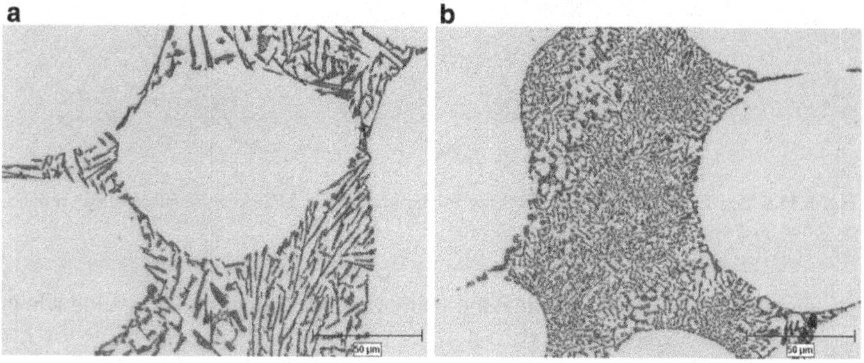

Fig. 6.111 Effect of Sr addition in the deformed billets: (**a**) non-modified, (**b**) 164 ppm Sr addition [64]

master alloy. By strontium addition, surface tension of the liquid decreases [61, 62] and this is believed to be the main reason for improving in the slope of the modified curve. Combined treatment shows itself as a lower limit of the refined curve in the strain–time graphs.

Comparing the results here with A356 alloy, Fig. 6.98, it appears that both alloying systems have the same trend in strain values before and after melt treatment. However, it should be emphasized that the strain values for untreated alloy are higher in the case of commercial 356 alloy which is associated to the

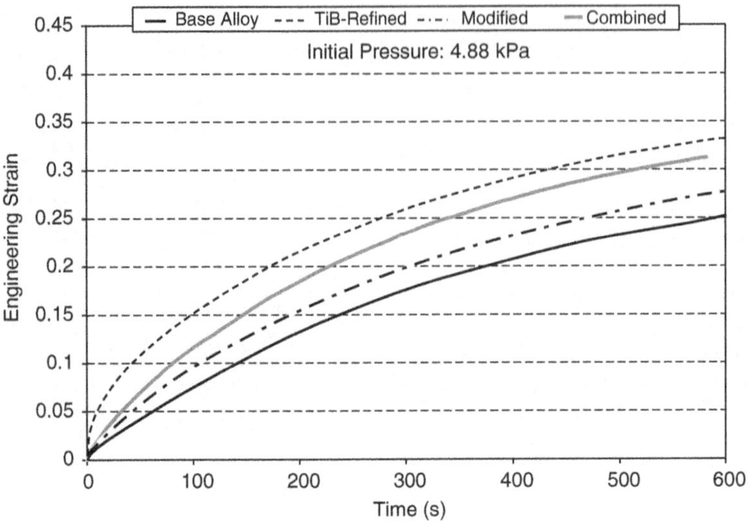

Fig. 6.112 Strain–time graphs for entire treatments [64]

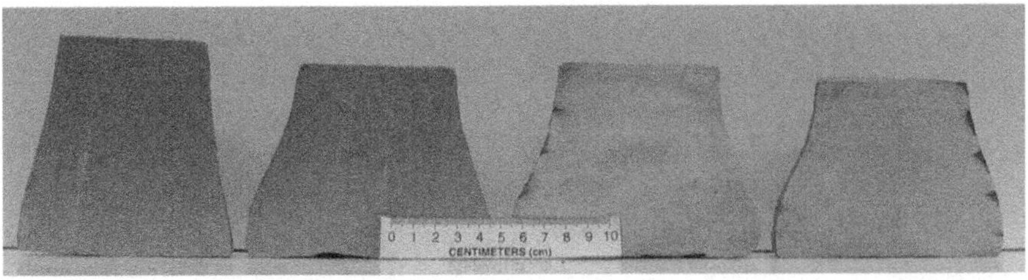

Fig. 6.113 Sectional view of the deformed billets under 4.88 kPa pressure (base alloy, refined, modified, and combined from *left* to *right*) [64]

pre-dissolved titanium in the alloy and its role on increasing the GRF value which was already discussed in Sect. 6.2.1.4.

Figure 6.113 compares longitudinal sections of the compressed billets with different levels of deformation under 4.88 kPa pressure. The trend in the level of deformation in these samples matches the strain–time graphs, Fig. 6.112.

Viscosity

Figure 6.114 and Table 6.16 show the viscosity measurements. Comparing the viscosity values with A356, the values are lower for commercial 356 for similar treatments. As noted, this is directly related to the effectiveness of titanium in restricting the globules growth. By lower growth rate of primary α-Al particles, the particles become smaller and they can flow easier.

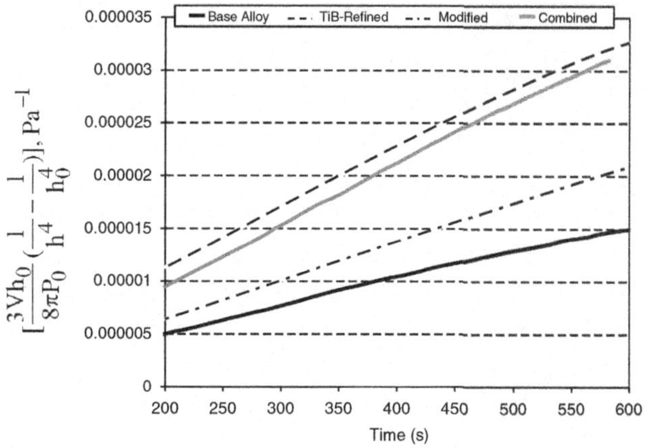

Fig. 6.114 Steady state part of the graphs, where the left hand side of equation 4.30 (Chap. 4) is plotted against time within 200–600 s after the beginning of each test to calculate the viscosity

Table 6.16 Viscosity values for various treatments

	Base alloy	TiB-refined	Modified	Combined
Log η (Pa.s)	7.5–7.55	7.3–7.32	7.39–7.4	7.15–7.22

Liquid Segregation

The microstructural changes from center to the wall of the deformed zones of the billets are shown in Fig. 6.115. The central parts are almost unchanged and there is no difference regarding the treatment. Normally in the deformed billets, the liquid segregates adjacent to the wall due to the application of vertical force. Liquid segregation is unavoidable for the base alloys both in A356 and Commercial 356 groups. This zone in the refined case appears to have the same segregation pattern as the base alloy with slightly more uniform pattern which is believed relates to the overshadowing of the refiner effect by the pre-dissolved titanium in the melt. The same pattern also exists for the modified sample, however in the combined case, the segregation pattern has disappeared to some extent.

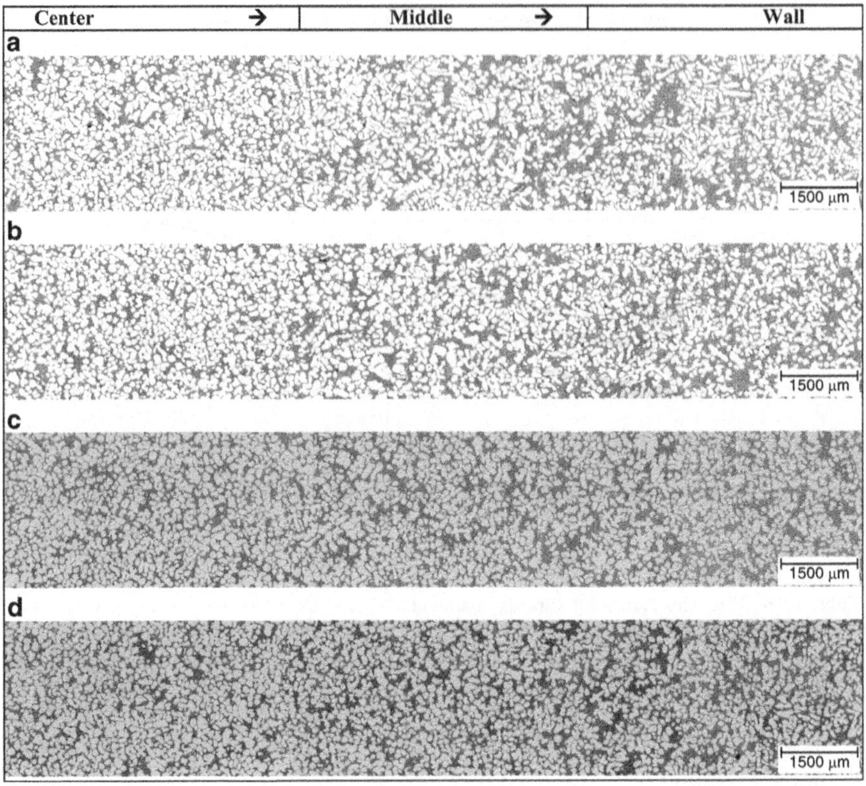

Fig. 6.115 Microstructure of deformed SSM billets from center to the wall; (**a**) base alloy, (**b**) refined, (**c**) modified, and (**d**) combined [64]

References

1. D.G. McCartney, Grain refining of Aluminum and its alloys using inoculants. Int. Mater. Rev. **34**(5), 247–260 (1989)
2. J.P. Gabathuler, D. Barras, Y. Krahenbuhl, Evaluation of various processes for the production of Billet with thixotropic properties. in *2nd International Conference on Semi-Solid Processing of Alloys and Composites* (MIT, Cambridge, 1992), 33–46
3. G. Wan, T. Witulski, G. Hirt, Thixoforming of Al alloys using modified chemical grain refinement for Billet production. La Metallurgia Italiana **86**, 29–36 (1994)
4. G. Wan, T. Witulski, G. Hirt, Thixoforming of Al alloys using modified chemical grain refinement for Billet production. in *Conference on Aluminum Alloys: New Process Technologies* (Italy, June 1993), 129–141
5. H.P. Mertens, R. Kopp, T. Bremer, D. Neudenberger, G. Hirt, T. Witulski, P. Ward, D.H. Kirkwood, Comparison of different feedstock materials for thixocasting, EUROMAT 97. in *Proceedings of the 5th European Conference on Advanced Materials and Processes and Applications* (1997), 439–444
6. S.C. Bergsma, M.C. Tolle, M.E. Kassner, X. Li, E. Evangelista, Semi-solid thermal transformations of Al-Si alloys and the resulting mechanical properties. Mater. Sci. Eng. **A237**, 24–34 (1997)

7. S.C. Bergsma, Casting, thermal transforming and semi-solid forming aluminum alloys. U.S. Patent 5571346, 5 Nov 1996
8. K. Tahara, H. Tezuka, T. Sato, A. Kamio, Semi-solid solidification in grain refined Al-7%Si-3%Cu alloy. in *6th International Conference on Aluminum Alloys* (1998), 303–308
9. H. Wang, C.J. Davidson, J.A. Taylor, D.H. St. John, Semisolid casting of AlSi7Mg0.35 alloy produced by low-temperature pouring. Mater. Sci. Forum **396–402**, 143–148 (2002)
10. H. Wang, C.J. Davidson, D.H. St. John, Semisolid microstructural evolution of AlSi7Mg alloy during partial remelting. Mater. Sci. Eng. **A368**, 159–167 (2004)
11. H. Wang, Semisolid processing of aluminium alloys. Ph.D. Thesis, The University of Queensland, Australia, Sep 2001
12. Q.Y. Pan, M. Arsenault, D. Apelian, M.M. Makhlouf, SSM processing of AlB2 grain refined Al-Si alloys. AFS Trans. (2004), Paper 04-053
13. Q.Y. Pan, D. Apelian, M.M. Makhlouf, AlB_2 grain refined Al-Si alloys: rheocasting/thixocasting applications. in *8th International Conference on Semi-Solid Processing of Alloys and Composites* (Limassol, Cyprus, 2004)
14. R.S. Rachmat, H. Takano, N. Ikeya, S. Kamado, Y. Kojima, Application of semi-solid forming to 2024 and 7075 wrought Aluminum Billets fabricated by the EMC process. Mater. Sci. Forum **329–330**, 487–492 (2000)
15. T. Grimmig, J. Aguilar, M. Fehlbier, A. Bührig-Polaczek, Optimization of the rheocasting process under consideration of the main influence parameters on the microstructure. in *8th International Conference on Semi-Solid Processing of Alloys and Composites* (Limassol, Cyprus, 2004)
16. R. Shibata, T. Kaneuchi, T. Souda, H. Yamane, Formation of spherical solid phase in die casting shot sleeve without any agitation. in *5th International Conference on Semi-Solid Processing of Alloys and Composites* (Golden, 1998), 465–469
17. K. Sukumaran, B.C. Pai, M. Chakraborty, The effect of isothermal stirring on an Al-Si alloy in the semisolid condition. Mater. Sci. Eng. **A369**, 275–283 (2004)
18. L. Yu, X. Liu, The relationship between viscosity and refinement efficiency of pure aluminum by AlTiB refiner. J. Alloys Compd. **425**, 245–250 (2006)
19. R. Elliot, *Eutectic Solidification Process* (Butterworth, London, 1983)
20. N. Fat-Halla, Microstructure and mechanical properties of modified and nonmodified stir-cast Al-Si hypoeutectic alloys. J. Mater. Sci. **23**, 2419–2423 (1988)
21. W.R. Loué, M. Suéry, Microstructural evolution during partial remelting of Al-Si7Mg alloys. Mater. Sci. Eng. **A203**, 1–13 (1995)
22. B.I. Jung, C.H. Jung, T.K. Han, Y.H. Kim, Electromagnetic stirring and Sr modification in A356 alloy. J. Mater. Proc. Tech. **111**, 69–73 (2001)
23. R.J. Kissling, J.F. Wallace, Grain refinement of Aluminum castings. Foundry (June 1963), 78–82
24. S. Nafisi, R. Ghomashchi, Grain refining of conventional and semi-solid A356 Al-Si alloy. J. Mater. Proc. Tech. **174**, 371–383 (2006)
25. S. Nafisi, R. Ghomashchi, Boron-based refiners: implications in conventional casting of Al-Si alloys. Mater. Sci. Eng. **A452–453**, 445–453 (2007)
26. ASTM International Standard Worldwide, Volume 02.02, Aluminum & Magnesium Alloys (2004), 78–92
27. L. Backerud, G. Chai, J. Tamminen, *Solidification Characteristics of Aluminum Alloys, Volume 2, Foundry Alloys* (American Foundry Society, Des Plaines, 1990)
28. D. Apelian, G.K. Sigworth, K.R. Whaler, Assessment of grain refinement and modification of Al-Si foundry alloys by thermal analysis. AFS Trans. **92**, 297–307 (1984)
29. P.A. Tøndel, G. Halvorsen, L. Arnberg, Grain refinement of hypoeutectic Al-Si foundry alloys by addition of Boron containing silicon metal, in *Light Metals*, ed. by S.K. Das (TMS, Denver, 1993), 783–790
30. J.A. Marcantonio, L.F. Mondolfo, Grain refinement in Aluminum alloyed with titanium and Boron. Metal. Trans. B **2**, 465–471 (1971)

31. P.S. Mohanty, J.E. Gruzleski, Mechanism of grain refinement in Aluminum. Acta Mater. **43** (5), 2001–2012 (1995)
32. D. Apelian, Semi-solid processing routes and microstructure evolution. in *7th International Conference on Semi-Solid Processing of Alloys and Composites* (Tsukuba, Japan, 2002), 25–30
33. BORAL, Aluminum Boron Master Alloy, Information sheet, KBAlloys, Inc. (www.kballoys.com)
34. W.C. Setzer, G.W. Boone, The use of Aluminum/Boron master alloys to improve electrical conductivity, in *Light Metals*, ed. by E.R. Cutshall (TMS, San Diego, 1992), 837–844
35. I. Barin, F. Sauert, E.S. Rhonhof, W.S. Sheng, *Thermochemical Data for Pure Substances* (VCH Verlagsgesellschaft mbH, Germany, 1993)
36. A.E. Karantzalis, A.R. Kennedy, Nucleation behavior of TiB2 particles in pure Al and effect of elemental additions. J. Mater. Sci. Tech. **14**, 1092–1096 (1998)
37. S. Nafisi, R. Ghomashchi, Boron-based refiners: advantages in semi-solid-metal casting of Al-Si alloys. Mater. Sci. Eng. **A452–453**, 437–444 (2007)
38. B. Chalmers, *Principles of Solidification* (Wiley, New York, 1964)
39. T.E. Quested, A.T. Dinsdale, A.L. Greer, Thermodynamic modeling of growth restriction effects in Aluminum alloys. Acta Mater. **53**, 1323–1334 (2005)
40. I. Maxwell, A. Hellawell, A simple model for grain refinement during solidification. ACTA Metal **23**, 229–237 (1975)
41. M. Johnsson, L. Backerud, Z. Metallkde. **87**, 216–220 (1996)
42. L. Backerud, M. Johnsson, The relative importance of nucleation and growth mechanisms to control grain size in various aluminum alloys. TMS Light Metals (1996), 679–685
43. M.A. Easton, D.H. St. John, A model of grain refinement incorporating alloy constitution and potency of heterogeneous nucleant particles. Acta Mater. **49**, 1867–1878 (2001)
44. S. Nafisi, R. Ghomashchi, The effect of dissolved Ti on the primary α-Al grain and globule size in the conventional and semi-solid casting of 356 Al-Si alloy. J. Mater. Sci. **41**, 7954–7963 (2006)
45. W. Kurz, D.J. Fisher, *Fundamental of Solidification* (Trans Tech, Switzerland, 1989)
46. S. Nafisi, R. Ghomashchi, Effects of modification during conventional and semi-solid metal processing of A356 Al-Si alloy. Mater. Sci. Eng. A **415**, 273–285 (2006)
47. S. Nafisi, D. Emadi, R. Ghomashchi, Impact of Mg addition on solidification behaviour of Al-7%Si alloy. J. Mater. Sci. Technol. **24**(6), 718–724 (2008)
48. FactSage. http://www.crct.polymtl.ca/factsage
49. A. Joenoes, J. Gruzleski, Magnesium effects on the microstructure of unmodified and modified Al-Si alloys. Cast Metals **4**, 62–71 (1991)
50. A.P. Bates, D.S. Calvert, Refinement and foundry characteristics of hypereutectic aluminum-silicon alloys. Br. Foundry Man **59**, 113–119 (1966)
51. S. Shankar, Y.W. Riddle, M.M. Makhlouf, Nucleation mechanism of the eutectic phases in Aluminum-Silicon hypoeutectic alloys. Acta Mater. **52**, 4447–4460 (2004)
52. S. Nafisi, R. Ghomashchi, S.M.A. Boutorabi, J. Hedjazi, New approaches to melt treatment of Al-Si alloys: application of thermal analysis technique. AFS Trans. **112** (2004), Paper 04-018
53. S.-Z. Lu, A. Hellawell, The mechanism of silicon modification in Aluminum-Silicon alloys: impurity induced twinning. Metal. Trans. A **18A**, 1721–1733 (1987)
54. L.M. Hogan, M. Shamsuzzoha, Crystallography of the flake-fiber transition in the Al-Si eutectic. Mater. Forum **10**, 270–277 (1987)
55. S. Argyropoulos, B. Closset, J.E. Gruzleski, H. Oger, The quantitative control of modification of Al-Si foundry alloys using a thermal analysis technique. AFS Trans. **91**, 351–357 (1983)
56. R. DasGupta, C.G. Brown, S. Marek, Analysis of overmodified 356 Aluminum alloy. AFS Trans. **92**, 297–310 (1984)
57. J. Charbonnier, Microprocessor assisted thermal analysis of Aluminum alloys structure. AFS Trans. **92**, 907–922 (1984)

58. K. Nogita, A.K. Dahle, Eutectic solidification in hypoeutectic Al–Si alloys: electron backscatter diffraction analysis. Mater. Charact. **46**, 305–310 (2001)
59. G.K. Sigworth, Theoretical and practical aspects of the modification of Al-Si alloys. AFS Trans. **91**, 7–16 (1983)
60. L. Heusler, W. Schneider, Influence of alloying elements on the thermal analysis results of Al-Si cast alloys. J. Light Metals **2**, 17–26 (2002)
61. J.P. Anson, J.E. Gruzleski, M. Stucky, Effect of strontium concentration on microporosity in A356 Aluminum alloy. AFS Trans. **108**, 01-009 (2001)
62. D. Emadi, J.E. Gruzleski, M. Toguri, The effect of Na and Sr modification on surface tension and volumetric shrinkage of A356 alloy and their influence on porosity formation. Metal. Trans. B **24B**, 1055–1063 (1993)
63. S. Nafisi, R. Ghomashchi, Combined grain refining and modification of conventional & Rheo-Cast A356 Al-Si alloy. Mater. Charact. **57**, 371–385 (2006)
64. S. Nafisi, Effects of grain refining and modification on the microstructural evolution of semi-solid 356 alloy. Ph.D. Thesis, University of Quebec, 2006
65. S. Nafisi, O. Lashkari, R. Ghomashchi, F. Ajersch, A. Charette, Microstructure and rheological behavior of grain refined and modified semi-solid A356 Al-Si slurries. Acta Mater. **54**, 3503–3511 (2006)
66. G. Chai, T. Roland, L. Arnberg, L. Backerud, Studies of dendrite coherency in solidifying aluminum alloy melts by rheological measurements. in *2nd International Conference, Semi-Solid Processing of Alloys and Composites* (Cambridge, 1992), 193–201
67. V. Laxmanan, M.C. Flemings, Deformation of semi-solid Sn-15%Pb alloy. Metal. Trans. A **11A**, 1927–1937 (1980)
68. J.A. Yurko, M.C. Flemings, Rheology and microstructure of semi solid aluminum alloys compressed in drop forge viscometer. Metal. Trans. A **33A**, 2737–2746 (2002)
69. L. Azzi, F. Ajersch, Development of aluminum-base alloys for forming in semi solid state. in *TransAl Conference* (Lyon, France, June 2002), 23–33
70. A. Beaulieu, L. Azzi, F. Ajersch, S. Turenne, Numerical modeling and experimental analysis of die cast semi-solid A356 alloy. in *Proceeding of M. C. Flemings* (TMS, 2001), 261–265
71. O. Lashkari, R. Ghomashchi, F. Ajersch, Deformation behavior of semi-solid A356 Al-Si alloy at low shear rates: the effect of sample size. Mater. Sci. Eng. **A444**, 198–205 (2007)
72. S. Nafisi, O. Lashkari, R. Ghomashchi, J. Langlais, B. Kulunk, The SEED technology: a new generation in rheocasting. Light Metals, CIM (2005), 359–371
73. A. Figueredo, *Science and Technology of Semi-Solid Metal Processing* (North American Die Casting Association, Rosemont, 2001)

Chapter 7
Thixocasting

Abstract Following a very detailed analysis of rheocasting, this chapter concentrates on thixocasting as another alternative route for SSM processing of metallic alloys. The effect of reheating parameters on the resulted microstructure of SSM billets is studied in detail and the effect of grain refining on the final structure of thixocast alloys is highlighted. The issue of liquid entrapment is discussed with respect to grain refinement.

7.1 Introduction

The previous chapters have dealt with the process of SSM casting directly from the melt, that is, rheocasting. The slurry is prepared by stirring the superheated molten alloy as it cools down to the mushy zone. This chapter describes thixocasting process where reheating of the alloy to temperatures above solidus and holding it isothermally within the mushy zone induces the desired structure for SSM casting.

One of the important issues in deciding whether to switch to one specific SSM fabrication route is the supply of feedstock. The term Slurry on Demand (SoD) describes the slurry-making operations to provide a constant supply of slurry for shaping operations. This may include the billets (slugs) produced via conventional casting or rheocasting, stored and used later by simple and short reheating to mushy zone. This in fact prescribes the SoD procedure based on a combination of rheocasting and thixocasting. Such an approach could be beneficial to establish manufacturing plants in regions with restricted environmental regulations while the feedstock can be produced in specialized sites and shipped to the manufacturing plants to fulfill the SoD requirements. This is similar to mini mills for steel industry where the feed stock is shipped to mills to be reheated and rolled to plates and sheets. Therefore, it is necessary to discuss the process of thixocasting to provide a basis for comparison.

S. Nafisi, R. Ghomashchi, *Semi-Solid Processing of Aluminum Alloys*,
DOI 10.1007/978-3-319-40335-9_7

7.2 Thixocasting of the Refined Conventional Cast Specimens

Limited series of tests were performed by reheating the specimens prepared for thermal analysis (Sect. 4.1). The main objective was to verify the effectiveness of inoculation during the reheating stage. Samples were chosen according to the results obtained in Sect. 6.2.1. During the entire experiments, three critical parameters were investigated: effect of partial remelting time, solidification cooling rate, and type of the refiner. These include 5 and 10 min reheating time, two different cooling rates during solidification process, and addition of sole B and Ti–B via Al5B and Al5Ti1B master alloys, respectively.

7.2.1 Effect of Reheating Time on SSM Structure

Figure 7.1 shows microstructural evolution of as-cast solidified samples in graphite cup which partially remelted at 583 ± 3 °C (Sect. 4.1.1.4). At this temperature, there is about 38–40 % fraction solid according to ThermoCalc calculations. During partial remelting and subsequent isothermal holding, there are several mechanisms active as follows:

- Remelting of low melting points constituents, eutectic mixture
  At this stage, the α-Al portion of the eutectic phase gradually precipitates on the α-Al primary phase leading to their growth. Numerous examples will be provided in this chapter (it is important to consider that the majority of eutectic phase is α-Al).
- Grain coarsening and ripening
  Dendrite fragmentation and coalescence could happen at the same time. Roots of some dendrite arms may get narrow, detach, and form separate particles. Also coalescence may result in liquid entrapment.
- Globularization and further particle coarsening
  Diffusion from areas with higher radius of curvature to areas with lower radius of curvature to decrease the interfacial energy between solid–liquid.

As shown in Fig. 7.1a, the as-cast untreated structure has a dendritic morphology. Other than coarsening of dendrites, reheating for 5 min seems had no major structural variation. After 10 min isothermal holding, the initial dendritic structure becomes thicker and rounder but still far from globular structure. In fact, the final microstructure is highly dependent on the original as-cast structure. Nevertheless, at the initial stage of holding, the main development is remelting of the eutectic regions which were indeed the last region to form during solidification process. For dendritic solidification, the eutectic already formed between the secondary or tertiary arms which by reheating process, this liquid is more likely to be entrapped within the arms. Meanwhile other mechanisms are in operation as well.

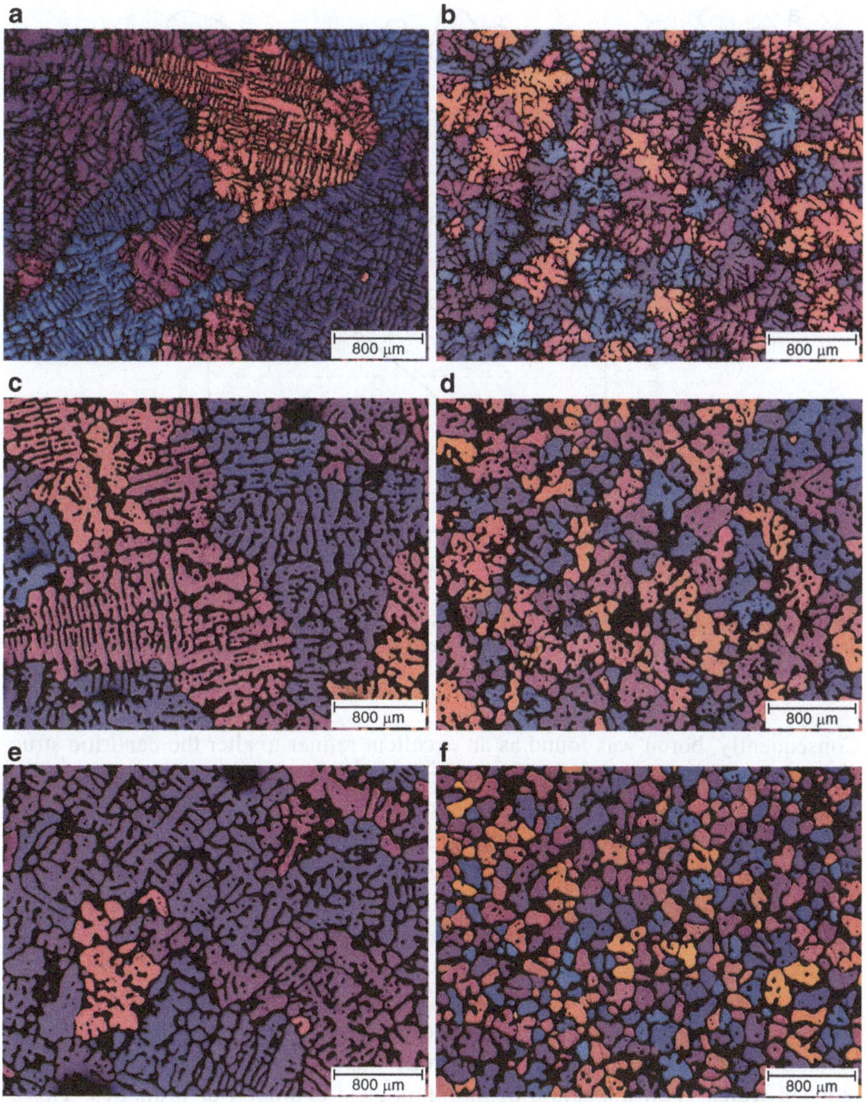

Fig. 7.1 Polarized light micrographs; (**a**), (**c**), (**e**) A356, untreated, (**b**), (**d**), (**f**) A356 with 226 ppm B (as-cast "(**a**), (**b**)", 5 min "(**c**), (**d**)", and 10 min "(**e**), (**f**)" reheated samples and isothermally held at 583 °C)

Grain coarsening leads to thicker secondary arms. By ripening, the smaller grains remelt and the larger ones grow. Ripening of secondary dendrite arms is quite evident which is due to the reheating time/temperature and tendency of the structure to reach the lower energy level, specifically rounder particles.

In Sect. 6.2, various refiners were used to evaluate their effectiveness and it was found that the addition of sole boron results in minimum average grain size.

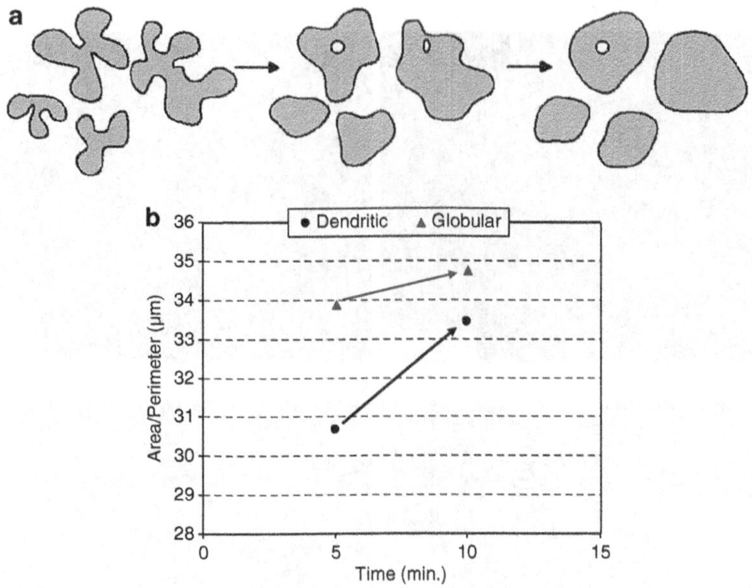

Fig. 7.2 (**a**) Schematic presentation of evolution toward globules formation, (**b**) solid–liquid interfacial evolution as a function of holding time (*circles* for untreated A356 and *triangles* for A356 refined with boron)

Consequently, boron was found as an excellent refiner to alter the dendritic structure even with high superheat casting process. After 5 min isothermal heating at 583 °C, the structure had lost the initial dendritic structure and evolved into more chunky and isolated particles resembling rosette/globule morphology. By 10 min reheating, the structure becomes almost globular with nearly uniform particle size. The liquid entrapped within the α-Al particles is reduced in timely manner and most of it well distributed between α-Al particles (Fig. 7.1).

The driving force for the evolution toward globularization is the reduction of the eutectic-primary particles interface area. This could be interpreted in terms of A/P ratio (area over perimeter) for both dendritic and globular structures. By longer holding time, the interfacial area between the liquid and primary particles is reduced which is an indication of morphological evolution of primaries. This is well depicted in Fig. 7.2. The as-cast structure has a great influence on the kinetics of evolution and as shown, the evolution rate is greater for dendritic structure. The highly branched dendritic structure has a large solid–liquid interface which is equivalent to greater driving force toward area reduction, greater reaction kinetic.

As reported by other researchers [1–3], thixocast SSM structure is directly affected by the as-cast microstructure. This is to say that for a fully developed dendritic structure, the transformation to globular morphology is very unlikely and requires very long reheating period. The effect of isothermal holding time within the mushy zone on globularization is shown in B-treated samples in Fig. 7.3. In addition, there is a tendency for the primary particles to thicken through

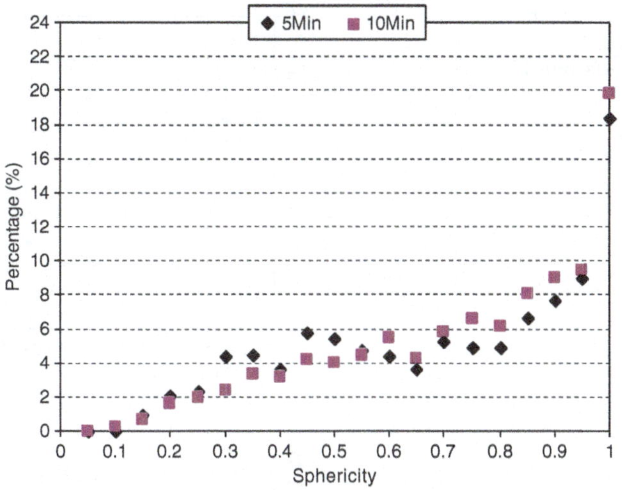

Fig. 7.3 Comparison of sphericity values for B-refined samples with different holding time (5 and 10 min)

mechanisms such as "Ostwald ripening" to reduce their interfacial area, that is, a combined process of coarsening and globularization. This is indicated by sphericity measurement (image analysis) where 10 min isothermal holding results in formation of rounder globules.

During reheating process and as a result of grain growth, parts of the liquid may be encapsulated within the coarsened grains. This entrapped liquid adversely affects the deformability of material which is associated with a reduction in the interconnected liquid phase. In fact, the percentage of the residual liquid plays an important role on the rheological behavior of the slurries. It may be regarded as a lubricant between the primary particles to ease up their movement. As for instance, by Sr addition, the surface tension of the liquid decreases and therefore easier sliding of primary particles would result in better die filling (Sect. 6.2.2).

It is also interesting to point out that the percentage of entrapped liquid decreases with increasing reheating time (Fig. 7.4). Such phenomenon may be interpreted as further segregation of aluminum onto the existing aluminum particles. The formation of entrapped liquid may be regarded as a reduction in the level of liquid within the overall 3D-interconnected liquid pool, but this cannot be with great certainty considering the difference between what is seen in 2D comparing to the real case in 3D.

7.2.2 Effect of Solidification Cooling Rate on SSM Structure

As stated in the Sect. 4.1, the entire thermal analysis cup samples solidified within the cooling rates of 1.5–2 $°Cs^{-1}$. Therefore, to study the effect of solidification cooling rate on SSM structure, new samples were solidified with much higher

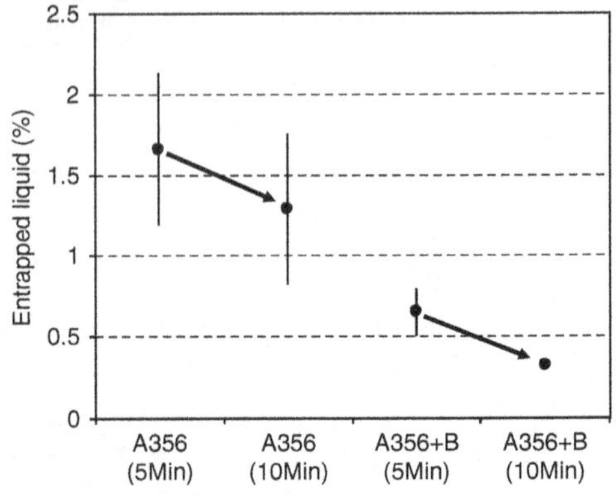

Fig. 7.4 Percentage of entrapped liquid as a function of partial remelting time

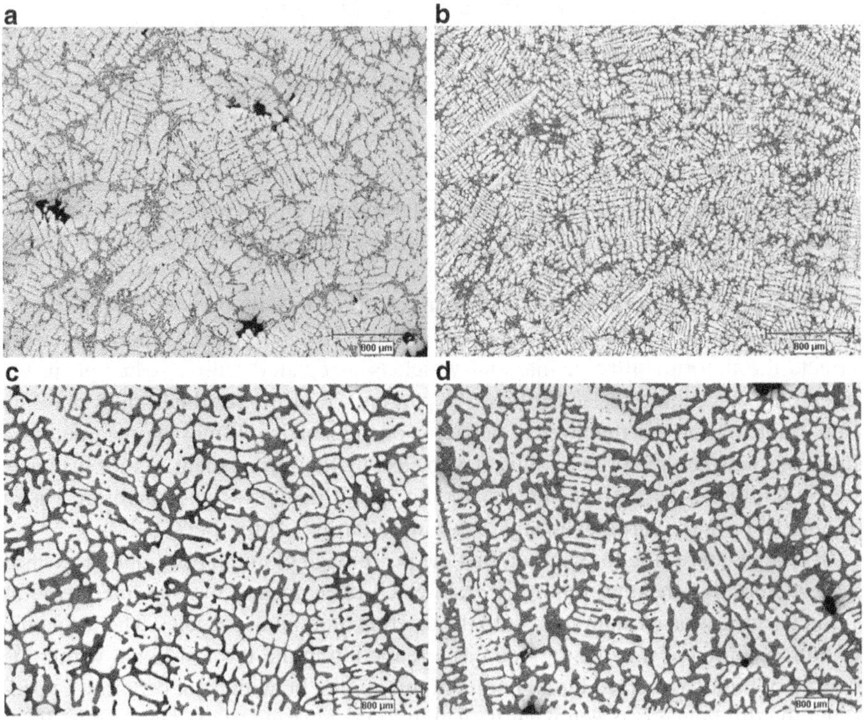

Fig. 7.5 Evolution of semi-solid structure as a function of cooling rate, A356 alloy (**a**) as-cast, ~1.5 °Cs^{-1}, (**b**) as-cast, ~11 °Cs^{-1}, (**c**) results from 10 min reheating sample "(**a**)" at 583 °C, and (**d**) results from 10 min reheating sample "(**b**)" at 583 °C

cooling rate of about 11 °Cs^{-1} with the application of compressed air. The as-cast structures are shown in Fig. 7.5a, b where higher cooling rate resulted in more compacted dendritic structure with obvious lower dendrite arm spacing "DAS" as expected. Figure 7.5c, d shows the structural evolution after 10 min holding time at

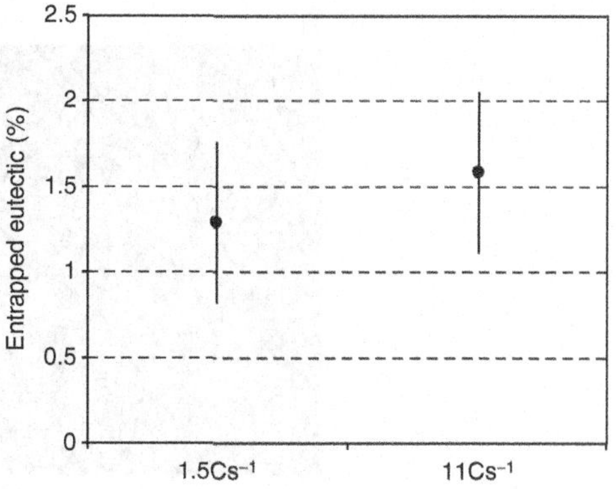

Fig. 7.6 Percentage of entrapped liquid in different cooling rates (10 min reheating at 583 °C)

about 583 °C. Both samples show dendritic structure and there is no sign of globularization. It is also difficult to ascertain if a finer as-cast billet structure has any effect on the kinetics of morphological changes as it is expected to be the case due to higher driving force initiated from larger surface area for the finer structures. Such conclusion required a full quantitative analysis which is beyond the scope of this chapter. The measured area of the entrapped liquid is shown in Fig. 7.6. Interestingly, the higher cooling rate here resulted in higher percentage of entrapped liquid. This may be attributed to finer dendritic structure having larger interfacial area which eventually results in higher number of liquid pools for faster cooled sample for this specific reheating time/temperature.

7.2.3 Addition of Various Refiners

A comparison between unrefined and refined samples is shown in Fig. 7.7. Partial remelting of B-refined alloy leads to the formation of α-Al globules with the added bonus of uniform distribution of eutectic mixture. Polarized light micrographs confirm that in the untreated and to some extent in Ti–B-treated samples, the individual particles do not essentially initiate independently from different nucleants and the concept of globule size measurement is not reasonable. Consequently, the grain size measurement was done in the entire experiments as depicted later in a graph, Fig. 7.8.

Untreated alloy has the largest grain size with nonhomogeneous size distribution considering the span of its standard deviation bar in Fig. 7.8. The unique size distribution relates to the B-refined alloy which shows the smallest standard deviation. Economically, longer reheating time has shortcomings such as higher energy consumption as well as nonuniform liquid distribution within the billet due to the gravitational force. On a microstructural stand point, increasing the duration of partial remelting leads to coarsening of primary α-Al globules as shown in Fig. 7.9 where the average circular diameter of globules was increased. However, it is believed that for this experimental condition, increasing the sphericity and

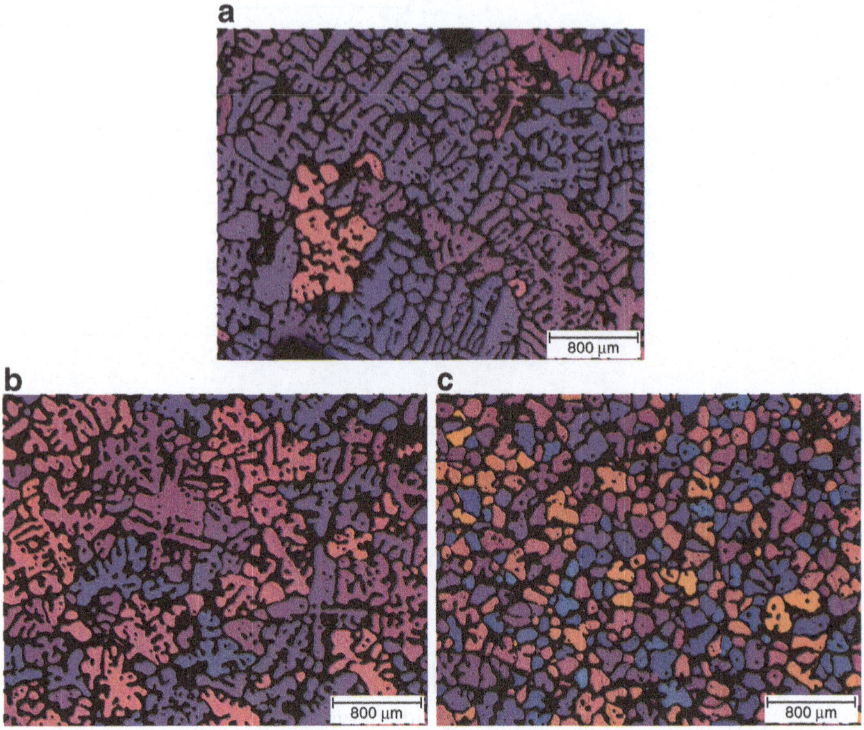

Fig. 7.7 Polarized light micrographs showing the effectiveness of refiners (10 min reheating time, 583 °C) (**a**) A356, untreated, (**b**) A356 + 622 ppm Ti + 110 ppm B, and (**c**) A356 + 226 ppm B

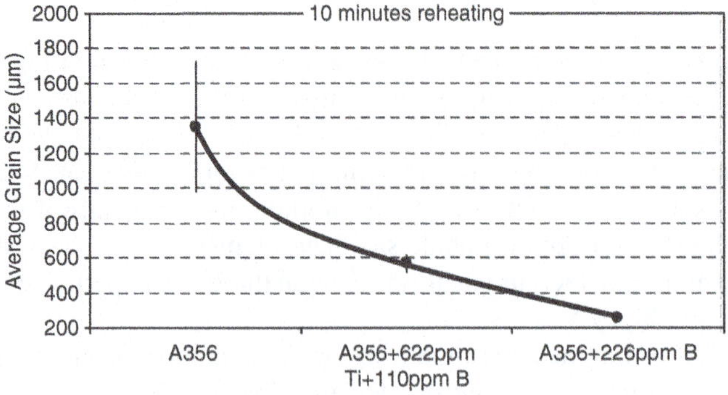

Fig. 7.8 Grain size variation according to treatment

globularity is much more important than the slight increase in the size of particles. By having more spherical primary particles, the flowability improves and consequently the viscosity decreases, Sect. 4.2.

Figure 7.10 compares the typical globule morphology of Ti/B and B-refined samples. As confirmed by quantitative metallography, the entrapped liquid content of B-refined sample is reduced by 4–5 times compared to the other samples (Fig. 7.11).

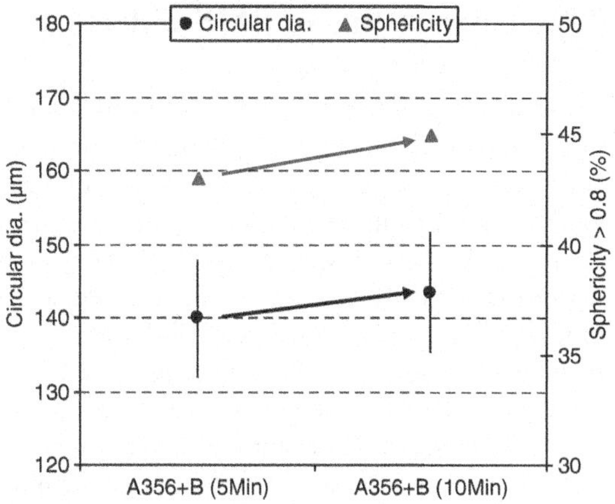

Fig. 7.9 Average circular diameter and percentage of particles having sphericity >0.8 (%)

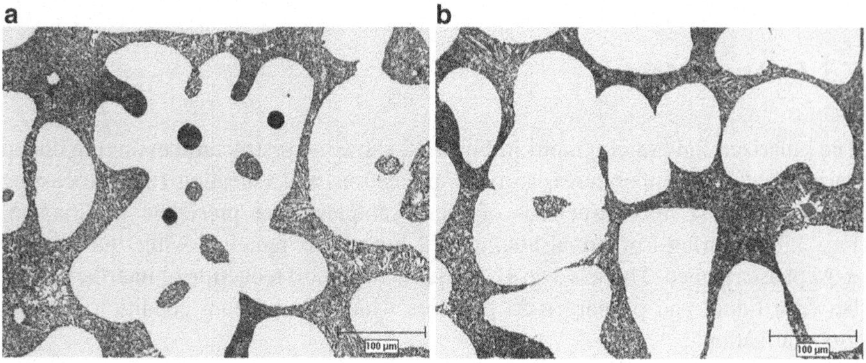

Fig. 7.10 Entrapped eutectic with different treatment (reheating time: 10 min) (**a**) A356 + 622 ppm Ti + 110 ppm B and (**b**) A356 + 226 ppm B (water quenching produces very fine Si in the eutectic)

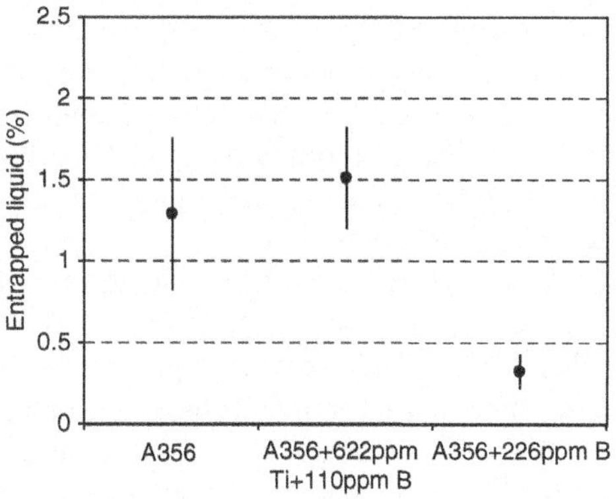

Fig. 7.11 Percentage of the entrapped liquid (reheating time: 10 min)

7.3 Thixocasting of EMS Billets

Binary Al–7 % Si alloys (6.7–6.9 % Si and 0.8–0.81 % Fe) were prepared and cast at two different cooling rates by pouring the molten metal in copper and sand molds at various pouring temperatures. After EMS application, samples were cooled to room temperature. For the experiments without stirring, the liquid was poured into the same molds and allowed to air cool. For thixocasting (reheating to semi-solid region), samples were cut from the transverse sections (200 mm from the bottom of EMS billets), in areas between the billet center and wall, and were reheated in an induction furnace. The reheating cycle included 2–3 min of heating up to 583 ± 3 °C and 10 min holding time at this temperature followed by water quenching (about 38–40 % fraction solid according to ThermoCalc calculations). Temperature variation during the tests was monitored by attaching thermocouples to both the billet center and the wall. More details were explained in Sects. 4.1.1.3 and 5.4.

7.3.1 Sand Mold

The polarized light micrographs in Fig. 7.12 show microstructural evolution due to variant pouring temperatures, stirring application, and reheating for thixocasting process (as-cast microstructures of these samples were presented in Chap. 5, Fig. 5.31). During isothermal holding, the eutectic is remelted while the primary α-Al phase ripened. There is also a driving force toward reduction of interfacial area between liquid and primary α-Al particles which in optimum condition leads to globularization.

Conventional samples with high pouring temperatures preserved their coarse dendritic structure even after 10 min reheating time. The very coarse structure seems to form a solid network with full 3D interconnections with a relatively high proportion of intragranular liquid. Reducing the superheat results in the evolution of the α-Al particles to rosette/equiaxed which elaborated in Sect. 5.4. By EMS application and consequent fragmentation of primary dendrites, the as-cast structure is well suited for additional morphological changes (spheroidization) through reheating process and encapsulation of the liquid phase by becoming mostly intragranular. Globule size is a bit larger within the high superheated samples but its size is reduced by lowering the superheat which is associated with the higher number of favorite nucleation sites, thermal and solute convection, and fragmentation of primary dendrites.

As described in Sect. 4.3, there is a difference between globule and grain size. Globules are primary particles which are apparently detached from each other however by applying a polarized light; it is evident that the neighbouring individual particles might have a connection from underneath of the plane of polished surface. As a result, similar color adjacent globules specify a particular grain. By this method, grains could be differentiated from globules and

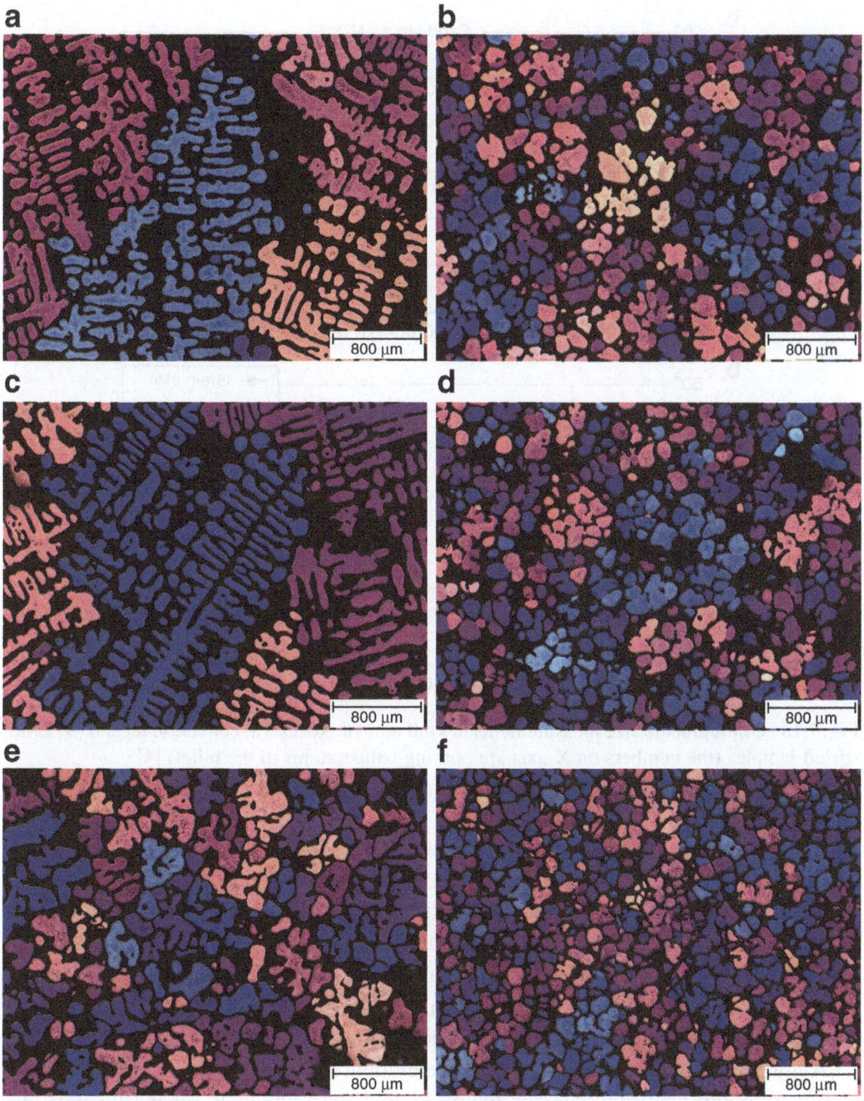

Fig. 7.12 Polarized light micrographs showing the effect of pouring temperature (690 °C "**a, b**", 660 °C "**c, d**", and 630 °C "**e, f**") and stirring on the grain and globule size variations in the sand mold casting (samples reheated at 583 °C for 10 min); (**a**), (**c**), and (**e**) conventional and (**b**), (**d**), and (**f**) EMS-stirred

Fig. 7.13 shows measured particle size of conventional and EMS reheated samples (particle size measurement by image analysis excluded all the entrapped eutectic areas). It is worth noting that the concept of globule size measurement in conventional cast samples is not completely valid which is associated with the errors related to the sectioning of dendritic branches as described in Sect. 4.3.

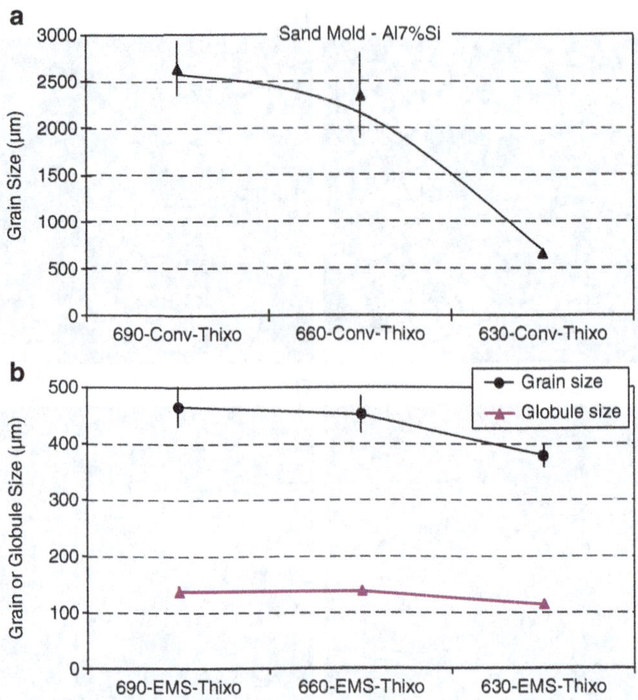

Fig. 7.13 Grain/globule size measurements in sand mold thixocast (**a**) conventional and (**b**) EMS-stirred samples (the numbers on X-axis are pouring temperatures of the billet) [4]

For conventional thixocast samples, the primary α-Al morphology is identical to the conventional cast samples (Figs. 5.31 and 7.12), although growth is noticeable due to reheating process. As a result, there is not much difference between the average grain size values (Figs. 5.34 and 7.13). In the EMS thixocast graph, Fig. 7.13b, both the grain and the globule sizes are presented. In contrast to the conventional samples, there is no sudden reduction in grain size and the values of globule size have a considerable difference from grain size values.

Agglomeration of primary particles are evident by examining the microstructure at higher magnifications, see for example Fig. 7.14. As it is clear, the particles become interconnected by solid necks and spheroidized by material transport specifically through the neck area which has a negative radius of curvature. These agglomerates are formed due to the sintering processes being activated with prolonged holding time (may be referred as coalescence ripening). Longer holding time and resultant agglomeration may encapsulate the eutectic mixture inside and form entrapped liquid area.

Selected results of image processing are presented in Fig. 7.15. By lowering the pouring temperature from 660 to 630 °C, average globule size reduced by ~15 %. Superheat reduction promotes the equiaxed particles formation with the added

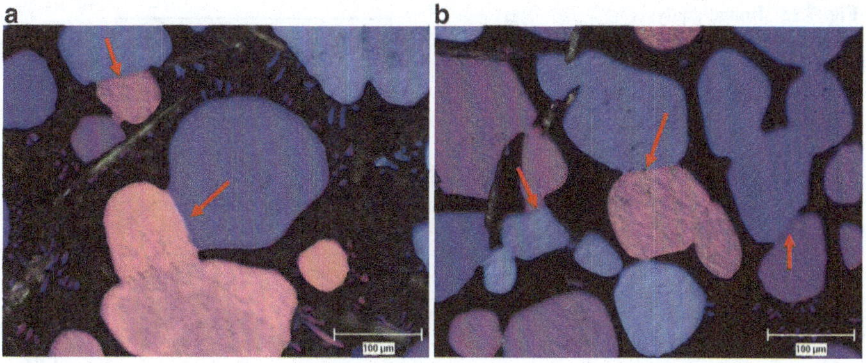

Fig. 7.14 Sintering of globules, EMS billets poured at (**a**) 690 °C and (**b**) 630 °C (sand mold, samples reheated at 583 °C for 10 min)

bonus of more nucleation sites as proven by the primary α-Al number density which is increased abruptly at 630 °C (Fig. 7.15a).

The as-cast structure is a function of pouring temperature and the greater the superheat, the structure is more dendritic. This concept is quite clear in the form of percentage of the primary α-Al particles having certain aspect ratio. By lowering the pouring temperature in conventional samples, the concentration of particles with an aspect ratio >2 decreases (Fig. 7.15b). The lowest amounts of primary particles with aspect ratio >2 were observed in EMS billets with a reasonable difference to that of the reheated conventionally cast samples.

By decreasing the pouring temperature, particles become more spherical and as a result, percentage of particles having sphericity values greater than 0.8 increases. Comparing the conventional and EMS-treated samples, it is clear that the percentage of particles having sphericity >0.8 is higher for EM stirred samples. In rheological studies, more spherical particles results in lower viscosity (Sect. 4.2) which is expected to induce better flowability and filling of the die cavity during high pressure diecasting.

Entrapped liquid has an impact on the viscosity of the semi-solid slurries and less entrapped liquid results in better fluidity [5–7]. Conventionally cast samples have a tendency to entrap the liquid which is normal in dendritic structure. Casting the billets with high superheat leads to a complex and massive dendritic structure which could encapsulate a portion of the liquid formed during reheating. Accordingly by superheat reduction, the structure tends to form rosette/equiaxed primary α-Al particles with less probability of liquid entrapment. In the case of EMS samples, the structure not only transforms to globules but also contains the least entrapped liquid, Fig. 7.16.

It is observed that the majority of entrapped liquid in the EMS samples initiates from liquid encapsulation by the surrounding globules, while for conventional samples, the main source is the liquid between secondary or tertiary dendrite arms. This notion is depicted in Fig. 7.17 for billets poured at 690 °C. This may have some effect on the homogeneity of entrapped liquid where a more uniform

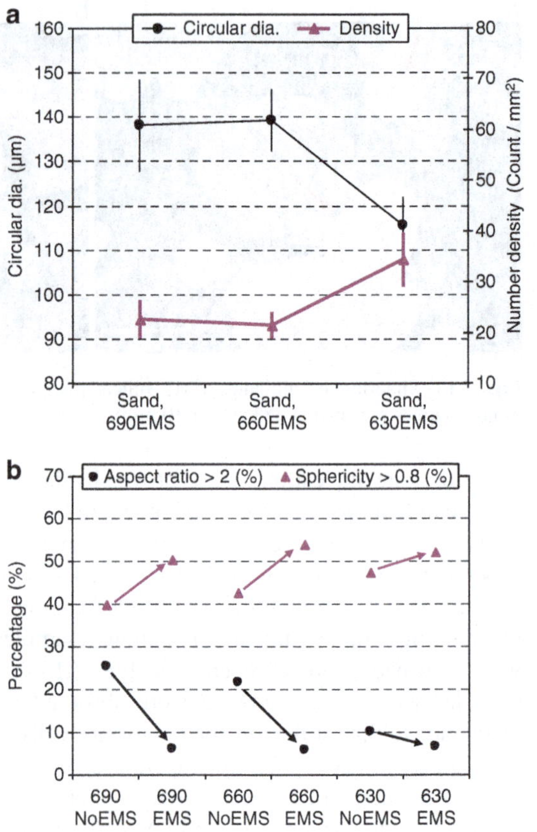

Fig. 7.15 Image analysis results from various pouring temperatures/application of EMS, sand mold thixocast [4]

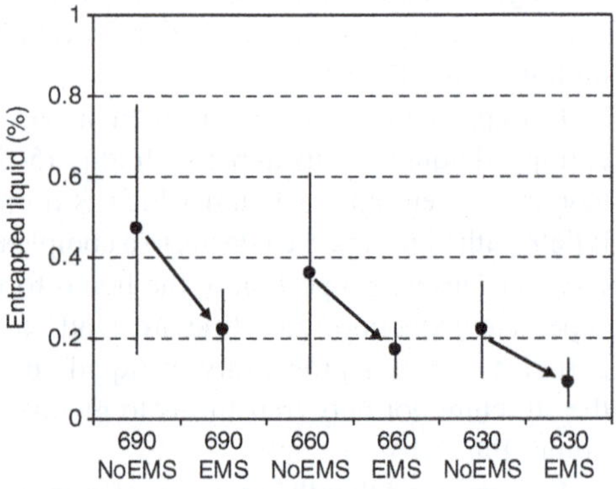

Fig. 7.16 Entrapped liquid measurement, sand thixocast [4]

composition is expected for the entrapped liquid for EMS thixocast billets. Conventionally cast thixo billets have entrapped liquid that is more characteristics of the interdendritic segregation.

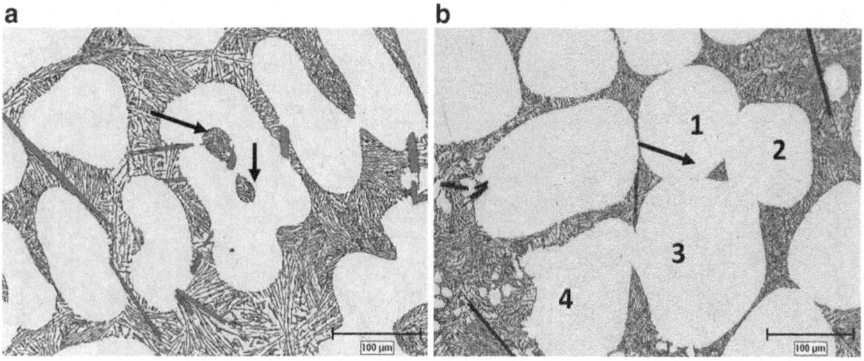

Fig. 7.17 Entrapped liquid in billets poured at 690 °C (**a**) conventional and (**b**) EMS (numbers show the globules and arrows show the liquid pockets)

7.3.2 Copper Mold

Figure 7.18 shows the microstructural evolution of copper mold cast samples by reheating at ~583 °C. Conventional thixo structures have dendritic morphology characterized by continuous solid network and liquid pockets. In contrast to the sand mold cast billets, casting in copper mold leads to thinner dendrite branches and smaller dendrite arm spacing (DAS). Lowering the pouring temperature results in rounder and more isolated particles with least entrapped liquid (Fig. 7.18a, c, e). EMS-reheated structures, however consist of almost all globules with an average size of about 100 µm even at higher superheats. The globules are well distributed within the eutectic network and nearly all the eutectic liquid pools are intergranular (Fig. 7.18b, d, f).

The trend of grain size values for the reheated conventional samples is similar to that of the as-cast structure (Fig. 5.40) including the direct relation to the pouring temperature (Fig. 7.19a). Comparing the grain/globule size values for sand and copper mold billets shows that by stirring, the grain/globule size values are getting closer, indicating almost independent relationship to the casting condition. In other words, the application of EMS diminishes the importance of the cooling rate and pouring temperature which could control the globule and grain size during reheating.

As mentioned in Sect. 4.3.1, sintering and coalescence of particles are parts of isothermal holding. Figure 7.20 shows typical sintering incidents in copper cast samples.

The process of globularization and refinement of the primary α-Al particles are shown in Fig. 7.21. In EMS samples, the average globule size reduces by lowering the pouring temperature. Interestingly, the rate of globule size reduction is lower than the sand mold cast specimen while the increasing rate of number density is higher indicating the importance of the as-cast structure.

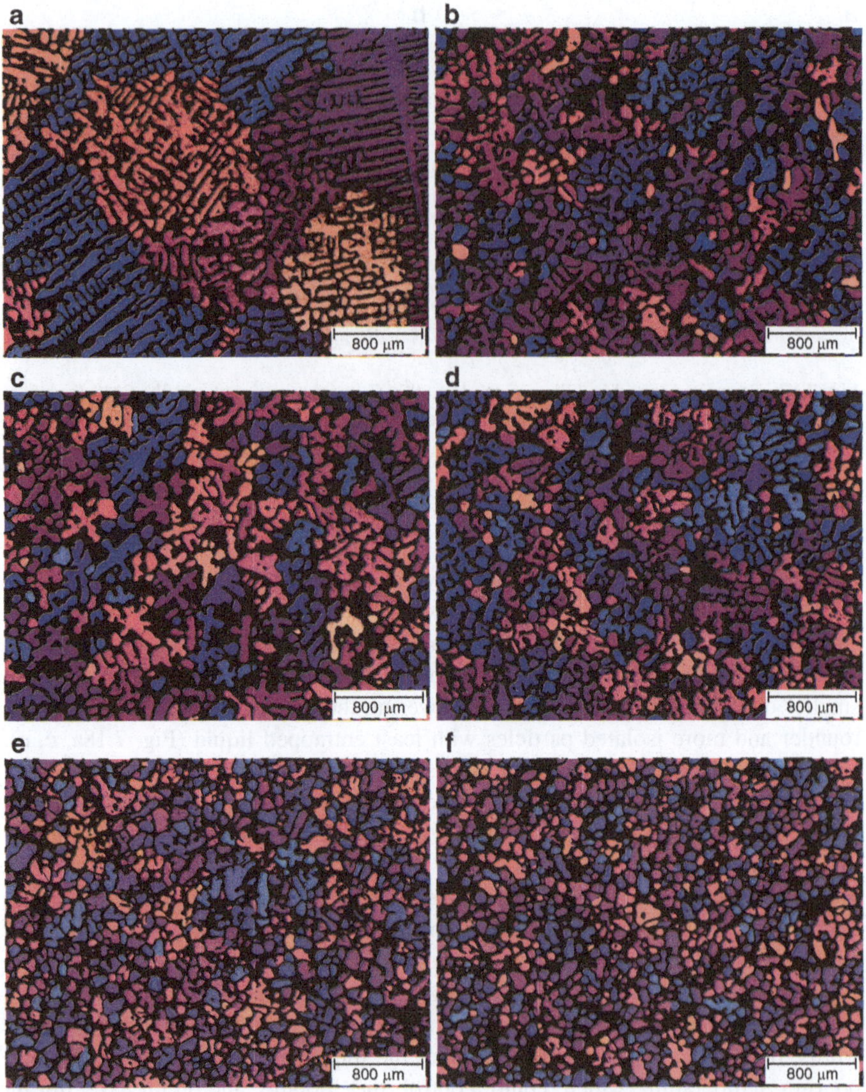

Fig. 7.18 Polarized light micrographs showing the effect of pouring temperature (690 °C "**a, b**", 660 °C "**c, d**", and 630 °C "**e, f**") and stirring on the grain and globule size variations in the copper mold casting (samples reheated at 583 °C for 10 min) (**a**), (**c**), and (**e**) conventional and (**b**), (**d**), and (**f**) EMS-stirred

The aspect ratio exhibits the same trend as sand cast samples, and the percentage of particles with aspect ratio >2 decreases with decreasing pouring temperature in both conventional and EMS samples; however, the reduction rate is higher for conventional samples. In fact, during the reheating process, an initial higher cooling rate (copper mold) results in the billets having greater potential for microstructural

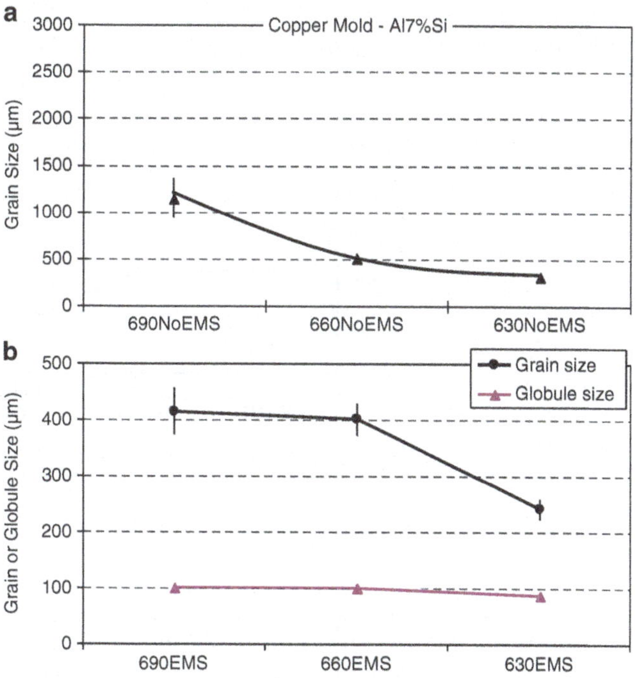

Fig. 7.19 Grain/globule size measurements in copper mold thixocast (**a**) conventional and (**b**) EMS-stirred samples (the numbers on X-axis are pouring temperatures of the billet) [4]

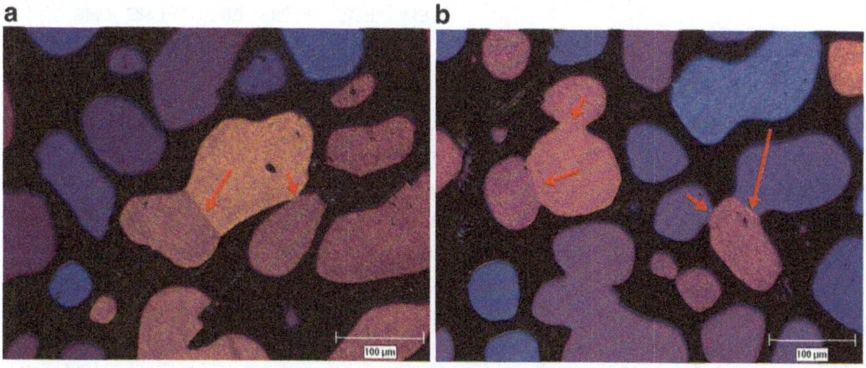

Fig. 7.20 Sintering of globules, EMS billets poured at (**a**) 690 °C and (**b**) 630 °C (copper mold, samples reheated at 583 °C for 10 min)

evolution, that is, larger interfacial area, and this is evident by comparing the results, Figs. 7.15 and 7.21. By lowering the superheat, the percentage of particles with sphericity value greater than 0.8 increases and the maximum value belong to EMS and lower superheated samples. In addition, the sphericity values are greater in comparison to the sand mold cast samples.

Fig. 7.21 Image analysis results from various pouring temperatures/application of EMS, copper mold thixocast [4]

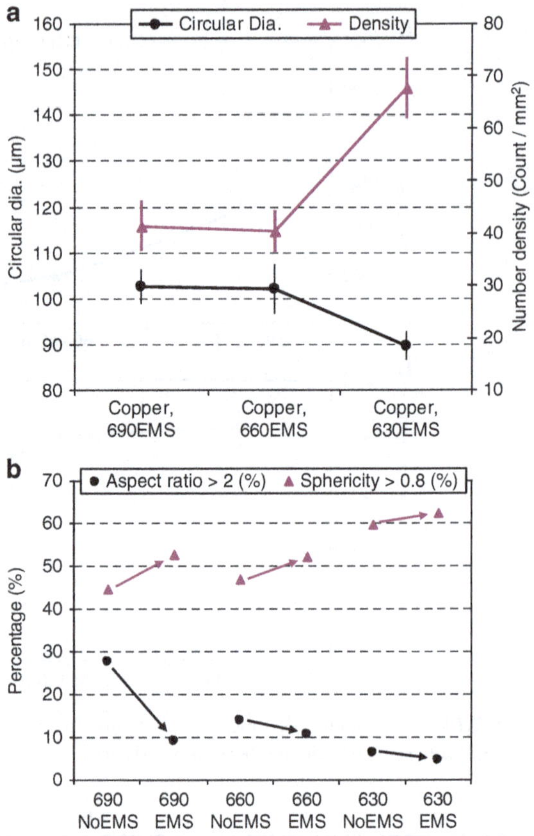

According to Loué and Suéry [1], during partial remelting, coarsening first proceeds predominantly through coalescence of dendrite arms. As the dendrite arms of the same grain have a perfectly matching crystallographic orientation, this results in high quantity of intragranular liquid which depends on the cooling rate. In the case of EMS samples, the coalescence of short dendrite arms leads to almost globular structure with a smaller amount of liquid pocket. The formation of spherical particles could be assisted by lowering the pouring temperature and thus having higher probability of equiaxed grains formation. Figure 7.22 illustrates the concept within the copper cast structures with high superheat value.

Comparing the micrographs with those of sand mold structures suggests that the percentage of liquid encapsulation in the conventional castings is also dependent on the cooling rate and it is higher for lower cooling rate. Quantitative results also reveal that cooling rate has a key role to play on the liquid encapsulation and its percentage is higher for lower cooling rates. In general, higher cooling rate during solidification leads to more uniform (less standard deviation) liquid entrapment throughout the sample. It is worth noting that the efficiency of EMS in refining the structure becomes more noticable at lower cooling rate during solidification (Figs. 7.16 and 7.23).

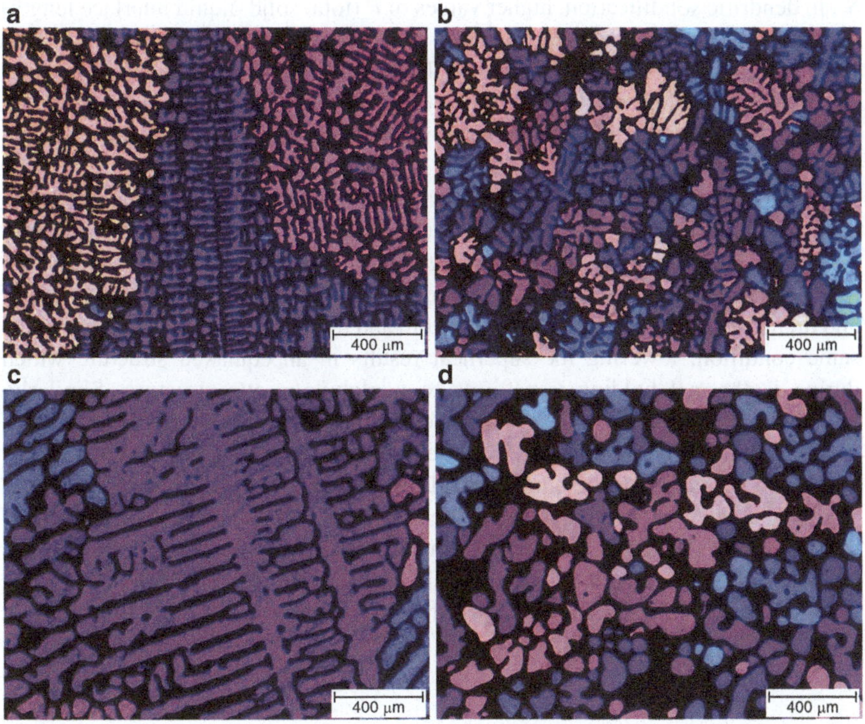

Fig. 7.22 Copper mold samples, poured @690 °C (**a**) as-cast, conventional, (**b**) as-cast, EM stirred, (**c**) thixocast, sample "(**a**)", and (**d**) thixocast, sample "(**b**)" [4]

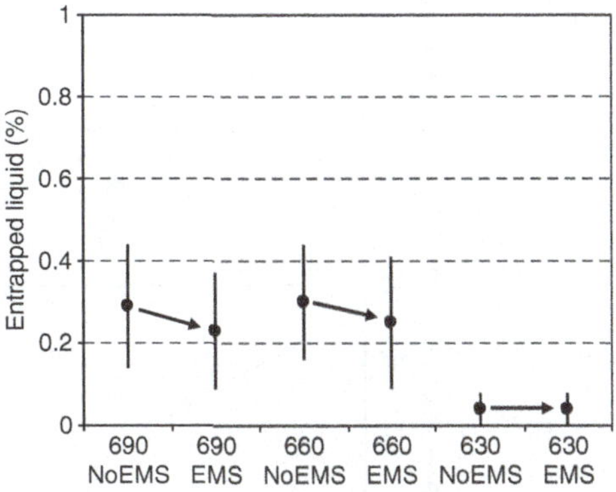

Fig. 7.23 Entrapped liquid measurement, copper thixocast [4]

Solidification time has an impact on partial remelting procedure. Basically, the driving force for the microstructural evolution within the mushy zone is the reduction of the interfacial area between liquid and solid. This could be estimated by the area to perimeter ratio. In fact, as mentioned in Sect. 4.3, this factor has an inverse relationship with the specific surface area per unit volume of the particles,

S_v. In dendritic solidification, higher values of P (total solid–liquid interface length) resemble a structure with more dendrite branches. Solid–liquid interfacial length mainly depends on the solidification of the alloy. For instance, the higher cooling rate of copper mold samples leads to lower dendrite arm spacing and finer secondary and tertiary branches and therefore the interface boundary of the eutectic-primary particles increases.

The effectiveness of A/P ratio in expressing the morphological changes in the as-cast structure during thixocasting (reheating at 583 °C for 10 min) for different cooling rates and application of EMS is given in Fig. 7.24. In conventional casting method, copper cast samples have a shorter solidification time which leads to a finer and more highly branched dendritic structure and therefore the A/P is smaller. In the same condition, lowering the superheat results in an equiaxed structure which during isothermal holding transforms to a globule/rosette structure. Similarly, higher cooling rate results in finer particles with smaller value of A/P (Fig. 7.24a).

Electromagnetic stirred billets exhibited a similar trend. The higher cooling rate of the copper mold leads to smaller dendrite size, and when coupled with stirring, results in greater percentage of fragmented dendrites with better distribution within the bulk liquid. A shorter solidification time also results in a limited growth with the eventual structure containing globules with smaller size compared to those of the sand mold billets (Fig. 7.24b). It is worth noting that the correlation between A/P

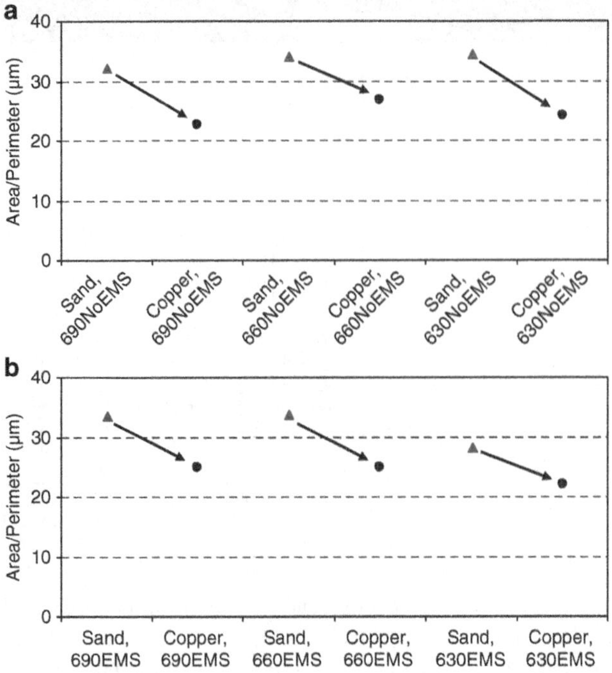

Fig. 7.24 Evolution of A/P ratio as a function of pouring temperature and stirring (*triangles* and *circles* for sand and copper molds cast samples, respectively) [4]

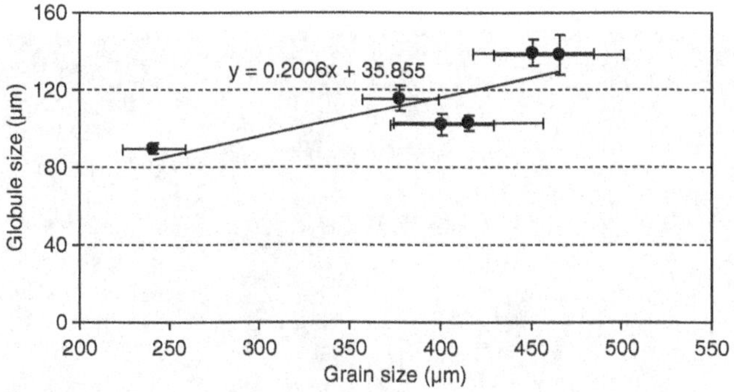

Fig. 7.25 Correlation between grain and globule size in EMS samples [4]

and application of EMS is not significant for either sand or copper molds at higher casting temperatures, 660 and 690 °C, but when temperature drops the effect of EMS on A/P ratio becomes more pronounced which is an indication of smaller particle size.

There is a direct correlation between grain/globule sizes and macro/microstructural evolution. The trend is shown in Fig. 7.25 where the smaller the grain size, the smaller is the globule size. This concept becomes complicated if one fails to differentiate between the globule and grain distinction as mentioned in Sect. 4.3.

7.3.3 Grain Refining/EMS: the Premium Choice

In Sect. 5.4, it was confirmed that electromagnetic stirring leads to fragmentation of primary phase and consequently homogeneous distribution of broken α-Al particles within the billets. A question may be raised on the influence of the refiner and vigorous agitation by EMS and whether it is favorable or not. In order to examine the benefit of grain refinement during EMS process, boron was selected as the refining agent and two series of tests were performed within the sand mold, with and without EMS application as explained in Sect. 4.1.1.3. Different pouring temperatures of 660 and 630 °C were examined and for refining objective, about 220 ppm boron was added using Al-5 % B master alloy with the final composition of Al (balance), 6.87 % Si, 0.84 % Fe, and 0.022 % B.

Figures 7.26 and 7.27 compare the resulting morphological evolution due to the boron addition/EM stirring. The primary α-Al phase has a fully columnar (dendritic) structure in the untreated sample and transforms to equiaxed morphology with the boron addition. This is evident in samples cast at higher superheat, for example, comparing Fig. 7.26a, b. However, when pouring temperature is decreased, the structure transforms to equiaxed, but finer and more spherical particles are achieved by refining process (comparing Fig. 7.27a, b). By stirring, the refining process is improved as revealed in the polarized light micrographs of

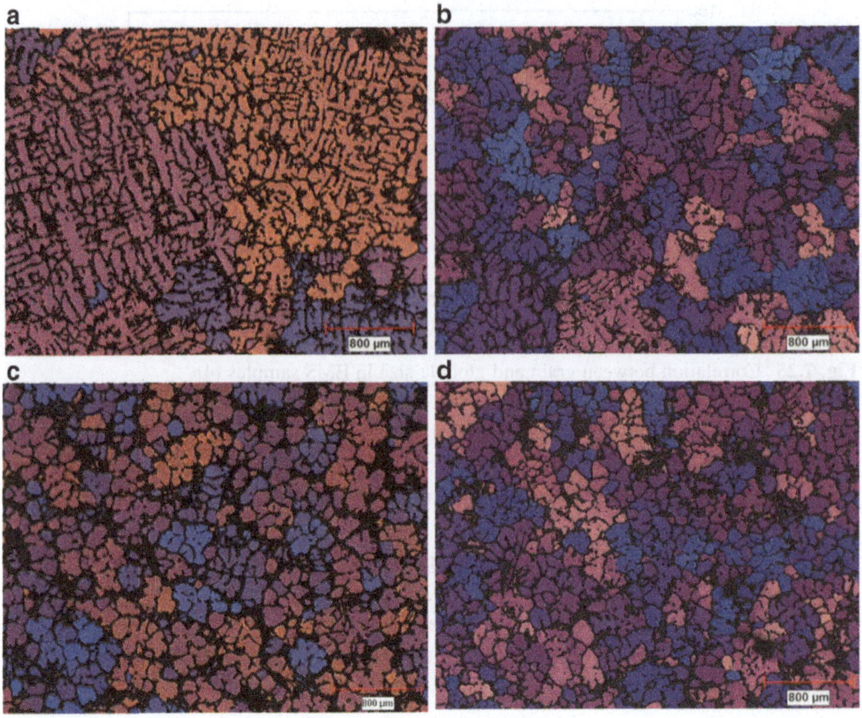

Fig. 7.26 Effects of boron addition/stirring on the formation of primary particles (sand cast, poured at 660 °C) (**a**) unstirred, unrefined, (**b**) unstirred, refined with ~220 ppm B, (**c**) EM stirred, unrefined, and (**d**) EM stirred, refined with ~220 ppm B

Figs. 7.26d and 7.27d for both pouring temperatures. The EM stirred samples show smaller and more compacted (denser) primaries.

As the pouring temperature decreases, the rate of heat extraction from the mold walls reduces and therefore a shallow temperature gradient within the bulk liquid is established (Sect. 5.2). Such temperature gradient encourages the formation of smaller primary particles with better distribution and thus more equiaxed particles will form. This concept is clearly shown by grain size measurement in Fig. 7.28a. By decreasing the pouring temperature from 660 to 630 °C, average grain size is reduced by a factor greater than 3. On the other hand by boron refining, nucleation sites increases drastically, so-called copious nucleation mechanism, and therefore smaller and more equiaxed particles were formed. It is worth mentioning that the refining impact is more pronounced in the case of higher superheats while by lowering the superheat, the shallow temperature gradient has a main contribution and as a result, the reduction in size consists of two elements which one overshadows the effect of the other (comparing ~72 % to ~56 % reduction in size for 660 °C and 630 °C, respectively).

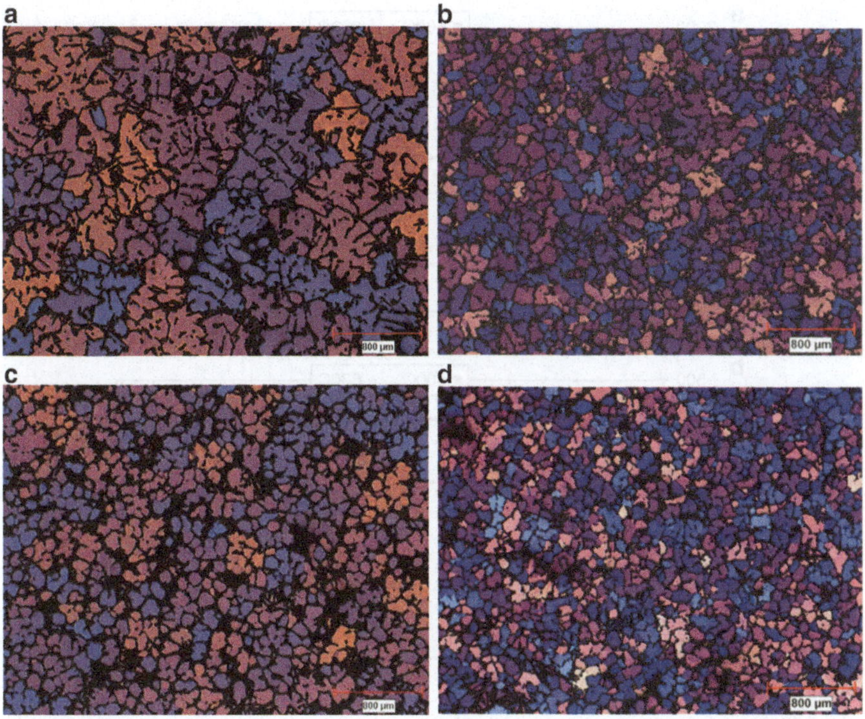

Fig. 7.27 Effects of boron addition/stirring on the formation of primary particles (sand cast, poured at 630 °C) (**a**) unstirred, unrefined, (**b**) unstirred, refined with ~220 ppm B, (**c**) EM stirred, unrefined, and (**d**) EM stirred, refined with ~220 ppm B

As described in Sect. 5.4, vigorous agitation leads to fragmentation of primary particles and the final structure is fully equiaxed. Grain size measurement (Fig. 7.28b) confirms that in the case of stirring, refiner addition is less prominent in comparison with the conventional casting, that is, stirring overshadows grain refinement.

7.3.3.1 Thixocasting of the Refined EMS Billets

Effects of isothermal holding on the aforementioned samples (Sect. 7.3.3) are shown in Figs. 7.29 and 7.30. The following points are noticeable:

- Generally, the lower the pouring temperature, the better is the results of thixocasting process. Better distribution (isolated particles) and equiaxed morphology before isothermal holding are the controlling parameters for rounder and smaller particles. This is clear through comparison of Figs. 7.29 and 7.30.
- Refining process results in the formation of more uniform equiaxed particles having less energy barrier for conversion to spherical particles through isothermal holding. During thixocasting, particles tend to reduce their energy level by

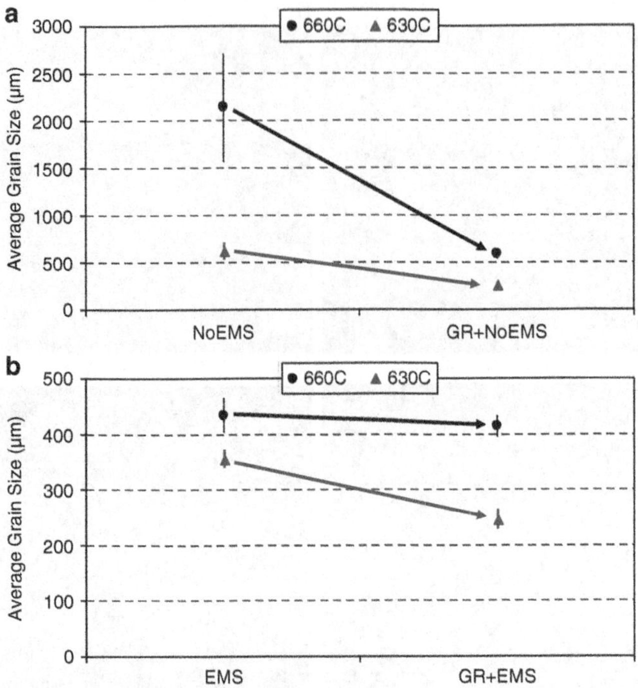

Fig. 7.28 Correlation between grain size variation and process parameters (grain refiner/stirring) (**a**) without stirring and (**b**) EM stirred

decreasing their surface area through transformation to most appropriate form (morphology) which is sphere (globule).

- Addition of refiner results in more potent sites for nucleation process rendering a larger number of particles per unit volume (number density of primary particles increases). This is apparent both in unstirred and stirred samples and it is more discernible at the lower superheats.

Figure 7.31 demonstrates the grain size evolution during the entire thixocasting experiments. Results are in line with conventional cast and/or stirred samples. Lower superheat and vigorous agitation lead to smaller and more isolated particles which are then transform to globules during isothermal holding (coarsening of primary particles is evident which is typical during isothermal holding).

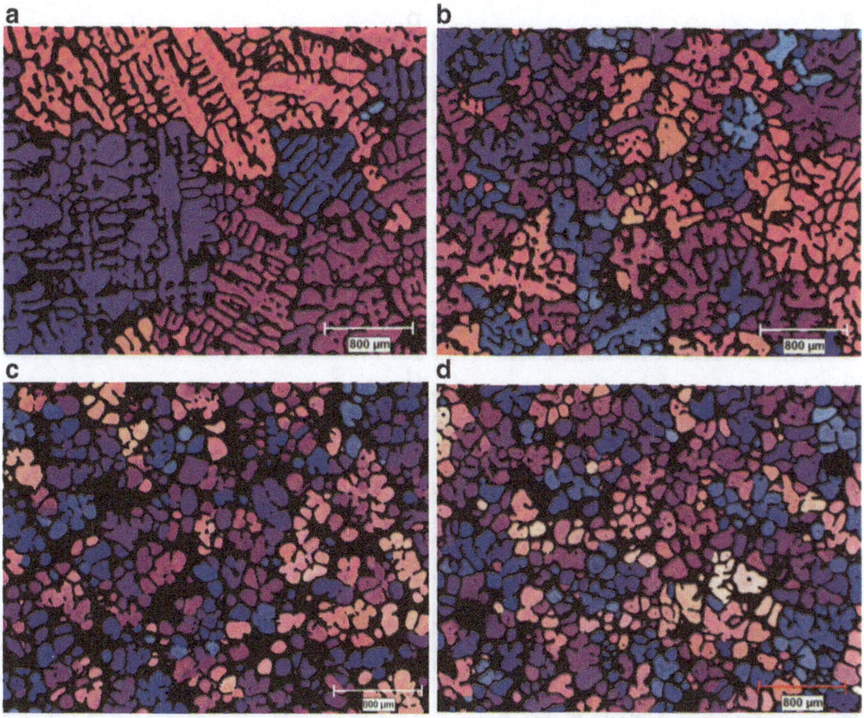

Fig. 7.29 Effects of boron addition/stirring on the formation of primary particles (sand mold cast, poured at 660 °C, reheated at 583 °C for 10 min) (**a**) unstirred, unrefined, (**b**) unstirred refined with ~220 ppm B, (**c**) EM stirred, unrefined, and (**d**) EM stirred, refined with ~220 ppm B

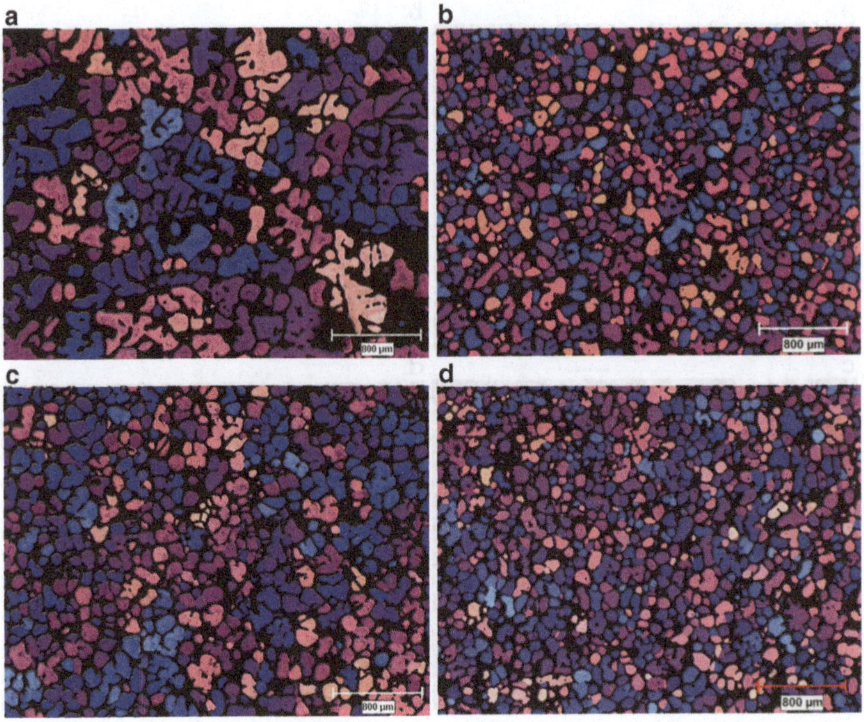

Fig. 7.30 Effects of boron addition/stirring on the formation of primary particles (sand mold cast, poured at 630 °C, reheated at 583 °C for 10 min) (**a**) unstirred, unrefined, (**b**) unstirred refined with ~220 ppm B, (**c**) EM stirred, unrefined, and (**d**) EM stirred, refined with ~220 ppm B

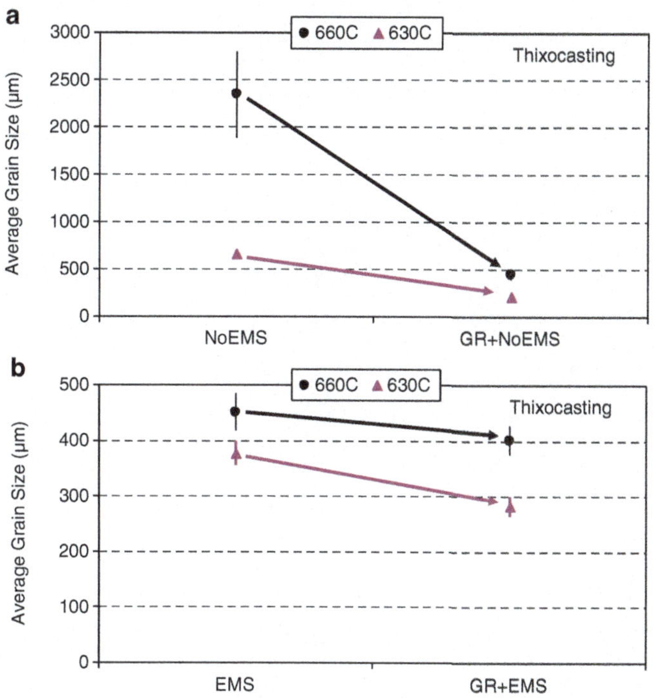

Fig. 7.31 Correlation between grain size variation and process parameters in thixocasting process (sand mold cast, reheated at 583 °C for 10 min) (**a**) without stirring and (**b**) EM stirred

References

1. W.R. Loue, M. Suery, Microstructural evolution during partial remelting of Al-Si7Mg alloys. Mater. Sci. Eng. A **A203**, 1–13 (1995)
2. Q.Y. Pan, M. Arsenault, D. Apelian, M.M. Makhlouf, SSM processing of AlB2 grain refined Al-Si alloys. AFS Trans. **112**, (2004), Paper 04-053
3. H. Wang, C.J. Davidson, D.H. St John, Semisolid microstructural evolution of AlSi7Mg alloy during partial remelting. Mater. Sci. Eng. **A368**, 159–167 (2004)
4. S. Nafisi, R. Ghomashchi, Microstructural evolution of electromagnetically stirred feedstock SSM billets during reheating process. J. Metallogr. Microstruct. Anal. **2**(2), 96–106 (2013)
5. D.H. Kirkwood, Semi-solid metal processing. Int. Mater. Rev. **39**(5), 173–189 (1994)
6. Z. Fan, Semisolid metal processing. Int. Mater. Rev. **47**(2), 49–85 (2002)
7. M.C. Flemings, Behavior of metal alloys in the semi-solid state. Metal. Trans. A **22A**, 952–981 (1991)

Chapter 8
Commercial Applications of Semi-Solid Metal Processing

Stephen P. Midson

Abstract The commercial applications of SSM processes are detailed in this chapter. The lower energy consumption along with longer die life and less expensive tooling make the SSM processes quite attractive to manufacturing industry. In this chapter, few applications in fabrication of a range of engineering components are introduced to highlight the practicality and cost-effectiveness of SSM processes. The findings are expanded into thixomolding to further justify the suitability of SSM processes for light metals such as Al and Mg alloys.

8.1 Introduction

The commercial appeal of semi-solid casting is that it combines many of the advantages of conventional high pressure die casting (high productivity, low cost, thin walls, excellent surface finish, and close dimensional tolerances) with a combination of excellent mechanical properties and outstanding pressure tightness. This allows semi-solid castings to be used in safety-critical and pressure-sensitive applications. As described in previous chapters of this book, semi-solid castings achieve this high level of performance by minimizing casting defects through a combination of utilizing a viscous semi-solid feed material that provides high levels of control during die filling, with very high pressures (>15,000 psi, in excess of 100 MPa) that are applied to minimize (or eliminate) solidification shrinkage as the castings solidify.

Over the past 20 years or more, there have been numerous commercial applications of semi-solid castings around the world, utilized in a wide range of markets such as automotive, aerospace, motorcycles, bicycles, electronics, defense, and sporting goods. These applications of semi-solid castings can generally be classified into the following categories:

(a) High quality aluminum castings
(b) Improved quality aluminum die castings
(c) Magnesium alloy castings

S.P. Midson
The Midson Group, Denver, Colorado, USA

S. Nafisi, R. Ghomashchi, *Semi-Solid Processing of Aluminum Alloys*,
DOI 10.1007/978-3-319-40335-9_8

Although there has been extensive research on the semi-solid casting of the higher melting temperature metals such as copper alloys and steels, there have been few commercial applications, mainly due to the limited die life obtained with these types of alloys.

The objective of this chapter is to provide a brief review of the range of commercial applications for semi-solid castings, list typical mechanical properties for both aluminum and magnesium alloys, and show some examples of commercial semi-solid castings. Similar to all fabrication processes, components produced by semi-solid casting shift in and out of commercial production as manufacturers continually revise and modify their product ranges. Therefore, rather than attempting to list semi-solid cast components that are currently in production, which is frequently difficult due to the proprietary concerns of both casters and end users, this chapter reviews notable components that have been produced by semi-solid casting over the past 20 years or so.

8.2 High Quality Castings

These types of high quality semi-solid castings are normally produced from primary aluminum alloys such as A356, 357, and 319S, and are typically produced using high solid fraction semi-solid slurries, utilizing a solid content of about 50 %. The nominal chemical compositions of these three alloys are listed in Table 8.1.

Handbook mechanical property data for the three alloys after heat treating to either the T5 or T6 tempers are listed in Tables 8.2 and 8.3, respectively [1]. T5 heat treating involves water quenching the castings immediately after ejection from the

Table 8.1 Nominal composition (wt%) of aluminum alloys commonly used for semi-solid casting [1]

Element*	A356	357	319S
Silicon	6.5–7.5	6.5–7.5	5.5–6.5
Iron	0.20	0.15	0.15
Copper	0.20	0.05	2.5–3.5
Manganese	0.10	0.03	0.03
Magnesium	0.25–0.45	0.45–0.60	0.30–0.40
Titanium	0.20	0.20	0.20
Strontium	0.01–0.05	0.01–0.05	0.01–0.05
Other (each)	0.03	0.03	0.03
Other (total)	0.10	0.10	0.10

Table 8.2 Handbook mechanical properties data for two semi-solid cast alloys after heat treating to the T5 temper (note that alloy 319S semi-solid castings are not normally used in the T5 temper) [1]

Alloy	0.2 % YS MPa (ksi)	UTS MPa (ksi)	Elongation (%)
357	200 (29)	283 (41)	8
A356	179 (26)	248–269 (36–39)	7–10

Table 8.3 Handbook mechanical properties data for three semi-solid cast alloys after heat treating to the T6 temper [1]

Alloy	0.2 % YS MPa (ksi)	UTS MPa (ksi)	Elongation (%)
357	283–290 (41–42)	345 (50)	7–9
A356	228–234 (33–34)	303–310 (44–45)	12–13
319S	317 (46)	400 (58)	5

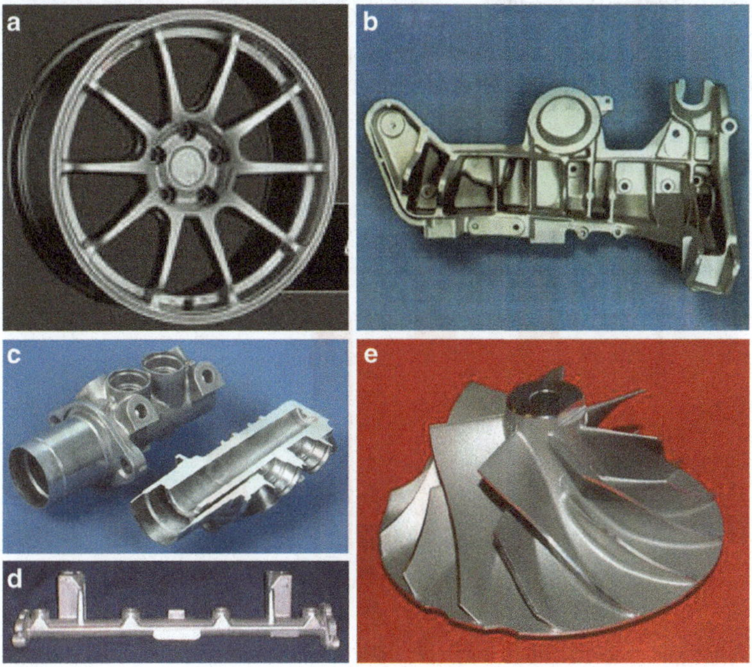

Fig. 8.1 Examples of high quality semi-solid castings. (**a**) Aftermarket automobile wheel produced by SSR [3], (**b**) suspension component produced by Stampal for Alfa Romeo [3], (**c**) master brake cylinder [3], (**d**) fuel rail [4], (**e**) turbocharger impeller produced by ACC for Cummins Turbo Technologies [5]

casting die, followed by a low temperature aging treating. Historically, many semi-solid castings have been used in the T5 temper, as unlike many other casting processes, the T5 temper provides a good combination of strength and ductility for semi-solid casting, especially when considering the low cost and simplicity of this heat treatment. However, in recent years, more and more semi-solid castings are being given the full T6 heat treatment (solution heat treatment, water quench, and low temperature age) as this maximizes mechanical performance (see Table 8.3). Since surface blistering of the castings during the high temperature solution heat treatment can be an issue during T6 heat treating, the caster has to take care to optimize their process to minimize air entrapment during die filling, and to utilize sleeve and die lubricants that do not promote blistering [2]. Figure 8.1 shows some examples of high quality semi-solid castings.

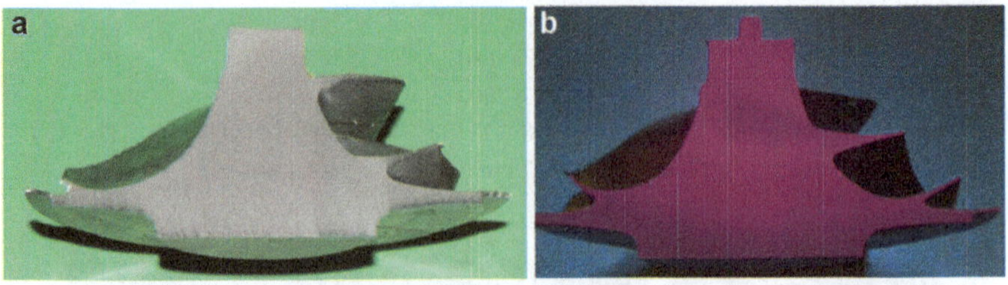

Fig. 8.2 Sections through impellers showing lack of porosity [5]; (**a**) Machined surface, (**b**) Machined and penetrant tested (viewed under ultraviolet light)

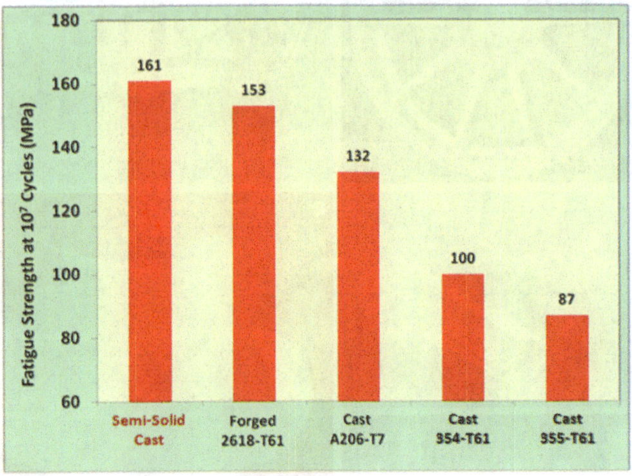

Fig. 8.3 Fatigue data for castings, forgings, and semi-solid castings [5]

Information has been recently published on the production of the alloy 319S semi-solid cast impeller shown in Fig. 8.1e [5]. Fatigue performance is especially important for turbocharger impellers, and porosity as small as 50 μm can reduce fatigue life [6], and so to ensure that the impellers were indeed porosity free, testing involved sectioning the impellers and abrading the machined surfaces using 600-grit paper followed by macroetching to remove aluminum smeared on the machined surface. The prepared surfaces were then penetrant tested to ensure that porosity as small as 50 μm were not present. Figure 8.2a shows the sectioned surface, whereas Fig. 8.2b shows a photograph viewed under ultraviolet light of a penetrant tested surface, showing the lack of porosity or other defects in the semi-solid cast impeller.

The fatigue life of the 319S alloy impellers was measured on samples machined from semi-solid castings and compared with samples produced by conventional casting or produced by forging [5]. Figure 8.3 shows that the fatigue life of the semi-solid casts material was significantly higher than conventional castings (including high strength alloy 206), and in fact was comparable to the wrought aluminum alloy 2618.

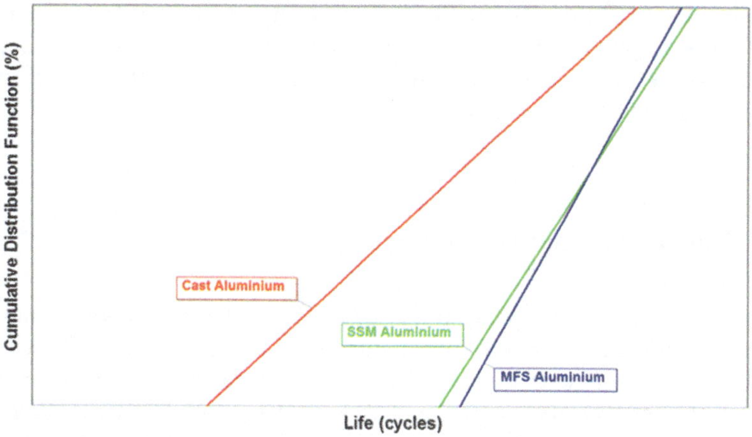

Fig. 8.4 Comparative fatigue life of cast, SSM, and forged + machined (MFS) impellers [7] (reprinted by permission of Taylor & Francis Ltd)

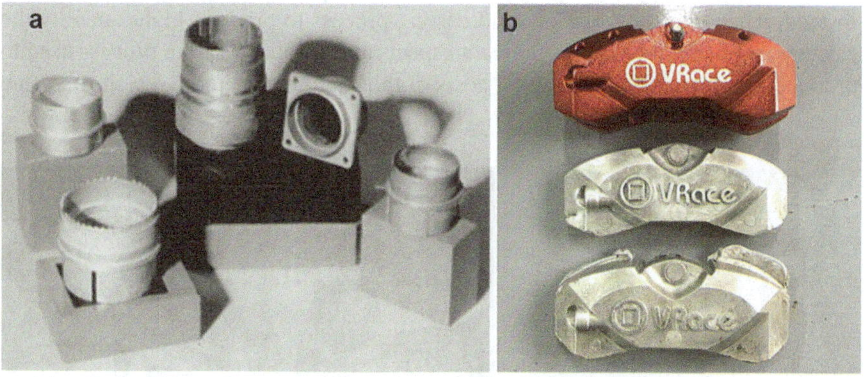

Fig. 8.5 (**a**) Electrical connectors produced from wrought alloy 6262 [8], (**b**) Brake caliper for motorcycle, cast from 6061-T6, color anodized (courtesy of GISSCO Co., Ltd, Thailand)

Results from a functional test were also reported [7]. The actual fatigue life is proprietary, but the data in Fig. 8.4 shows that semi-solid cast impellers were significantly better than castings produced via conventional casting methods, and again the fatigue life of these semi-solid castings was comparable to the wrought alloy 2618.

Another potential approach for the production of high quality semi-solid castings is through the use of silicon-free casting alloys (2xx) and wrought alloys (2xxx, 6xxx), as these types of alloys generally have better mechanical properties than can be achieved with conventional foundry alloys. These alloys are only infrequently produced by conventional casting as they suffer from poor resistance to hot tearing, but it is well known that semi-solid casting significantly reduces the tendency for hot tearing [5], and some commercial semi-solid castings have been produced from these types of alloys. Figure 8.5a shows electrical connectors semi-solid cast from wrought alloy 6262 [8], whereas Fig. 8.5b shows a brake caliper for a motorcycle, cast from wrought alloy 6061, heat treated to the T6 temper, and color anodized. One advantage of the low-silicon wrought alloys is that excellent colors can be produced by anodizing after casting.

8.3 Improved Quality Die Castings

The semi-solid casting processes can also be used for the production of improved quality die castings. These are typically produced using a lower solid fraction of 20–30 % (70–80 % liquid). Conventional die castings can contain significant levels of residual porosity, but castings produced using low solid fraction rheocasting processes have been reported to reduce (but generally not eliminate) this residual porosity [3]. These types of low solid fraction castings generally employ the same turbulent die filling conditions as used by conventional die castings, so the generation of an extremely high quality semi-solid slurry is not of primary importance. Instead, when choosing a semi-solid casting process to produce higher quality die castings, probably the most significant aspects are simplicity of the process used to produce the slurry, and low capital costs of the equipment. Figure 8.6 shows some

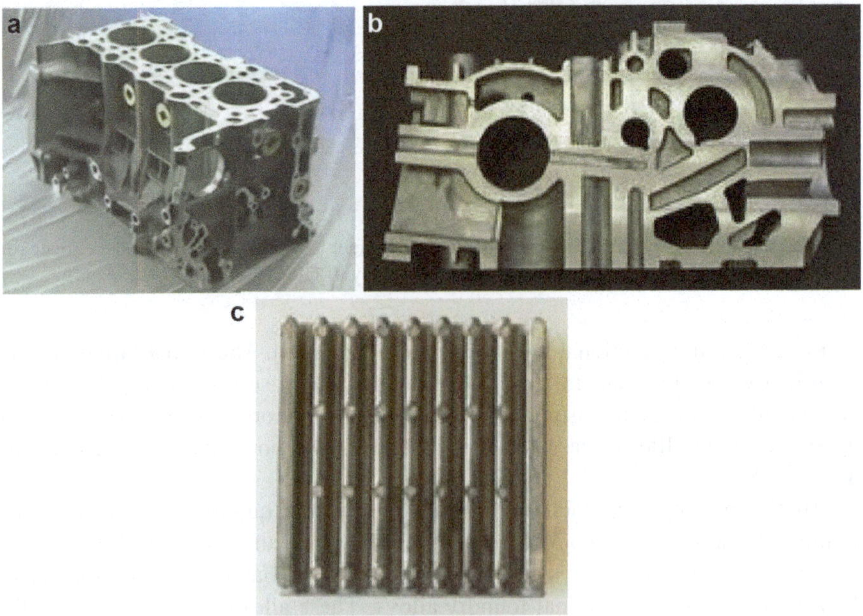

Fig. 8.6 Examples of semi-solid castings used to replace conventional die castings. (**a**) Diesel engine block produced by Honda [9]; (**b**) oil pump filter housing [10], and (**c**) heat sink produced by AMAX [3]

Table 8.4 Parameters for the oil pump filter housing produced by semi-solid rheocasting [10]

Parameter	Unit	Die casting	SSR
Casting shot weight	kg	6.5	6.5
Furnace temperature	°C	720	640
Pouring temperature	°C	720	588
Cycle time	s	120	86–90
Solidification time	s	30	28
First phase velocity	ms^{-1}	0.25–0.30	0.36
Second phase velocity	ms^{-1}	2.03–2.54	N/A
Final metal pressure (intensification)	Bar (MPa)	1000 (100)	800 (80)

examples of low solid fraction semi-solid castings used to replace conventional die castings.

The oil pump filter housing shown in Fig. 8.6b was originally designed as a gravity casting (which accounts for the thick walls), but was converted to a die casting. The thick walls created problems with porosity when produced by conventional high pressure die casting, and the machined castings required impregnation. Porosity was significantly reduced (but not eliminated) after conversion to a low solid fraction semi-solid casting process (semi-solid rheocasting-SSR), eliminating the need for impregnation. Table 8.4 compares process data for the oil pump filter housing produced by both die casting and semi-solid casting.

8.4 Thixomolding

Thixomolding is a semi-solid process developed by Thixomat [11] and is used exclusively for the production of castings from magnesium alloys [1]. Over the past 20 years or so, it has probably been the most commercially successful of all the semi-solid processes. Thixomolded components are generally produced from the three magnesium alloys listed in Table 8.5, and the casting are normally used in the as-cast (unheat treated) condition. Published mechanical properties for thixomolded components produced from the three magnesium alloys are listed in Table 8.6.

Thixomolded components have been used in a range of markets, including automotive, consumer electronics and consumer hardware applications as varied as automotive shift cams, notebook computer cases, digital camera bodies, and chain saw housings. Figure 8.7 shows a number of commercial examples, and many of these examples include extremely thin-walled components used in electronic or electrical applications.

Some additional thixomolded magnesium components are described below. Figure 8.8a shows an example of a drive gear. Six separate components thixomolded for a projector are shown in Fig. 8.8b—three of the parts make up the external case, whereas the other three are internal structural members.

Table 8.5 Nominal compositions (wt%) of magnesium alloys used for thixomolding [1]

Element	AZ91D	AM-50	AM-60
Aluminum	8.3–9.7	4.4–5.4	5.5–6.5
Zinc	0.35–1.0	0.22 (max)	0.22 (max)
Manganese	0.15–0.50[a]	0.26–0.6[a]	0.24–0.6[a]
Silicon	0.10 (max)	0.10 (max)	0.10 (max)
Iron	0.005[a]	0.004[a]	0.005[a]
Copper (Max)	0.030	0.010	0.010
Nickel (Max)	0.002	0.002	0.002
Others (Total)	0.02	0.02	0.02
Magnesium	Bal.	Bal.	Bal.

[a]If either of the minimum manganese limit or the maximum iron limit is not met, then the iron/manganese ratio shall not exceed 0.032 for AZ91D, 0.015 for AM-50 and 0.021 for AM-60

Table 8.6 Handbook data for thixomolded magnesium castings in the as-cast condition [1]

Alloy	0.2 % YS MPa (ksi)	UTS MPa (ksi)	Elongation (%)
AZ91D	159 (23)	234 (34)	3–6
AM-60	124 (18)	221 (32)	6–13
AM-50	131 (19)	221 (32)	6–9

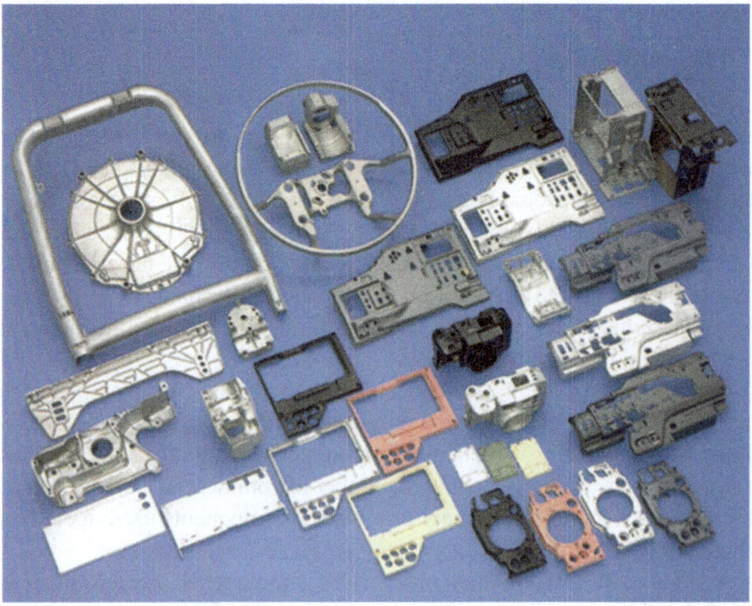

Fig. 8.7 Examples of magnesium alloy thixomolded components (courtesy of Thixomat Inc., and Japan Steel Works)

Fig. 8.8 Thixomolded components produced for several different applications. (**a**) Drive gear, (**b**) mini-projector, and (**c**) components thixomolded for a motorcycle, AM60 (courtesy of Thixomat Inc., and Japan Steel Works), (**d**) plate replaced a stamped part (courtesy of Phillips-Medisize Corp.)

Thixomolding was used to provide control of critical dimensions, dimensional stability, and alignment of the various components. Figure 8.8c shows parts thixomolded for a US-manufactured motorcycle. Figure 8.8d is a thixomolded magnesium part which replaced a stamped component.

References

1. NADCA, *Product Specification Standards for Die Castings Produced by the Semi Solid and Squeeze Casting Processes*, 5th edn. (NADCA, Wheeling, 2009)
2. Y.F. He, X.J. Xu, F. Zhang, D.Q. Li, S.P. Midson, Q. Zhu, Impact of die and plunger lubricants on blistering during T6 heat treatment of semi solid castings. in *Trans 2013 NADCA Die Casting Congress and Tabletop*, paper number T13-012
3. S.P. Midson, Industrial applications for aluminum semi solid castings. in *13th International Conference on Semi solid Processing of Alloys and Composites* (Muscat, Oman, 2014) (published in Solid State Phenomena, vol. 217–218, 2015, 487–495)

4. C.S. Rice, P.F. Mendez, Slurry based semi solid diecasting. Adv. Mater. Process. **159**(10), 49–53 (2001)
5. G. Wallace, A.P. Jackson, S.P. Midson, Q. Zhu, High-quality aluminum turbocharger impellers produced by thixocasting. Trans. Nonferrous Met. Soc. China **20**, 1786–1791 (2010)
6. F.J. Major, Porosity control and fatigue behavior in A356-T61 Aluminum Alloy, AFS 1997 Transactions, paper no. 97–94
7. G. Wallace, A.P. Jackson, S.P. Midson, Novel method for casting high quality aluminum turbocharger impellers SAE paper number 2010-01-0655. SAE Int. Mater. Manuf. **3**, 405–412 (2010)
8. M.P. Kenny, J.A. Courtois, R.D. Evans, G.M. Farrior, C.P. Kyonka, A.A. Koch, K.P. Young, Semisolid metal casting and forging, ASM Handbook, vol. 15, Casting (1992)
9. K. Kuroki, T. Suenaga, H. Tanikawa, T. Masaki, A. Suzuki, T. Umemoto, M. Yamazaki, Establishment of a manufacturing technology for the high strength aluminum cylinder block in diesel engines applying a rheocasting process. in *Eighth International Conference on Semi Solid Processing of Alloys and Composites* (Limassol, Cyprus, 2004)
10. J. Yurko, R. Boni, SSR™ semi solid rheocasting. La Metallurgia Italiana (March 2006), 35–41
11. L. Pasternak, R. Carnahan, R. Decker, R. Kilbert, Semi-solid production processing of magnesium alloys by thixomolding. in Second International Conference on Semi-Solid Processing of Alloys and Composites (MIT, Cambridge, 1992), 159–169

Appendix A
Mathematical Solutions For Non-Newtonian and Newtonian Fluids [1]

The deformation behavior of non-Newtonian fluids squeezed between two large parallel plates, when considered in cylindrical coordinates, are expressed by the following equation [2]:

$$\tau_{rz} = -m\left|\frac{\partial v_r}{\partial z}\right|^{n-1}\left(\frac{\partial v_r}{\partial z}\right) \tag{A.1}$$

where v_r is the radial velocity and m and n are the material constants. Assuming steady state deformation, $\frac{\partial v_z}{\partial t} = 0$, and the fact that the reduction in height is rather negligible during steady state deformation in contrast to deformation in radial direction, the normal velocity of flow v_z, can be neglected in comparison with the radial velocity, v_r. Also assuming circular symmetry, v_θ is zero. With these assumptions, the basic equations for the flow field become,

$$\text{Continuity}\quad \frac{1}{r}\frac{\partial}{\partial r}(rv_r) + \frac{\partial v_z}{\partial z} = 0 \tag{A.2}$$

$$\text{Momentum}\quad \frac{\partial P}{\partial r} = -m\frac{\partial}{\partial z}\left(-\frac{\partial v_r}{\partial z}\right)^n \tag{A.3}$$

$$\frac{\partial P}{\partial z} = 0 \tag{A.4}$$

With the boundary conditions

$$z = 0: \qquad \frac{\partial v_r}{\partial z}(r, o) = 0 \tag{A.5}$$

$$z = \frac{h}{2}: \qquad v_z\left(r, \frac{h}{2}\right) = \frac{\mathrm{d}h}{\mathrm{d}t}, \quad v_r\left(r, \frac{h}{2}\right) = 0 \tag{A.6}$$

S. Nafisi, R. Ghomashchi, *Semi-Solid Processing of Aluminum Alloys*,
DOI 10.1007/978-3-319-40335-9

$$r = 0: \qquad v_r(o, z) = 0 \tag{A.7}$$

$$r = R \qquad P(R) = P_0 \tag{A.8}$$

Equation (A.4) means that P (momentum) is independent of z. Hence integrating (A.3) twice with respect to z, yields:

$$v_r = (-1)^{1/n}\left(\frac{n}{n+1}\right)\left(\frac{1}{m}\frac{\mathrm{d}P}{\mathrm{d}r}\right)^{1/n} \times \left[\left(\frac{h}{2}\right)^{n+1/n} - z^{(n+1)/n}\right] \tag{A.9}$$

With integrating the continuity (A.2) over z:

$$\frac{\mathrm{d}h}{\mathrm{d}t} = -\frac{2}{r}\frac{\partial}{\partial r}\left(r\int_0^{h/2} v_r \mathrm{d}z\right) \tag{A.10}$$

Integrating (A.10) over (r), results:

$$\frac{r}{2}\frac{\mathrm{d}h}{\mathrm{d}t} = -2\int_0^{h/2} v_r \mathrm{d}z \tag{A.11}$$

Substituting (A.9) in (A.11) and integrating the pressure gradient so obtained gives:

$$P - P_0 = \left(\frac{m}{n+1}\right)(-1)^{n+1} \times \left\{\left(\frac{2n+1}{4n}\right)\left(\frac{h}{2}\right)^{-[(2n+1)/n]}\frac{\mathrm{d}h}{\mathrm{d}t}\right\}^n \left(r^{n+1} - R^{n+1}\right) \tag{A.12}$$

Here, R is radius of the specimen at time, t. Integrating over the surface of the plates gives:

$$F = (-1)^n\left(\frac{m}{n+3}\right)\pi R^{n+3} \times \left\{\left(\frac{2n+1}{4n}\right)\left(\frac{h}{2}\right)^{-[(2n+1)/n]}\frac{\mathrm{d}h}{\mathrm{d}t}\right\}^n \tag{A.13}$$

Since the fluid does not completely fill the space between the plates, R varies with time, t. However since the volume of the work piece remains unchanged during deformation, that is, $R^2h = R_o^2h_o$, and letting $d_o = 2R_o$, (A.13) may be rewritten as

$$-\frac{dh}{dt} = \left\{\left(\frac{2n}{2n+1}\right)^n \frac{4F}{\pi d_o^{n+3} m}(n+3)\right\}^{1/n} h_o^{-[(n+3)/2n]} h^{[5(n+1)/2n]} \tag{A.14}$$

If

$$k = \left\{\left(\frac{2n}{2n+1}\right)^n \frac{4F}{\pi d_o^{n+3} m}(n+3)\right\}^{1/n} \tag{A.15}$$

Thus

$$-\frac{dh}{dt} = k h_o^{-[(n+3)/2n]} h^{[5(n+1)/2n]} \tag{A.16}$$

Integrating from $h = h_o$ at $t = 0$

$$\left(\frac{1}{h}\right)^{(3n+5)/2n} - \left(\frac{1}{h_o}\right)^{(3n+5)/2n} = \left(\frac{3n+5}{2n}\right) k h_o^{-[(n+3)/2n]} t \tag{A.17}$$

If $n = 1$, m becomes equal to η according to the power law equation and (A.17) reduces to equation (4.29) in the text.

Rewriting (A.17) yields:

$$\frac{h_o}{h} = \left\{1 + \left(\frac{3n+5}{2n}\right) h_o^{(n+1)/n} kt\right\}^{2n/(3n+5)} \tag{A.18}$$

This is a modified form of the condition where the fluid does not fill the space between the plates.

Assuming the engineering strain $e = \left(1 - \frac{h}{h_o}\right)$ results:

$$e = 1 - \left\{1 + \left(\frac{3n+5}{2n}\right) h_o^{(n+1)/n} kt\right\}^{-[2n/(3n+5)]} \tag{A.19}$$

Equation (A.19) is the theoretical strain–time curve for a power law fluid. From this equation, the following is achieved:

$$\log(1-e) = -\left(\frac{2n}{3n+5}\right) \times \log\left\{1 + \left(\frac{3n+5}{2n}\right) h_o^{(n+1)/n} kt\right\} \tag{A.20}$$

For $1 \ll \left(\frac{3n+5}{2n}\right) h_o^{(n+1)/n} kt$, that is, at steady state deformation; after appreciable deformation-long times, the (4.34) in the text is given as

$$\log(1-e) = -\left(\frac{2n}{3n+5}\right)\log t - \left(\frac{2n}{3n+5}\right)\log\left(\frac{3n+5}{2n}h_o^{(n+1)/n}k\right) \quad (A.21)$$

The shear rate, $\dot{\gamma}$, at any point (r, z) within the fluid is given by [3]:

$$\dot{\gamma} = -\frac{\partial v_r}{\partial z} = -\frac{2(\mathrm{d}h/\mathrm{d}t)}{h^2}\left(\frac{2n+1}{2n}\right)\left(\frac{z}{h/2}\right)^{1/n} r \quad (A.22)$$

For $n = 1$, this reduces to the following:

$$\dot{\gamma} = -\frac{6rz}{h^3}\frac{\mathrm{d}h}{\mathrm{d}t} \quad (A.23)$$

The average shear rate, $\dot{\gamma}_{av}$, obtained by integrating throughout the volume is given as

$$\dot{\gamma}_{av} = \frac{2\pi}{V}\int_0^R \int_0^{h/2} \dot{\gamma}\, r\mathrm{d}r\,\mathrm{d}z \quad (A.24)$$

Combining (A.23) and (A.24) yields:

$$\dot{\gamma}_{av} = -\frac{R}{2h^2}\frac{\mathrm{d}h}{\mathrm{d}t} \quad (A.25)$$

Rewriting (A.25), using $V = \pi R^2 h$ gives the average shear rate quoted in the text, (4.31).

$$\dot{\gamma}_{av} = -\sqrt{\frac{V}{\pi}}\left(\frac{\mathrm{d}h/\mathrm{d}t}{2h^{2.5}}\right) \quad (A.26)$$

References

1. O. Lashkari, R. Ghomashchi, Deformation behavior of semi-solid A356 Al-Si alloy at low shear rates: effect of fraction solid. Mater. Sci. Eng. A **486**, 333–340 (2008)
2. V. Laxmanan, M.C. Flemings, Deformation of semi-solid Sn-15%Pb alloy. Metall. Trans. A **11A**, 1927–1937 (1980)
3. J. F. Stefan, Versuche Uber Die Scheinbare Adhasion SitzberM th. Naturw. Kl. Bagar Akad. Wiss Munchen, **69**, part 2 (1874)

Appendix B
Lattice Disregistry Between Solid Al and Inoculant Particle

As mentioned in Chap. 6, one of the main requirements for a particle to be an active refiner is its atomic match with the new phase to form. This is defined by lattice disregistry which is regarded as a controller for the effectiveness of a substrate for heterogeneous nucleation. Lattice disregistry is defined as:

$$\delta = \frac{d_p - d_m}{d_m} \times 100 \tag{B.1}$$

where d_p and d_m are the interatomic spacings of the particles and matrix, respectively. The interatomic spacing is related to lattice parameter and type of crystal structure. For aluminum with FCC crystal structure and a unit cell of $a = 4.0494$ A° at 25 °C, an effective grain refiner should match with this lattice parameter. Table B.1 shows some physical characteristics for the most common grain refiners and their lattice disregistry. Porter and Easterling [1] divided the interfacial boundary nature into three classes: coherent, semi-coherent, and incoherent. A coherent interface arises when the two crystals match perfectly at the interface plane so that the two lattices are continuous across the interface (Fig. B.1). With partial atomic misfit across the interfacial area, it becomes more favorable to replace the interface with a semi-coherent interface with a specified lattice disregistry (Fig. B.2). The last case occurs when the interfacial plane has a very different configuration at the interface. This condition results when the patterns are totally different or interatomic distances differ more than 25 % (Fig. B.3).

S. Nafisi, R. Ghomashchi, *Semi-Solid Processing of Aluminum Alloys*,
DOI 10.1007/978-3-319-40335-9

Table B.1 Some physical properties of grain refiner particles [2]

Particle	Lattice parameter (A°)	Density (g/cm^3)	Bravais system	Disregistry at 25 °C (%)
Al	$a = 4.049$	2.697	fcc	0
TiB_2	$a = 3.030$	4.495	hcp	5.8
	$c = 3.229$			
$TiAl_3$	$a = 3.846$	3.35	Tetragonal	5.02
	$c = 8.594$			
AlB_2	$a = 3.005$	3.1	hcp	4.96
	$c = 3.25$			
AlB_{12}	$a = 10.17$	2.55	Tetragonal	151
	$c = 14.28$			
SrB_6	$a = 4.193$	3.43	Cubic	3.546

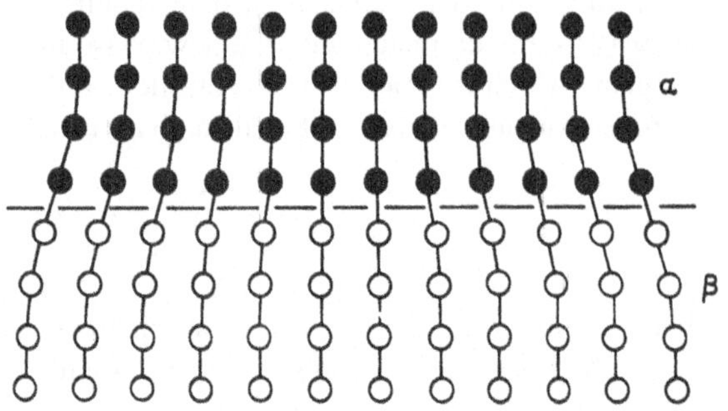

Fig. B.1 A coherent interface with slight mismatch leads to coherency in the adjoining lattices [1]

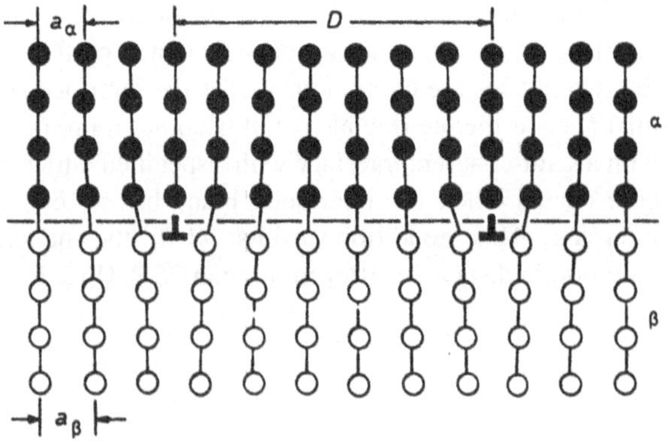

Fig. B.2 A semi-coherent interface. The misfit parallel to the interface is accommodated by a series of edge dislocations [1]

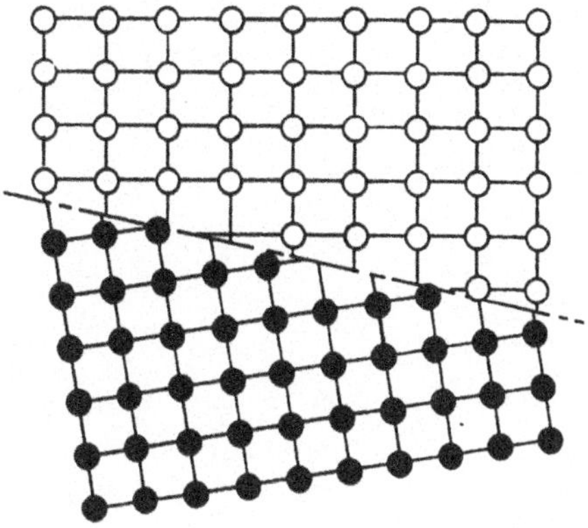

Fig. B.3 An incoherent interface [1]

References

1. D.A. Porter, K.E. Easterling, *Phase Transformations in Metals and Alloys* (Van Nostrand Reinhold, England, 1981)
2. International Center of Diffraction Data, www.ICDD.com

Index

S. Nafisi, R. Ghomashchi, *Semi-Solid Processing of Aluminum Alloys*,
DOI 10.1007/978-3-319-40335-9

C

D

E

F